LCA Compendium – The Complete World of Life Cycle Assessment

Series editors
Walter Klöpffer, LCA Consult & Review, Frankfurt am Main, Germany
Mary Ann Curran, BAMAC, Ltd., LCA & Sustainability Consultant, Rock Hill, SC, USA

Aims and Scope

Life Cycle Assessment (LCA) has become the recognized instrument to assess the ecological burdens and human health impacts connected with the complete life cycle (creation, use, end-of-life) of products, processes and activities, enabling the assessor to model the entire system from which products are derived or in which processes and activities operate. Due to the steady, world-wide growth of the field of LCA, the wealth of information produced in journals, reports, books and electronic media has made it difficult for readers to stay abreast of activity and recent developments in the field. This led to the realization of the need for a comprehensive and authoritative publication.

LCA Compendium – The Complete World of Life Cycle Assessment will discuss the main drivers in LCA (SETAC, UNEP/SETAC Life Cycle Initiative, etc.), the strengths and limitations of LCA, the LCA phases as defined by ISO standards, specific applications of LCA, Life Cycle Management (LCM) and Life Cycle Sustainability Assessment (LCSA). Further volumes, which are closely related to these themes will cover examples of exemplary LCA studies ordered according to the importance of the fields of application. They will also present new insights and new developments and will keep the whole work current. The aim of the series is to provide a well-structured treatise of the field of LCA to give orientation and guidance through detailed descriptions on all steps necessary to conduct an LCA study according to the state of the art and in full agreement with the standards.

LCA Compendium – The Complete World of Life Cycle Assessment anticipates publishing volumes on the following themes:

- Background and Future Prospects in Life Cycle Assessment (published in March 2014)
- Goal and Scope Definition in Life Cycle Assessment (published in August 2016)
- Life Cycle Inventory Analysis (LCI)
- Life Cycle Impact Assessment (LCIA) (published in March 2015)
- Interpretation, Critical Review and Reporting in Life Cycle Assessment
- Applications of Life Cycle Assessment
- Special Types of Life Cycle Assessment (published in July 2016)
- Life Cycle Management (LCM) (published in August 2015)
- Life Cycle Sustainability Assessment (LCSA)
- Life Cycle Assessment Worldwide

More information about this series at http://www.springer.com/series/11776

Matthias Finkbeiner
Editor

Special Types of Life Cycle Assessment

Springer

Editor
Matthias Finkbeiner
Chair of Sustainable Engineering
Institute of Environmental Technology
Technische Universität Berlin
Berlin, Germany

ISSN 2214-3505 ISSN 2214-3513 (electronic)
LCA Compendium – The Complete World of Life Cycle Assessment
ISBN 978-94-017-7608-0 ISBN 978-94-017-7610-3 (eBook)
DOI 10.1007/978-94-017-7610-3

Library of Congress Control Number: 2016947060

Printed on acid-free paper

This Springer imprint is published by Springer Nature
The registered company is Springer Science+Business Media B.V. Dordrecht

Preface

The title of this volume of the *LCA Compendium* book series reads *Special Types of Life Cycle Assessment*. This may raise immediate questions about what is actually meant by special types of LCA and which types of LCA are supposed to be covered by this term. Let me give you some background first, before I try to answer this question more concretely than the usual LCA experts' answer: "…it depends…."

When the *LCA Compendium* was conceived and as a follow-up to the introductory volume *Background and Future Prospects in Life Cycle Assessment* (published 2014), the series editors, Walter Klöpffer and Mary Ann Curran, planned an overall structure of individual topics/volumes that could be clustered into two main categories:

1. Volumes that basically address the classical four phases of LCA as defined by the ISO:

   Goal and scope definition (published 2016)
   Inventory analysis
   Impact assessment (published 2015)
   Interpretation

2. Volumes that focus on applications of LCA that go beyond the environment as the only dimension and that include new developments and approaches:

   - LCA application
   - Special types of LCA (published 2016)
   - Life cycle management (published 2015)
   - Life cycle sustainability assessment

While the allocation of specific topics and methods is rather straightforward for the first category, it gets a bit trickier for the latter. When I started to populate the outline of the present volume *Special Types of Life Cycle Assessment*, I tried to cover basically some main developments of "new" approaches that go beyond or were built on the basis of classical, product-related, attributional, process-based

LCA. Several of them were driven by the trend toward – mainly communication or "footprint"-driven – simplification; others were the result of the trend toward sustainability-driven sophistication.

Following this approach, synergies and overlaps with other volumes of the *LCA Compendium* series became obvious. While life cycle sustainability assessment (LCSA) and the associated methods of life cycle costing (LCC) and social LCA (SLCA) would clearly fall under this generic working definition of special types of LCA, they are actually covered as a separate *LCA Compendium* volume on LCSA edited by Alessandra Zamagni, Tomas Ekvall, and Michael Martin. As a consequence, LCC, SLCA, and LCSA are not included here.

Other topics like water footprint or resource efficiency are partly addressed by the volume on impact assessment edited by Michael Hauschild and Mark Huijbregts. While they are discussed there from the viewpoint of an individual impact category within LCA, they are still included in this volume, but more from the perspective of a stand-alone approach.

In that sense, all the topics covered in this volume fall under the broad definition given above, but they obviously do not provide a complete and exclusive coverage of it. This volume contains a rather interesting potpourri of topics from carbon footprinting, water footprinting, eco-efficiency assessment, resource efficiency assessment, input-output (IO) and hybrid LCA, and material flow analysis (MFA) to organizational LCA, which are not covered by the other volumes.

It is evident that these topics cannot comprehensively be covered in a single volume, and besides the development goes on. Further volumes, which are closely related to these themes, will present new insights and new developments and will keep the whole work current.

Because this *Special Types of Life Cycle Assessment* volume is therefore indeed special, we decided to go into a bit more detail on recent trends toward special types of LCA. This is part of Chap. 1 "Introducing Special Types of Life Cycle Assessment," which also provides a detailed overview of the contents, the authors, and the individual chapters of this volume.

Nora Roberts wrote once: "If something isn't special, then it's ordinary." In that sense, I sincerely hope that this volume lives up to the "special" expectations expressed by its title and that – most importantly of all – you enjoy reading it.

Berlin, Germany Matthias Finkbeiner
1 February 2016

Acknowledgments

It is rewarding to see this volume finally published after having worked on it for about 2 years. This would have not been possible without the outstanding contributions and tremendous efforts of many colleagues.

First and foremost, I have to acknowledge all the **authors** of this volume. I am grateful that these leading scholars for the different topics committed to provide a contribution for this book. It was a great pleasure to experience the competence, motivation, and engagement of all of you – despite the different challenges involved. Those of you who delivered your contributions early in the process had to patiently wait for publication, while others had to bear with me for pushing them to complete their chapters at the very end of the process. Atsushi Inaba has to be acknowledged in particular as he was not just the main author, but basically the editor of Chap. 2 "Carbon Footprint of Products." He brought together a team of nine coauthors from various organizations in different countries, who contributed individual sections to this comprehensive overview on the development, application, and impact of CFP in different parts of the world.

Next to the authors, the **referees** (Jeroen Guinée, Yasunari Matsuno, Llorenç Milà i Canals, Shinichiro Nakamura, Sangwon Suh) deserve sincere thanks for sharing their knowledge and experience. The extensive and valuable feedback provided by them helped the authors to improve their chapters in many respects.

The comments of the referees were complemented by the final revision of the **series editors** of the *LCA Compendium*, Walter Klöpffer and Mary Ann Curran. They handled the reviewed articles out of their long experience and contributed valuable suggestions and final advice. I personally like to thank Walter and Mary Ann for their trust in me as editor of this volume.

Last, but by no means least, I have to express my cordial gratitude to Almut B. Heinrich, the **managing editor** of the *LCA Compendium*, who supported me in a really substantial way by spending plenty of hours working with authors and referees, editing the articles, and controlling the production process.

I particularly like to thank my sustainable engineering group at Technische Universität Berlin for all their support. Their commitment and professionalism allows me to do such extra projects on top of our regular duties.

Finally, a sincere thanks is given to my family for their courtesy and patience.

Contents

Contributors

Vanessa Bach Chair of Sustainable Engineering – Institute of Environmental Technology, Technische Universität Berlin, Berlin, Germany

Markus Berger Chair of Sustainable Engineering – Institute of Environmental Technology, Technische Universität Berlin, Berlin, Germany

Sylvain Chevassus French Ministry for Environnent, Energy and Sea (MEEM), Paris, France

Tom Cumberlege The Carbon Trust, London, UK

Matthias Finkbeiner Chair of Sustainable Engineering – Institute of Environmental Technology, Technische Universität Berlin, Berlin, Germany

Eunah Hong Korea Environmental Industry & Technology Institute (KEITI), Seoul, Republic of Korea

Atsushi Inaba Department of Environmental and Energy Chemistry, Kogakuin University, Tokyo, Japan

Akira Kataoka Japan Environmental Association for Industry (JEMAI), Tokyo, Japan

David Laner Institute for Water Quality, Resource and Waste Management, Vienna University of Technology, Vienna, Austria

Christian Doppler Laboratory for Anthropogenic Resources, Vienna University of Technology, Vienna, Austria

Pongvipa Lohsomboon Thailand Greenhouse Gas Management Organization (TGO), Bangkok, Thailand

Julia Martínez-Blanco Chair of Sustainable Engineering – Institute of Environmental Technology, Technische Universität Berlin, Berlin, Germany

Corinne Mercadie EMC Distribution Groupe Casino, Vitry sur Seine, France

Masaharu Motoshita Research Institute of Science for Safety and Sustainability, National Institute of Advanced Industrial Science and Technology (AIST), Tsukuba, Ibaragi, Japan

Thumrongrut Mungcharoen National Science & Technology Development Agency and Faculty of Engineering, Kasetsart University, Bangkok, Thailand

Shinichiro Nakamura Faculty of Political Science and Economics, Waseda University, Tokyo, Japan

Keisuke Nansai National Institute for Environmental Studies, Center for Material Cycles and Waste Management Research, Ibaraki, Japan

Stephan Pfister Institute of Environmental Engineering (IfU), ETH Zurich, Zurich, Switzerland

Klaus Radunsky Umweltbundesamt GmbH, Vienna, Austria

Helmut Rechberger Institute for Water Quality, Resource and Waste Management, Vienna University of Technology, Vienna, Austria

Peter Saling Sustainability Strategy, BASF SE, Ludwigshafen, Germany

Laura Schneider Chair of Sustainable Engineering – Institute of Environmental Technology, Technische Universität Berlin, Berlin, Germany

Acronyms

AADP	Anthropogenic stock extended abiotic depletion potential
ACFN	Asia Carbon Footprint Network
ADP	Abiotic depletion potential
AHG	Ad hoc group
AoP	Area of protection
AOXs	Adsorbable organic halogens
AP	Acidification potential
APC	Air pollution control
ASUE	Arbeitsgemeinschaft für sparsamen und umweltfreundlichen Energieverbrauch e.V. Germany: Comparison of heating costs in new developments, 2009
AWS	Alliance for Water Stewardship
BOD	Biological oxygen demand
BSI	British Standards Institution
CBO	Carbon business office
CDP	Carbon disclosure project
CED	Cumulative energy demand
CFs	Characterization factors
CFCs	Chlorofluorocarbons
CFL	Compact fluorescent lamp
CFP	Carbon footprint of products
CGE	Computable general equilibrium
CML-IA	CML impact assessment
COD	Chemical oxygen demand
CSI	Cement sustainability initiative
CSR	Corporate social responsibility
CTA	Ratio of water use to consumption availability
CV	Coefficient of variation
DALY	Disability-adjusted life year
DMC	Domestic material consumption

DMI	Domestic material input
DTA	Demand to availability
ECF	Elemental chlorine-free
ECM	Eco-care matrix
EDP	Ecosystem damage potential
EEA	Eco-efficiency analysis
EEIO	Environmentally extended IO
EF	GHG emission factor
EIA	Environmental impact assessment
EMAS	Eco-management and audit scheme
EMC	Environmentally weighted material consumption
EMS	Environmental management system
EoL	End of life
EPDs	Environmental product declarations
EPR	Extended producer responsibility
ESP	Economic resource scarcity potential
EWR	Environmental water requirement
EWS	European Water Stewardship
EXIOPOL	Externality data and input-output tools for policy analysis
GDP	Gross domestic product
GE	General equilibrium model
GefStoffV	Hazardous Substances Regulation Act
GHGs	Greenhouse gases
GLS	General lighting standard
GRI	Global reporting initiative
GWP	Global warming potential
HAL	Halogen lamp
HCl	Hydrogen chloride
HCs	Hydrocarbons
HF	Hydrogen fluoride
HMs	Heavy metals
ICP	International comparison program
IO-LCA	Input-output and hybrid LCA
IOT	IO table
IPCC	Intergovernmental Panel on Climate Change
ISIC	International Standard Industrial Classification of All Economic Activities
ISO	International Organization for Standardization
LCA	Life cycle assessment
LCC	Life cycle costing
LCIA	Life cycle impact assessment
LCSA	Life cycle sustainability assessment
LSFs	Life support functions
MFA	Material flow analysis

MIOT	Monetary IO table
MRIO	Multiregional input-output table
MSW	Municipal solid waste
MSY	Maximum sustainable yield
NACE codes	Nomenclature Générale des Activités Economiques dans les Communautés Européennes
NPP	Net primary production
NPV	Net present value
NSF	National Sanitation Foundation
OBIA	Overall business impact assessment
ODP	Ozone depletion potential
OEF	Organizational environmental footprint
OEFSRs	Organization environmental footprint sector rules
OLCA	Organizational LCA
OSHA	Occupational safety and health act
OSR	Old scrap ratio
PAS	Publicly available specification
PCRs	Product category rules/environmental product declarations
PDF	Potentially disappeared fraction
PEF	Product environmental footprint
PEF guide	Organization environmental footprint guide (PEF guide)
PEMFC	Proton exchange membrane fuel cell
PGMs	Platinum group metals
PIOT	Physical IOT
PLCA	Process-based LCA
POCP	Photochemical oxidation creation potential
PPP	Purchasing power parity
PUR	Polyurethane
RDF	Residue-derived fuel
REEs	Rare earth elements
ROW	Rest of the world
SCP	Sustainable consumption and production
SD	Sustainable development
SE	Statistical entropy
SEA	Statistical entropy analysis
SEEBALANCE	Social eco-efficiency analysis, trade name SEEBALANCE
SETAC	Society of Environmental Toxicology and Chemistry
SFA	Substance flow analysis
SKUs	Stock keeping units
SLCA	Social LCA
SMEs	Small- to medium-sized enterprises
TGO	Thailand Greenhouse Gas Management Organization
TMI	Total material requirement
UNCED	United Nations Conference on Sustainable Development

UNESCAP	United Nations Economic and Social Commission for Asia and the Pacific
UNFCCC	United Nations Framework Convention on Climate Change
USLP	Unilever Sustainable Living Plan
VOCs	Volatile organic compounds
WAVE	Water accounting and vulnerability evaluation
WBCSD	World Business Council for Sustainable Development
WF	Water footprint
WFN	WaterStat database
WIO-MFA	Waste input-output material flow analysis
WRI	World Resources Institute
WSI	Water stress index
WTA	Ratio of water use to water availability
WTO CTE	Committee on Trade and Environment of the World Trade Organization
WULCA	Water use in LCA

Chapter 1
Introducing "Special Types of Life Cycle Assessment"

Matthias Finkbeiner

Abstract Based on the classical methods and standards of life cycle assessment (LCA), there is recently a trend toward the diversification and proliferation of "new" life cycle-based assessment approaches. They are summarized in this volume/book under the heading of *Special Types of Life Cycle Assessment* and include:

- Carbon footprinting
- Water footprinting
- Eco-efficiency assessment
- Resource efficiency assessment
- Input-output (IO) and hybrid LCA
- Material flow analysis (MFA)
- Organizational LCA

The nature and scope of these special types of LCA are rather different. Some represent specific impact categories in the form of seemingly simplified stand-alone footprinting methods; some represent complementary modeling approaches like MFA, IO-LCA, and hybrid LCA; and some represent a broader application context (organizational LCA, OLCA) or a broader sustainability scope of the assessment (resource and eco-efficiency assessment). This volume contains state-of-the-art contributions for each of these special types of LCA by leading experts in the respective field.

Keywords Carbon footprint • CLCA • Consequential LCA • Eco-efficiency assessment • Hybrid LCA • IO-LCA • Input-output LCA • ISO 14040 • ISO 14044 • LCA • LCC • LCSA • Life cycle assessment • Life cycle costing • Life cycle sustainability assessment • Material flow analysis • MFA • OEF • Organizational environmental footprint • Organizational LCA • OLCA • PAS 2050 • PEF • Product environmental footprint • Resource efficiency assessment • SLCA • Social LCA • Water footprint

---

M. Finkbeiner (✉)
Chair of Sustainable Engineering – Institute of Environmental Technology, Technische Universität Berlin, Office Z1, Strasse des 17. Juni 135, 10623 Berlin, Germany
e-mail: matthias.finkbeiner@tu-berlin.de

M. Finkbeiner (ed.), *Special Types of Life Cycle Assessment*, LCA Compendium – The Complete World of Life Cycle Assessment,
DOI 10.1007/978-94-017-7610-3_1

Acronyms

IO-LCA	Input-output and hybrid LCA
LCC	Life cycle costing
LCSA	Life cycle sustainability assessment
MFA	Material flow analysis
OEF	Organizational environmental footprint
OLCA	Organizational LCA
PEF	Product environmental footprint
SLCA	Social LCA
WBCSD	World Business Council for Sustainable Development
WRI	World Resources Institute

1 Introduction

Life cycle assessment (LCA) is a well-established and widely used environmental management tool. The history and several milestones of its development were comprehensively described in the first volume called *Background and Future Prospects in Life Cycle Assessment* (Klöpffer 2014) of this book series *LCA Compendium – The Complete World of Life Cycle Assessment* (series editors Walter Klöpffer and Mary Ann Curran).

LCA started to boom around 20 years ago. Part of this growth came from increased application and implementation of LCA in both private and public decision-making, which was supported by well-accepted international standards (Finkbeiner et al. 2006; Finkbeiner 2013, 2014b; ISO 14040 2006; ISO 14044 2006). However, a significant additional momentum was generated by the development of "new" approaches built on the basis of classical LCA. This volume acknowledges these special types of LCA by dedicating the whole book to some of the most prominent species of this emerging trend:

- Carbon footprinting (see *Chap*. 2)
- Water footprinting (see *Chap*. 3)
- Eco-efficiency assessment (see *Chap*. 4)
- Resource efficiency assessment (see *Chap*. 5)
- Input-output (IO) and hybrid LCA (see *Chap*. 6)
- Material flow analysis (MFA) (see *Chap*. 7)
- Organizational LCA (see *Chap*. 8)

Life cycle sustainability assessment (LCSA) and the associated methods of life cycle costing (LCC) and social LCA (SLCA) are further important species of special types of LCA. However, there will be a separate volume on LCSA within the *Compendium* book series edited by Alessandra Zamagni, Tomas Ekvall, and Michael Martin, and so LCC, SLCA, and LCSA are not covered here. More recently, the EU Commission proposed methods called Product and Organizational

Environmental Footprint (PEF and OEF, respectively). They are still under development and currently undergo a pilot testing phase. If this process is completed and turns out to be successful, future volumes on special types of LCA may include contributions on PEF and OEF. As of now, the opportunities and threats of PEF and OEF are still under debate (Finkbeiner 2014a; Galatola and Pant 2014; Lehmann et al. 2015).

This introductory *Chap.* 1 describes some of the recent trends toward special types of LCA in Sect. 2. Section 3 contains an overview of the individual chapters of this volume. Section 4 concludes the chapter with a summary and outlook.

2 Recent Trends Toward Special Types of LCA

The LCA community has a fairly long tradition of methodological debates. Discussions about the best impact assessment method, the best allocation procedures for coproducts and end of life, the best database, the best uncertainty and data quality assessment, proper normalization, weighting or not, and many more are daily business for LCA practitioners. As a consequence, the first species of special types of LCA were mainly driven and inspired by methodological issues. IO-LCA (input-output LCA) and also consequential LCA (CLCA), therefore, still refer to LCA, but proposed a different methodological setting. Unfortunately, both of them were originally often oversold as a better alternative or substitute of good old standard LCA, i.e., attributional, process-based LCA in modern terms. Nowadays, it is recognized that they are more complementary than competitive. In real-world application for decision-making, standard LCA is the established tool, while IO-LCA can help as macroeconomic scale-up or screening tool, and consequential LCA can be used to simulate some future scenarios as additional sensitivity analysis.

More recently, special types of LCA were driven by different developments and trends. On the one hand, there is a trend toward – mainly communication-driven – simplification; on the other hand, there is a trend toward sustainability-driven sophistication. On the simplification side, the term “footprint” started to emerge.

Carbon footprinting was a huge driver for the market expansion of simplified “LCA.” The carbon footprint discussions led to a huge proliferation of different guidelines and standards including ISO/TS 14067 on carbon footprint of products (ISO/TS 14067 2013), the Greenhouse Gas Protocol Product Life Cycle Accounting and Reporting Standard of the World Business Council for Sustainable Development (WBCSD), and the World Resources Institute (WRI), PAS 2050, and further national guidelines from, e.g., Japan, Korea, the European Union, France, Germany, and New Zealand (Finkbeiner 2009; Finkbeiner et al. 2006).

The second generation of footprint was the *water footprint*. Next to stand-alone methods, such as virtual water, the method of the Water Footprint Network, the Global Water Tool, or the Corporate Water Gauge, many methods were developed in an LCA context (Berger and Finkbeiner 2013). Both the increasing relevance of water footprinting and the diverse methods were the drivers to develop the ISO 14046 as an international water footprint standard (ISO 14046 2014). The footprint trend is not over yet. Nowadays, there are "whatever" footprints for almost everything from nitrogen over biodiversity to bulky organizational environmental footprints.

The sophistication trend goes in the opposite direction. Rather than selecting specific environmental impacts and aspects out of the comprehensive set of impacts to be studied in classical LCA, the underlying methodology is either adapted for other sustainability dimensions like in LCC and SCLA and/or brought into the context of further sustainability dimensions like in resource and eco-efficiency assessments.

A schematic typology of these tools and developments is presented in Fig. 1.1 (Finkbeiner et al. 2010) according to an adapted pyramid of needs from Maslow (1943). While the original pyramid of Maslow has the basic physiological needs like food and water at the bottom, followed by safety needs, love and belonging, and esteem until self-actualization at the very top, the adapted version starts with the basic approach of life cycle thinking, followed by single-issue methods like carbon or water footprinting, LCA, and resource or eco-efficiency assessment up to LCSA at the top of the pyramid (see Fig. 1.1).

The interpretation of the Maslow pyramid of environmental and sustainability assessment should be similar to the original pyramid on human needs. This means that the hierarchy does not imply any ranking of which tool is better than another. It rather addresses different levels of sophistication and therefore can be used to define development paths. If organizations or stakeholders have just started with integrating sustainability considerations into their processes and practices, life cycle thinking is a good starting point. If climate change is the most relevant issue in some parts of the globe and water scarcity in others, it is feasible and pragmatic to start the more quantitative assessments with the respective single-aspect tools. Once this is done, the next step for these organizations is rather obvious, and they can develop more complete and comprehensive environmental assessments in the form of real LCA. Once this level is reached, the other sustainability dimensions can be integrated (Finkbeiner et al. 2010).

In that sense, the two LCA megatrends introduced above – simplification and sophistication – go in different directions, but they are still complementary from an application perspective. The Maslow's pyramid for life cycle-based environmental and sustainability assessment approaches unites both trends in a common framework. Simplification approaches are good to reduce the entry barrier to start working with quantitative life cycle-based assessment tools, and the sophistication approaches open up new application fields for those organizations which already have implemented LCA.

Last, but not least, a more recent development which could develop a new trend is the application of LCA on the organizational level. This "new member of the LCA family" (Martinez-Blanco et al. 2015a, b, c) takes up ideas and concepts toward product- and organization-related environmental assessment and management tools developed in the 1990s (Finkbeiner et al. 1998). It seems that time has

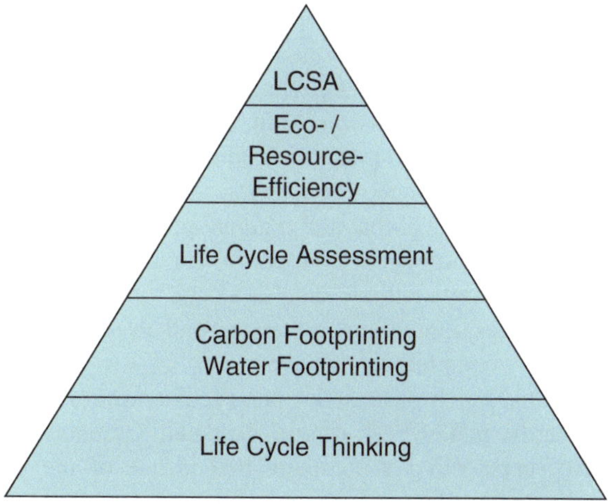

Fig. 1.1 Adaptation of Maslow's pyramid of human needs for life cycle-based environmental and sustainability assessment approaches (Reproduced from Finkbeiner et al. 2010)

come to expand the traditional product orientation of LCA toward an organizational perspective as the benefits and the potential of the life cycle approach are not limited to an application on products. The publication of ISO/TS 14072 on organizational LCA (ISO/TS 14072 2014) and the guidance document of the UNEP/SETAC Life Cycle Initiative are expected to support a growing application of LCA for organizations (UNEP 2015). Organizational LCA might also be a key catalyst for the application breakthrough of social LCA as most social issues and indicators are typically managed on the organizational level (Martinez-Blanco et al. 2015c).

3 Overview of the Volume

Chapter 2 describes the carbon footprint of products in several regions of the world. For this contribution, Atsushi Inaba from Kogakuin University in Tokyo did an outstanding job by providing an up-to-date status of this topic. He brought together a great team of coauthors representing key actors in this field including Sylvain Chevassus (French Environment Ministry, Paris, France), Tom Cumberlege (The Carbon Trust, London, United Kingdom), Eunah Hong (Korea Environmental Industry and Technology Institute, Seoul, Republic of Korea), Akira Kataoka (Japan Environmental Management Association for Industry, Tokyo, Japan), Pongvipa Lohsomboon (Thailand Greenhouse Gas Management Organization, Bangkok, Thailand), Corinne Mercadie (EMC Distribution Group Casino, Marne-la-Vallée, France), Thumrongrut Mungcharoen (National Science and Technology Development Agency and Kasetsart University, Bangkok, Thailand), and Klaus Radunsky (Umweltbundesamt, Vienna, Austria). The chapter starts with a historic overview of the developments and provides details of the carbon footprint approaches and programs in the UK, France, Japan, Korea, and Thailand.

Chapter 3 focuses on water footprinting and addresses the question of "how to count the drops and assess the impacts." The author team of Markus Berger (Technische Universität Berlin, Germany), Stephan Pfister (ETH Zurich, Switzerland), and Masaharu Motoshita (Agency of Industrial Science and Technology, Tsukuba, Japan) represents leading scholars in the field with complementary competencies. Their contribution starts with a status of water resources and demands from a global and regional perspective followed by a necessary clarification of terminology. A core part is the discussion and comparison of the different water footprint methods, databases, and tools. Lessons learned from water footprint case studies, remaining challenges, and an outlook including the consensus model WULCA complete this chapter.

Chapter 4 broadens the perspective toward eco-efficiency assessment. Peter Saling from BASF SE in Ludwigshafen, Germany, is one of the key actors in this field representing a company that is one of the pioneers and most prominent supporters of the eco-efficiency concept. In this chapter, a general introduction to the concept is followed by an introduction of the so-called sustainability assessment toolbox developed by BASF. This includes the BASF-specific type of eco-efficiency analysis plus adaptions of it like the so-called SEEBALANCE and AgBalance applications. Several case studies and examples are presented in order to demonstrate both the method as such and its application potential.

Chapter 5 addresses multidimensional LCA perspectives in the form of resource efficiency assessment. Laura Schneider and Vanessa Bach from my group at Technische Universität Berlin, Germany, provide a state-of-the-art review of this field. After a proper definition and classification of "resources" in the context of LCA, the current methods for the assessment of abiotic and biotic resource use in LCA are introduced. The shortcomings of the more established topic of abiotic resources with regard to the inherent property of materials, current reserves, and/or annual extraction rates as well as future consequences of resource extraction are discussed. The chapter concludes with an overview of research needs and proposed methodological developments for abiotic resource efficiency assessment and especially for the less developed area of biotic resources.

The fundamentals of input-output and hybrid LCA are covered in *Chap.* 6. Shinichiro Nakamura (Waseda University, Tokyo, Japan) and Keisuke Nansai (National Institute for Environmental Studies, Tsukuba, Japan) are leading contributors to this special type of LCA. Their comprehensive chapter introduces the basics of input-output analysis for LCA from a simple input-output model with one sector to complex multiregional extensions. Among others, the concepts of environmentally extended IO, different types of hybrid IO-LCA, and the waste IO (WIO) model are introduced. Tools and several databases for IO-LCA are presented and discussed. Several case studies demonstrate the potential and applicability of the approaches elaborated in this chapter.

Chapter 7 covers material flow analysis (MFA). David Laner and Helmut Rechberger from Vienna University of Technology, Austria, represent one of the leading groups on this topic established by Paul Brunner. They introduce the basic terms and procedures of MFA methodology from the selection of system boundaries, flows, substances, and processes to the presentation of results. Data reconciliation and uncertainty analysis are covered as well as the differences between static and dynamic MFA. Options for application including resource efficiency evaluation, identification of sources, sinks and final sinks, national materials accounting, or environmental impact assessment are introduced. Last, but not least, the combination of MFA and LCA is presented as a promising approach for environmental decision support.

The final *Chap.* 8 of this volume is dedicated to the life cycle assessment of organizations, such as organizational LCA. Julia Martínez-Blanco (Technische Universität Berlin, Germany; now at Inèdit, Barcelona, Spain), Atsushi Inaba (Kogakuin University, Tokyo, Japan), and myself represent the core leadership team of the flagship project of the UNEP/SETAC Life Cycle Initiative on this topic. As lead authors of the associated Guidance on Organizational LCA (UNEP 2015), we cover the topic also for this volume. After a brief sketch of the way toward LCA of organizations and an overview of existing initiatives for LCA of organizations, some of the main methodological issues of organizational LCA are discussed. They include the main differences with product LCA, reporting organization, reporting flow, system boundary, types of data, and prioritization of data collection efforts. For the practical implementation of organizational LCA, an approach of different implementation pathways is proposed. The chapter concludes with a presentation of early adopted case studies of Accor as a French international hotel group present in 92 countries with more than 3500 hotels and Unilever as an Anglo-Dutch multinational fast-moving consumer goods company with a wide-ranging portfolio in foods, household, and personal care products of around 400 brands.

4 Conclusion and Outlook

The establishment of the international standards of LCA (ISO 14040 series) was crucial for the broad acceptance of LCA all around the world and by all stakeholders. The standards contributed significantly to the transition of LCA from an academic toy or misused greenwashing machine toward a serious, robust, and professional tool to support decision-making in public and private organizations (Finkbeiner 2014b). At the same time, part and consequence of this success story is the increasing development of spin-off methods and tools built on LCA, which we call in this volume "Special Types of Life Cycle Assessment."

The special types of LCA address particular parts of LCA methodology or modeling, respectively, complementary methodologies (e.g., IO-LCA, MFA), new objects for the analysis (e.g., organizational LCA), simplified or single-issue types of LCA (e.g., carbon or water footprinting), or expanded versions of purely environmental LCA by addressing or including further sustainability dimensions (resource efficiency, life cycle costing, social LCA, life cycle sustainability assessment).

All these developments open up new opportunities for a credible and robust use of LCA for real-world decision-making in the sense of life cycle management and life cycle sustainability management (Finkbeiner 2011; Baitz et al. 2013). However, they may also pose certain threats to LCA as the core methodology if the proliferation trend leads to reduced consistency, scientific robustness, and information quality. More choice in the toolbox is a good opportunity for tailor-made solutions, but bears the risk of confusion, arbitrary choices, and inefficient competition.

Individualization and big data are among the megatrends of our societies. In that sense, it is more than likely that we will see a further blooming of special types of LCA. It is the challenging task of the global LCA community to support its growth by being as open, innovative, and market responsive as possible, but as conservative as necessary to safeguard the hard-earned credibility of LCA and its offspring. LCA and its special types are still more of a family business than an incorporated enterprise. Therefore, responsible growth and managing risks for the core product LCA appear more sustainable than jumping on every gravy train and selling out LCA to the max.

References

Baitz M, Albrecht S, Brauner E, Broadbent C, Castellan G, Conrath P, Fava J, Finkbeiner M, Fischer M, Fullana i Palmer P, Krinke S, Leroy C, Loebel O, McKeown P, Mersiowsky I, Möginger B, Pfaadt M, Rebitzer G, Rother E, Ruhland K, Schanssema A, Tikana L (2013) LCA's theory and practice: like ebony and ivory living in perfect harmony? Int J Life Cycle Assess 18:5–13

Berger M, Finkbeiner M (2013) Methodological challenges in volumetric and impact-oriented water footprints. J Ind Ecol 17:79–89

Finkbeiner M (2009) Carbon footprinting – opportunities and threats. Int J Life Cycle Assess 14 (2):91–94

Finkbeiner M (2011) Towards life cycle sustainability management. Springer, Berlin/Heidelberg. ISBN 978-94-007-1898-2

Finkbeiner M (2013) From the 40s to the 70s—the future of LCA in the ISO 14000 family. Int J Life Cycle Assess 18:1–4

Finkbeiner M (2014a) Product environmental footprint—breakthrough or breakdown for policy implementation of life cycle assessment? Int J Life Cycle Assess 19:266–271

Finkbeiner M (2014b) The international standards as the constitution of life cycle assessment: the ISO 14040 series and its offspring. In: Chapter 3 of Klöpffer W (ed) Background and future prospects in life cycle assessment. LCA compendium – the complete world of life cycle assessment, Springer, Dordrecht, pp 85–106, doi 10.1007/978-94-017-8697-3

Finkbeiner M, Wiedemann M, Saur K (1998) A comprehensive approach towards product and organisation related environmental management tools. Int J Life Cycle Assess 3:169–178

Finkbeiner M, Inaba A, Tan RBH, Christiansen K, Klüppel H-J (2006) The new international standards for life cycle assessment: ISO 14040 and ISO 14044. Int J Life Cycle Assess 11 (2):80–85

Finkbeiner M, Schau EM, Lehmann A, Traverso M (2010) Towards life cycle sustainability assessment. Sustainability 2:3309–3322

Galatola M, Pant R (2014) Reply to the editorial "Product environmental footprint—breakthrough or breakdown for policy implementation of life cycle assessment?" by Finkbeiner M. Int J Life Cycle Assess 19(2):266–271. Int J Life Cycle Assess (2014) 19:1356–1360

ISO 14040 (2006) Environmental management – life cycle assessment – principles and framework. Geneva

ISO 14044 (2006) Environmental management – life cycle assessment – requirements and guidelines. Geneva

ISO 14046 (2014) Water footprint – requirements and guidelines. Geneva

ISO/TS 14067 (2013) Carbon footprint of products – requirements and guidelines for quantification and communication. Geneva

ISO/TS 14072 (2014) Environmental management – life cycle assessment – requirements and guidelines for organizational life cycle assessment. Geneva

Klöpffer W (2014) Background and future prospects in life cycle assessment. LCA compendium – the complete world of life cycle assessment. Springer, Dordrecht. doi:10.1007/978-94-017-8697-3, 362 pp

Lehmann A, Bach V, Finkbeiner M (2015) Product environmental footprint in policy and market decisions: applicability and impact assessment. Integr Environ Assess Manag 11:417–424

Martínez-Blanco J, Inaba A, Finkbeiner M (2015a) Scoping organizational LCA – challenges and solutions. Int J Life Cycle Assess 20:829–841

Martínez-Blanco J, Inaba A, Quiros A, Valdivia S, Milà-i-Canals L, Finkbeiner M (2015b) Organizational LCA: the new member of the LCA family – introducing the UNEP/SETAC Life Cycle Initiative guidance document. Int J Life Cycle Assess 20:1045–1047

Martínez-Blanco J, Lehmann A, Chang YJ, Finkbeiner M (2015c) Social organizational LCA (SOLCA) – a new approach for implementing social LCA. Int J Life Cycle Assess 20:1586–1599

Maslow AH (1943) A theory of human motivation. Psychol Rev 50:370–396

UNEP (2015) Guidance on organizational life cycle assessment. Life Cycle Initiative, United Nations Environment Programme and Society for Environmental Toxicology and Chemistry, Paris, France, http://www.lifecycleinitiative.org/wp-content/uploads/2015/04/o-lca_24.4.15-web.pdf. Accessed 07 July 2015

Chapter 2
Carbon Footprint of Products

Atsushi Inaba, Sylvain Chevassus, Tom Cumberlege, Eunah Hong, Akira Kataoka, Pongvipa Lohsomboon, Corinne Mercadie, Thumrongrut Mungcharoen, and Klaus Radunsky

Abstract According to ISO 14040:2006 and ISO 14044:2006, the carbon footprint of products (CFPs) is the system to calculate the category indicator of the targeted product for the global warming potential or "climate change" in life cycle assessment.

There are many LCA studies focusing on greenhouse emissions. However, it is quite new to show consumers the calculation results on the shelves of supermarkets.

CFP started in the UK, and many countries followed. In this chapter, the background of CFP, the aims, and the relation to type 3 label known as ISO 14025:2006 are described in Sect. 2, followed by the general procedures of CFP and methodological issues of CFP.

A. Inaba (✉)
Department of Environmental and Energy Chemistry, Kogakuin University, 163-8677 Tokyo, Japan
e-mail: a-inaba@cc.kogakuin.ac.jp; inaba3423@gmail.com

S. Chevassus
French Ministry for Environnent, Energy and Sea (MEEM), La Défense cedex, 92055 Paris, France

T. Cumberlege
The Carbon Trust, SE1 9NT London, UK

E. Hong
Korea Environmental Industry & Technology Institute (KEITI), 122-706 Seoul, Republic of Korea

A. Kataoka
Japan Environmental Association for Industry (JEMAI), 101-0044 Tokyo, Japan

P. Lohsomboon
Thailand Greenhouse Gas Management Organization (TGO), 10210 Bangkok, Thailand

C. Mercadie
EMC Distribution Groupe Casino, 94400 Vitry sur Seine, France

T. Mungcharoen
National Science & Technology Development Agency and Faculty of Engineering, Kasetsart University, 10900 Bangkok, Thailand

K. Radunsky
Umweltbundesamt GmbH, Spittelauer Lände 5, 1090 Vienna, Austria

M. Finkbeiner (ed.), *Special Types of Life Cycle Assessment*, LCA Compendium – The Complete World of Life Cycle Assessment,
DOI 10.1007/978-94-017-7610-3_2

As the consumers can now compare CFPs directly in the store, it is needed to clarify the rules for the calculation and communication of CFP. In order to develop the internationally harmonized methodology of CFP, ISO/TS 14072 was published in 2013. In Sect. 3, the main discussion points are introduced, which were compromised when ISO/TS 14067:2013 was published.

In Sects. 4, 5, 6, 7, 8, and 9, the experiences of CFP in the UK, France, Japan, Korea, and Thailand are introduced.

CFP is the evaluation tool of the product focusing only on global warming, designed as "single criteria." Recently, the evaluation tool of the organization focusing on global warming has been paid attention, which is called "Organizational CFP" (ISO/TR 14069:2013). Moreover, the tools to evaluate more environmental categories than only global warming of the product and/or the organization are called "multicriteria."

In this chapter, the current status of CFP in the world is overviewed, and the outlook of CFP in the future is discussed. One of the reasons why CFP obtained such high attention in the society is that the consumers can see and compare CFPs directly in the store. CFP is expected to be a communication tool between producers and consumers. This is true in the present and will be true in the future.

Keywords LCA of organizations (ISO 14072) • Carbon footprint of products • CFP • Climate change • CO_2 • Consumer acceptance • Eco-label • European Commission • GHG Protocol • Global warming • Greenhouse gases • ISO 14020:2000 • ISO 14025:2006 • ISO 14040:2006 • ISO 14044:2006 • ISO/TR 14069:2013 • ISO/TS 14067:2013 • ISO 14072 • Multicriteria • PAS 2050 • PCR • Product category rules • Publicly available specification • Single criteria • Single impact • UNEP/SETAC Life Cycle Initiative

Acronyms

ACFN	Asia Carbon Footprint Network
AHG	Ad hoc group
BSI	British Standards Institution
CBO	Carbon Business Office
CFP	Carbon footprint of products
CSR	Corporate social responsibility
EF	GHG emission factor
EPD	Environmental product declaration
GHG	Greenhouse gases
GWP	Global warming potential
IPCC	Intergovernmental Panel on Climate Change
PAS	Publicly available specification
PCRs	Product category rules/environmental product declarations
SCP	Sustainable consumption and production

SKU	Stock keeping units
TGO	Thailand Greenhouse Gas Management Organization
UNCED	United Nations Conference on Sustainable Development
UNESCAP	Economic and Social Commission for Asia and the Pacific
UNFCCC	United Nations Framework Convention on Climate Change
WBCSD	World Business Council for Sustainable Development
WRI	World Resources Institute
WTO CTE	Committee on Trade and Environment of the World Trade Organization

1 Introducing "Carbon Footprint of Products"

Atsushi Inaba
Department of Environmental and Energy Chemistry, Kogakuin University, 163-8677 Tokyo, Japan

1.1 Outline of the Chapter

This chapter overviews the status of the "carbon footprint of products" (CFPs). It is presented in six main sections. First, the features, the aims, and the general implementation method of CFP are introduced, followed by the methodological issues of CFP discussed in Sect. 2. In Sect. 3, the main discussion points in connection with the publication of ISO/TS 14067:2013 (ISO 2013a) are explained including arguments from several countries. In Sects. 4, 5, 6, 7, 8, and 9, the world activities on CFP are introduced in the UK, France, Japan, Korea, and Thailand.

1.2 What Is Carbon Footprint of Products?

The CFP displays the greenhouse gas emissions associated with the life cycle of commercial goods on the packages in supermarkets. This development started in the UK in 2007. PepsiCo was one of the first companies to calculate, certify, and display the product carbon footprint on their potato chips. Since 2007, over 28,000 individual stock keeping units followed, independently certified by Carbon Trust for 178 organizations around the world, which means in 27 different countries and on every continent (Fig. 2.1). Also in France, Casino, in one of the supermarket chains, started in 2008 the sales of some foods showing the greenhouse gases on their packages (Fig. 2.2).

Fig. 2.1 Carbon footprint declaration from Carbon Trust, UK, when CFP started in 2007–2008 (Photo from Tom Cumberlege, Carbon Trust, see Sect. 4)

Fig. 2.2 Carbon footprint declaration from Casino, France (Photo from Corinne Mercadie, see Sect. 6)

Following the activities in the UK and France (Sects. 4, 5, and 6), Japan, the Republic of Korea, and Thailand launched the CFP pilot project (Sects. 7, 8, and 9).

The CFP shows the greenhouse gases (GHGs) of all life cycle stages of the product "from cradle to grave." It is the method of "life cycle assessment (LCA)" shown by ISO 14040:2006 (ISO 2006a) and ISO 14044:2006 (ISO 2006b).

1.3 The Carbon Footprint of Products as Example of Type 3 Label

The environmental label to show the result of an LCA study is called "type 3 label." ISO 14020:2000 (ISO 2000) describes three types of the environmental label. First, type 1 is the label that the third party guaranteed that it has passed the specific criteria required by the program. Second, type 2 is the self-declaration label of the environmental information by the producer. Type 3 is the label showing the results of the LCA study certified by the third party, but it does not compare them with the results of other products. The implementation procedures of type 3 label are indicated in ISO 14025:2006 (ISO 2006c).

For example, the International System EPD (environmental product declaration) has been operated in Europe mainly in Sweden (EPD 2015). Also, the EcoLeaf Program has been operated in Japan since 2002, where the calculation results of not only GHGs but also SO_x, NO_x, COD, etc. using the LCA of about 477 industrial products such as copy machines are disclosed in the web of JEMAI (Japan Environmental Management Association for Industry) (JEMAI 2015a).

In 1999, the Global Environmental Declaration Network (GEDnet) was founded as an international nonprofit association of EPD program holders among organizations in Denmark, Germany, Norway, Sweden, Chinese Taipei, Japan, and Korea and an organization in the USA recently joined (GEDnet 2015). EPD program holders are also increasing outside the GEDnet.

The characteristics of type 3 label such as EPD and EcoLeaf are as follows:

(a) Their aim is to assess the holistic environmental impacts not only on the climate change but also on other environmental impacts such as ozone layer depletion, acidification, and eutrophication.
(b) Their target products are mainly industrial products from business to business.
(c) Their results are published mainly on the Internet.

In contrast, the CFPs are as follows:

(a) They focus only on the climate change by GHGs.
(b) Their targets are mainly daily goods and foods sold in the supermarket, i.e., the products from business to consumers.
(c) Their results are disclosed mainly on the packages of the products.

1.4 The Carbon Footprint of Products as an Application of Life Cycle Assessment

Previous LCA studies in the private sector have been conducted to disclose environmental product information in their environmental reports or their CSR (corporate social responsibility) reports. Due to the many restrictions for comparative assertions in ISO 14044:2006 (ISO 2006b), it is almost prohibited. The CFP is a tool to disclose the GHGs of the products calculated by LCA, using the procedure of the type 3 label by ISO 14025:2006 (ISO 2006c).

1.5 Aims of Carbon Footprint of Products

Although the reduction of GHGs has been conducted in the industrial sectors, the transportation sector and the business and domestic sector become more urgent because the GHGs in these sectors are increasing continually (IPCC 2014).

The GHGs in these sectors are directly related to the consumer's behavior. For example, although the reduction of fuel consumption of automobiles is the task of the companies, the consumer's behavior such as the "eco-drive" including the idling stop on the red traffic light and "car sharing" is more important as the users want to apply these measures.

The concept and tools to reduce GHGs by consumers are well known as "sustainable consumption," which is recognized in WSSD (World Summit on Sustainable Development) in 2002 at Johannesburg, South Africa (UN 2002), as part of the "sustainable development" endorsed in UNCED (United Nations Conference on Sustainable Development) in 1992 at Rio de Janeiro, Brazil, together with "sustainable production." The Marrakech Process (UNEP 2003), which is the concrete plan toward sustainable consumption, was authorized in the experts meeting at Marrakesh in 2003; then the results were implemented under the leadership of UNEP.

To realize sustainable consumption, it is important to show the consumers the GHGs of the consumer's behavior or action, which is called as "CO_2 visualization." The CFP is one of the tools of the "CO_2 visualization," showing the GHGs of the daily goods and foods to the consumers in the supermarket, making the consumers purchase them and then moving to the sustainable society. In other words, the CFP is a tool to move to the sustainable society by the change of the consumer's behavior. Also the CFPs are expected to make the producers develop the new products with less environmental impacts and then move to the sustainable production, which is the continuous cycle of "sustainable consumption and production" (SCP).

2 Main Methodological Issues of Carbon Footprint of Products

Atsushi Inaba
Department of Environmental and Energy Chemistry, Kogakuin University, 163-8677 Tokyo, Japan

This section illustrates the general implementation procedure of CFP.

2.1 Program Operator and Product Category Rules

As mentioned in Sect. 1.3, the aim of CFP is not the comparison, but it is easy for the consumers to compare the GHGs. Therefore, it is required to be a fair and transparent implementation procedure. According to ISO 14025:2006 (ISO 2006c), the GHGs of the products which the consumers might compare shall be calculated according to the "product category rules" (PCRs).

Also, according to ISO 14025:2006 (ISO 2006c), the implementation of CFP began with the creation of PCR.

For the GHG calculation, data such as the consumption of energy and materials, i.e., the steps that are directly involved in the life cycle of products (production, use, transportation, disposal, etc.), are collected by examining the actual factories or processes. These are called "primary data."

The data that are indirectly involved, such as the production of materials and energy used in the processes of primary data, are searched generally in the LCA databases or the literature and the statistics. These are called "secondary data."

In the conventional LCA, the primary data is often referred as "foreground data" and the secondary data as "background data." However, the terms "primary data" and "secondary data" have become more popular because they are used in PAS 2050 in the UK, which is the worldwide first standard for CFP (see Sect. 2.2).

The program holder of the CFP shall establish and manage the PCRs. When the CFP is implemented in a variety of products, it becomes necessary to consider the consistency and the relevance of each PCR. In addition, the program holder shall provide the secondary data to the practitioner of the CFP. When the CFP is trusted by the consumers and the practitioner can carry out the CFP conveniently, the maintenance of secondary data and the disclosure of the PCR are indispensable. These are the most important issues in the implementation of the CFP.

2.2 Standards of Carbon Footprint of Products and Quantification

Because the CFP is the system to display the GHGs of products, the comparison of GHGs of the different products can easily be done by consumers. Therefore, in order to ensure that the CFP is carried out in a fair and equitable way, the standardization for the calculation method and the operation procedure has been taken place in various organizations.

For example, in the UK, the technical specification, "PAS 2050," was developed and published in October 2008 by cooperation of Carbon Trust, the British Standards Institute (BSI), and the Department for Environment, Food, and Rural Affairs (DEFRA) (BSI 2008). It was revised in 2011 by BSI, DEFRA, and the Department of Energy and Climate Change (BSI 2011). This document is referenced in each country as the most preceding standard, as well as the discussion basis for ISO/TS 14067:2013 (ISO 2013a) (see Sect. 3).

The CFP in the UK is the system not only to disclose the GHGs but to target their reduction. A document that indicates the requirements for the display and the disclosure of the reduction targets has been published in 2008 (Carbon Trust 2008).

In the UK, Tesco, a large supermarket chain, had actively implemented the CFP. The calculation of the GHGs of more than 500 products was performed, and more than 100 of them had been displayed in the supermarkets. Carbon Trust has supported the carbon footprinting and labeling of products via its consultancy and certification business. In order to display the carbon label, a business must calculate the footprint in accordance with PAS 2050. It is not mandatory to display the specific calculation methods used for each product that has achieved the carbon label.

The structure of Japanese PCRs of CFP is shown in Fig. 2.3. In Japan, the CFP pilot project supported by the Ministry of Economy, Trade, and Industry (METI) was carried out from 2008 to 2011, when the documents "basic guidelines of the carbon footprint of products" and "guide of establishing product category rules" have been developed. They show the criteria to establish the PCRs for all commodities and to operate the CFP. These two documents were merged into the technical specification, TS Q 0010:2009 (METI 2009) "General Principles for the Assessment and Labeling of Carbon Footprint of Products" in April 2009. Since then, the individual PCR has been developed in accordance with this standard. As of May 2015, 105 PCRs have been published.

As of June 2015, Korea's carbon footprint labeling provides 50 PCRs in total: two for general goods and 48 for individual energy-using products. General PCRs are for general goods (durables, nondurables, production goods, and services) and for energy-using products. Individual PCRs are the use scenarios of each energy-using product.

The two major reasons why KEITI developed PCRs in this way are to serve the convenience of applicant companies of carbon footprint labeling. First, it usually takes at least 6 months to a year to develop a PCR for one product, which causes the

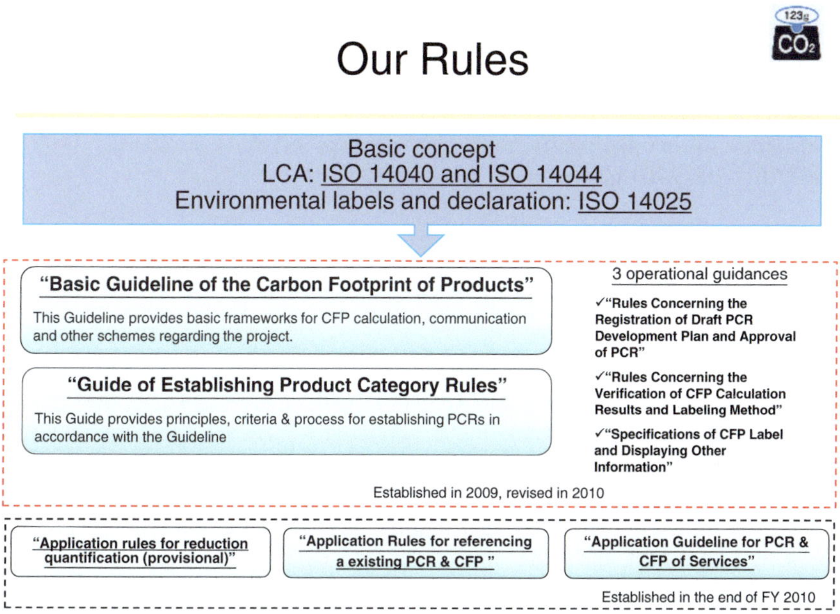

Fig. 2.3 Document structure of Japanese CFP in the trial and pilot project 2008–2012 (Two main documents in the *upper broken line box* were merged to TS Q 0010 in 2009)

complaints from the applicant companies. Second, when applying for certification of multiple products, applicants normally have to check numbers of PCRs equal to the number of products for certification; so KEITI collected all the common aspects of general products into two general PCRs and separated the specific characteristics of each energy-using product into individual PCRs. Applicants can now choose one suitable PCR among 48 different PCRs. When a product for certification shows special characteristics that are not included in general PCRs, additional internal guidelines are developed and provided to applicant companies.

In Thailand, there are two types of PCRs developed in different ways. Initially, PCRs were developed for each product by the company (or their consultant). The next PCR or the national PCR is developed by TGO with support from the National Technical Committee. As of May 2015, there are 161 approved PCRs.

The GHG Protocol, a collaboration of the World Business Council for Sustainable Development (WBCSD) and the World Resource Institute (WRI), issued the "The Product Accounting and Reporting Standard" in 2011, which is an international standard for companies to complete an LCA study, measured the GHGs of their products, and publicly reported the inventory results (WRI and WBCSD 2011). It differs slightly from the other standards which aim to disclose the GHGs of the product to the consumers. The decision on the final method of implementation has been left to the selection by the companies conducting the GHG inventory. This standard also provides guidance on target settings to reduce emissions from a product and track performance over time.

This standard was developed in parallel with ISO/TS 14067:2013 (ISO 2013a). Considerable efforts were made to harmonize these standards to meet business' needs for an international approach. These standards are generally aligned in their accounting approaches.

In order to internationally unify the concept of these standards, ISO/TS 14067:2013 (ISO 2013a) was published (see Sect. 3).

2.3 Methodological Issues for Quantification

Table 2.1 shows the main characteristics of ISO 14040:2006 (ISO 2006b), PAS 2050:2008 (BSI 2008) in the UK, JIS TS Q 0010 (METI 2009) in Japan, the Product Accounting and Reporting Standard:2011 (WRI and WBCSD 2011) by the GHG Protocol, and ISO/TS 14067:2013 (ISO 2013a).

In ISO 14044 (2006) (ISO 2006b), the LCA implementation method is determined. It means that the method of calculating the GHGs depends on the purpose of the LCA. In contrast, the main aim of the PAS 2050 is to ensure a consistent approach to GHG emission quantification that will improve the understanding and guide businesses to develop GHG mitigation initiatives. The concept of Japanese standards is to determine the calculation rules by each individual PCR because there are many different characteristics in each product category.

From Table 2.1, it is clear that the rules of cutoff, data quality, allocation, inclusion of emissions from capital goods and use phase, and GHG offset are almost the same among those standards due to the effort to calculate in detail as much as possible, but the quantification rules such as renewable electricity generation, land use change, carbon storage, and delayed emissions are different in each standard, which were discussed seriously when ISO/TS 14067 (2013) (ISO 2013a) was published.

For communication, PAS 2050 and Japanese TS Q 0010 are based on type 3 label of ISO 14025:2006, and the GHG Protocol focused mainly on external communication reports and performance tracking reports. All these communication methods are accepted in ISO/TS 14067:2013 (ISO 2013a), including type 1 shown in ISO 14024:1999 (ISO 1999a) but excluding type 2 shown in ISO 14021:1999 (ISO 1999b) because the consensus could not be established internationally regarding type 2 label, which is explained in Sect. 3.

2.4 Methodological Issues for Verification

According to ISO 14025:2006 (ISO 2006c), the calculation results of the GHGs are verified by the independent third-party reviewers facilitated by the program operator. However, as it is a time-consuming process and costly, the program operators in each country practically modified this process. For example, in Japan, the system

Table 2.1 Main features of each CFP standard

	ISO 14040/44 (LCA standard)	PAS 2050 (product carbon footprint standard)	TS Q 0010 (Japanese carbon footprint standard)	GHG Protocol (product accounting and reporting standard)	ISO14067 (product carbon footprint standard)
Life cycle approach	Basis	Adapted	Adapted	Adapted	Adapted
Environmental issue	Broad, i.e., GHG and many others	Only GHG	Only GHG (6 gases of Kyoto Protocol)	Only GHG	Only GHG (latest IPCC report)
Modeling approach	Consistent with goal and scope, mainly attributional	Attributional modeling	Attributional modeling	Attributional modeling	Attributional modeling
Partial GHG reporting	Consistent with goal and scope	Yes: requirements for passing information through the chain	Be shown as the adding information	For intermediate products, cradle-to-gate reporting is allowed, guidance given	Consistent with the scope, be justified
Functional unit	Consistent with goal and scope	Product category rules if available	Preference for product category rules	Consistent with goal and scope	Consistent with goal and scope
Cutoff rules	Consistent with goal and scope	Specific rules, 1 % of GHG emissions, 95 % complete	5 % of GHG emissions	No cutoff rule, all attributable processes included (estimated/proxy data allowed to reach 100 %)	Consistent with goal and scope
Data quality	Specific items	Specific data quality requirements	Specific rules	Pedigree matrix	ISO 14040/44
Data accepted	Consistent with goal and scope	Primary data: transparent process data Secondary data: governed by data quality rules (incl. IO)	Transparent process data, but IO data are acceptable if there is no process data	Primary data required for reporting-company operations, preference for transparent process data before	Consistent with goal and scope

(continued)

Table 2.1 (continued)

	ISO 14040/44 (LCA standard)	PAS 2050 (product carbon footprint standard)	TS Q 0010 (Japanese carbon footprint standard)	GHG Protocol (product accounting and reporting standard)	ISO14067 (product carbon footprint standard)
				IO but based on data quality assessment	
Allocation	1. System boundary expansion	Modified from ISO 14044:	Preference for product category rules (consistent with ISO/14040/44)	ISO 14040/44	ISO 14040/44
	2. Physical causality	1. Subdivide		Without the inclusion of avoided burden for system expansion (consequential)	
	3. Socio economic	2. Expand			
		3. Economic			
		Spec. rules for end of life			
Emissions from capital goods	Sensitivity and consistency analyses	Excluded	Excluded	Addressed through data quality rules, additional analysis may be included	Consistent with goal and scope
Use phase	Obligatory according to principle, but cradle-to-gate possible	Obligatory for final goods, disclose use profile. Inclusion in B-to-B measurement dependent on B-to-B boundary	Included	Cradle-to-grave obligatory for all products unless eventual fate of the product is unknown, then cradle-to-gate allowed	Consistent with goal and scope
GHG offset	Consistent with goal and scope	Cannot be included	Cannot be included	Cannot be included in inventory results	Shall not be included
Renewable electricity generation	Consistent with goal and scope	Rules to avoid double counting. Included if additionality and double counting is addressed	Cannot be included if it is not connected directly to the system	Are reported separately	To avoid double counting
Land use change	Consistent with goal and scope	Specifies procedure and provides default soil emissions per country	Not specified	Provides guidance for determining attributable impacts	Direct land use change shall be documented separately. In direct LUC should be considered

Carbon storage	Consistent with goal and scope	Included: provides calculation method	Not specified	Currently reported separately	Biomas: neutral, soil; should be included
Delayed emissions	Consistent with goal and scope	Included: describes calculation method	Not specified	Shall not be included	Shall not be included
Reduction	No guidance	Code of good practice: specific rules for calculating reduction	Could be included, if both CFPs are certified	Important goal of public reporting	Could be included, if both CFPs are certified
Verification	Peer review (independent verification)	Independent third party, other party, or self certification	ISO-14025 (third-party independent verification	Assurance required, self-assurance or third party accepted	Third-party verification or CFP disclosure report
Communication	No guidance	Requirements in code of good practice (separate from PAS 2050)	ISO-14025	External communication report required; if a target is set, the report must include progress against the target	External communication report, performance tracking report, type 1 and type 3 ISO 14025 for Type2
Comparative assertions	Possible but strict rules	Not possible	Not possible	Not possible	Not possible
Status	Published 2006	Published 2008	Published 2009	Published 2011	Published 2013

certification method has been introduced. It validates the quantification, verification, and disclosure of a CFP system set up within the company. As the system is third party verified, it enables the company to make a CFP declaration without going through the product-by-product verification process.

ISO/TS 14067:2013 (ISO 2013a) allows for communication of the CFP to the public also without third-party verification of the CFP study report provided that a much more comprehensive CFP disclosure report is made publicly available as well. It was introduced in this standard due to the restriction of ISO standardization that the third-party verification shall not be mandate. The program operator and the company to disclose CFP generally would like to have the third-party verification in order to assure the validity of the calculation.

2.5 *Methodological Issues for Communication*

As mentioned in Sect. 1.2, the CFP started in the UK by considering the many values of GHGs on the package of the product. The numerous values of GHGs should be expressed as a "functional unit," the typical function of the product or service, e.g., the GHGs per liter (volume) for orange juice and the GHGs per a package of potato chips.

The supermarket, Casino, in France introduced the CFP mainly for food, where the GHGs per 100 g were shown for all types of food.

In Japan, the amount of GHG emissions per package of product was tested when the CFP project started in 2008. Responding to the consumer survey results, different types of display methods, such as the amount of GHG emissions per wash for detergent, per a cup of coffee, per an hour for a fluorescent lamp, and per 100 g of vegetables, were accepted in 2011 during the pilot project period .

3 Discussing the Publication of ISO/TS 14067:2013 "Carbon Footprint" Greenhouse Gases: Carbon Footprint of Products – Requirements and Guidelines for Quantification and Communication (ISO 2013a)

Klaus Radunsky

Umweltbundesamt GmbH, Spittelauer Lände 5, 1090 Vienna, Austria

3.1 About the Process of Publishing ISO 14067

3.1.1 New Work Item Proposal

ISO technical committee ISO/TC 207, "environment management," and here subcommittee SC7, "greenhouse gas management and related activities", started the process on carbon footprint of products (CFPs) at its meeting in Beijing (July 2007). An important milestone was the agreement on a new work item proposal (November 2008) related to the quantification and communication of CFP.

ISO 14067 was designed as a two-part standard, part 1 "quantification" and part 2 "communication," to be developed in ISO/TC 207/SC 7/WG 2 "GHG in the value and supply chain." In developing ISO 14067, a range of standards is referred to, so ISO 14040, ISO 14044, ISO 14025, ISO 14064, and the experiences in their application.

The development of ISO 14067 faced various challenges. There was the need to combine methodological rigor with the need of broad application. The standard should be convenient due to the companies' need to monitor the CFP of a broad range of thousands of products including their progress toward decarbonization on a yearly basis. In addition, there was the challenge to combine concepts and terminology of a range of existing ISO standards. The agreement on a meaningful communication concerning the needs of purchasers and consumers turned out to be the greatest challenge, because of the diversity of products/markets/consumers.

3.1.2 Meetings of Working Group 2

ISO/TC 207 Working Group 2, "GHG management in the value or supply chain," had its first meeting in April 2008 in Vienna. The participants agreed to:

- Address "quantification" (part 1) and "communication" (part 2) of the upcoming standard
- Build on existing ISO standards such as ISO 14040 (LCA) (ISO 2006a), ISO 14044 (LCA) (ISO 2006b), ISO 14025 (environmental labels and declarations) (ISO 2006c), and ISO 14064-1 (quantification of greenhouse gas emissions of organizations) (ISO 2006d)
- Being largely consistent with GHG (greenhouse gases) Protocol, PAS 2050[1], and other national guidelines

The process under WG 2 was led by the conveners: Klaus Radunsky (Austria) and Daegun Oh (Korea) and supported by DIN, Germany, which hosted the

[1] Publicly Available Specification 2050:2008, specification for the assessment of the life cycle greenhouse gas emissions of goods and services addresses the single impact category of global warming to provide a standardized and simplified implementation of process LCA methods for assessing GHG emissions from products.

secretaries Katherina Wührl and Claudia Laabs. Overall, 107 experts from 35 member countries (including developing countries such as Argentina, Brazil, China, India, Indonesia, Malaysia, Mexico) participated.

The significant interest in ISO 14067 was demonstrated by about 50 countries participating in voting at the various stages of the ISO process. Furthermore, ISO was invited together with representatives from the GHG Protocol and from BSI (British Standards Institution) to inform delegates during the WTO CTE (Committee on Trade and Environment of the World Trade Organization) Information Session on "Carbon Footprint and Labeling Schemes" in February 2010 in Geneva on the progress of the carbon footprint standard.

Important messages emerging from the WTO session were:

- The upcoming standard will support the WTO mandate of facilitating international trade.
- The harmonization of the standards' requirements will be key to ensuring a trade-friendly standards' regime.

Strong efforts were made to achieve coherence among ISO, GHG Protocol, and BSI, in particular with respect to quantification.

Thanks to an initiative from the Swedish Standards Institute (SIS), ISO member for the country, and the Swedish International Development Authority (SIDA), the ISO process has gained significant engagement from developing countries, in particular from the Middle East and North Africa (MENA countries) and the East African Community (EAC countries).

From December 2008 to December 2009, WG 2 prepared four working drafts and the first committee draft by March 2010 after its sixth meeting in Tokyo, Japan. Due to the significant number of comments, WG 2 decided to move to ISO/DIS only after consideration of three additional committee drafts in January 2012, after its tenth meeting in Mississauga, Canada (carbon footprint of products – requirements and guidelines for quantification and communication, ISO/DIS 14067:2012). Finally, in *April 2013*, ISO/TS 14067 was approved, after the 12th and last meeting of WG 2 in Vienna, Austria.

Ten liaisons, some of them within TC 207 (e.g., SC3, SC5) and some with other TCs or with organizations outside ISO, helped to attract a broad range of experts.

Road Testing The number of companies prepared to assess their carbon footprints was growing during the period of developing the standard, as was reflected in the successful "road testing" exercises organized by WRI (World Resource Institute) and WBCSD (World Business Council for Sustainable Development) from January to June 2010. Forty-two companies representing various sectors and located in 17 countries participated in that exercise, and in doing so, they contributed to the finalization of the WRI/WBCSD's GHG Protocol.

The survey was completed by 1,018 respondents around the world, representing organizations of all sizes. The survey revealed the following:

- The majority of companies reported that the standard helped to achieve a better understanding of organizational processes.
- More than 40 % claimed to have achieved a reduction in greenhouse gas emissions.
- Thirty-two percent cited achievement of cost savings and efficiencies.

Leading companies such as Volkswagen, Sony Ericsson, Nokia, Unilever, Philips, and Timberland use advanced design tools to limit the carbon footprint of products under development. LCA Sustainable Product Design Europe, held in London in December 2010, examined strategies from these pioneers to cost-effectively incorporate life cycle thinking and sustainability into product design. The presentations showed that carbon footprint is a key parameter in designing new products.

The International Green Technology and Purchasing Conference in October 2010 in Kuala Lumpur addressed carbon footprint labels, sharing practical experiences from Japan (including also the ecopoint program), Thailand, Korea, and Taiwan and also the broader concept of eco-labeling by examples from Malaysia and Sweden (environmental product declaration) and reducing carbon footprint by life cycle performance strategies (SustainaLogic, USA).

Final Decision All these experiences contributed to the rich comments provided by member countries at the various stages of the development of ISO 14067; therefore, the final text represents a carefully drafted and well-balanced global view. The main difference, in particular to some carbon accounting rules included in PAS 2050, is that WG 2 decided to avoid subjective value judgments, e.g., with respect to discounting carbon emissions over time.

To improve user-friendliness and consistency, WG 2 decided to merge part 1, quantification, and part 2, communication. The working group allowed for a second round of balloting with the expectation that the standard would earn broad support in all countries.

Unfortunately, finally, the lack of support for ISO 14067 from some countries allowed its publication "only" as technical specification in *May 2013*. In particular, the guidance and requirements related to "communication" raised concerns.

3.2 *Main Discussion Points*

The following major challenges relate to both quantification and communication:

- Proper structure (two parts, one on quantification and another for communication or one standard)
- Coherence – building on independently developed ISO standards (related to GHG management and life cycle assessment), including diverse approaches related to verification
- Diverse interests and expectations (e.g., consumers versus industry)

- Linkage to "climate change," including potential trade barriers if CFP is introduced as policy to control GHG emissions at the national level

ISO/TS 14067 provides much more specific guidance than the underlying ISO 14044:2006 (environmental management – life cycle assessment – requirements and guidelines). The technical specification avoids excessively prescriptive language in order to effectively support carbon footprint measurement for all products and services.

ISO/TS 14067 calls for specific product category rules, including not only the specifications of ISO 14025:2006 (ISO 2006c) (environmental labels and declarations – type III environmental declarations – principles and procedures) but also other sector-specific standards or internationally agreed guidance documents related to material and product categories. ISO 14067 should therefore be seen as a framework standard. Such rules for more specific products could provide additional guidance, e.g., with respect to the functional unit, system boundary, data quality, use stage and use profile, end-of-life stage, and allocation.

ISO/TS 14067 also offers a range of communication options, including carbon footprint declarations, claims, labels, reporting, and performance tracking. The requirements on verification and the need for specific product category rules are partly dependent upon whether the communication is business to business or business to consumer oriented. When an organization decides to make a CFP communication publicly available, regardless of the chosen CFP communication option, that CFP communication shall either be verified by a third party in accordance with ISO 14025 or be supported by a CFP disclosure report which has to include all the results, data, methods, assumptions, and limitations in a transparent manner and sufficient detail to allow the reader to comprehend the complexities and trade-offs inherent in the CFP.

3.2.1 Structure of ISO/TS 14067

ISO/TS 14067 clarifies its application by including a clause specifying that it is not intended to create barriers to trade or to contradict any WTO requirements. In addition, ISO/TS 14067 clarifies that the CFP study shall not be used for a communication on overall environmental superiority because a CFP study *covers only a single impact category*, namely, "climate change." In addition, comparisons based on the CFP of different products shall not be made public unless the requirements of normative Annex D "comparisons based on the CFPs of different products," are fulfilled. This is due to the inherent limitations of the CFP approach that have been addressed in Annex B (normative), "Limitations of the carbon footprint of a product."

The final ISO/TS 14067 (2013) has the following structure:

Carbon Footprint of Products – Requirements and Guidelines for Quantification and Communication

3.3 Arguments Against the Acceptance of ISO 14067 as International Standard

As mentioned above, the limited support for ISO 14067 allowed its publication only as technical specification, because some countries, particularly developing countries, hesitated to accept it as international standard.

The main issues/concerns expressed by these countries were:

- CFP communication
  The main objective of the document should be to address climate change, not environmental communication. Communication is seen by those countries as a step toward regulation.
- Life cycle assessment (LCA)
  LCA is a discretionary subject that adds significant uncertainty to the CFP methodology, and consequently the document should be written as a guide. A standard could communicate the message that all parties concerned are in agreement, and so it could be used as a regulation reference.
- Product Category Rules (PCRs)
  PCRs are an important part of LCA regarding environmental product declaration methodology because they help to minimize market confusion by streamlining the procedures which products are evaluated for their environmental impacts and by ensuring globally consistent data collection and analysis.
  Some countries do not have the infrastructure to produce PCRs, nor does an international reference or repository exist. This process is very time-consuming and expensive and may put developing countries in a disadvantageous position.
- Trade barriers
  This point is dealt softly in the standard, and the sentence included in the document is weak; this might lead to the assumption that the document could be used as a trade barrier. Since LCA is used to assess the environmental impacts associated with all stages of a product's life, this could result in "favoring" products from certain countries or regions.
  ISO/TC207/SC7/WG2 made substantial efforts to address the trade barrier concerns by keeping the neutrality of the document and including the WTO disclaimer in clause 4; however, some members were convinced that the only way to ensure that the document is not improperly used was by not publishing it, ignoring the fact that the lack of a harmonized document would be the biggest trade barrier.

3.4 Current Situation

At the 20th ISO/TC 207 plenary meeting in Gaborone (Botswana) in June 2013, it was decided to carry out an international survey on CFP, in order to help identify the best path for the ISO/TS 14067 evolution, with the objective to:

- Investigate the diffusion of CFP practices
- Identify needs to support the development of the international standard

The first survey results have been presented in May 2014, during the ISO/TC 207 plenary meeting in Panama, showing:

- A good level of knowledge of the ISO/TS 14067, both in developed and developing countries.
- Many countries already accept ISO/TS 14067 as national TS.

- A strong interest to upgrade the technical specification to an international standard.
- Broad perception that this tool will have a high or medium level of interest for the market.

During the meeting in Panama, ISO/TC 207/SC7 decided to create an ad hoc group (AHG), in order to evaluate possible evolving scenarios for the CFP and to assess the possibility to upgrade the technical specification ISO/TS 14067 to an ISO standard. The AHG should also identify a possible path to do that.

By December 2014, the ISO 14067 survey/report has been completed, and the results, as well as those of the SC7 strategic plan survey, favor an upgrade of ISO/TS 14067 to a CFP standard. At the Sassuolo meeting, November 2014, the AHG members had consensus in recommending to convert the ISO/TS 14067 to a standard.

Meanwhile, it has been decided that CFP communication (ISO/AWI 14026) (ISO 2015) and PCR (ISO/DTS 14027) are being worked on in SC3 (ISO 14026 and 14027) and CFP verification in SC7/WG6 (ISO 14064-3).

3.5 Supporting Literature

Finkbeiner M (2009) Carbon footprinting – opportunities and threats. Int J Life Cycle Assess 14(2) 91–94

Finkbeiner M (2013) From the 40s to the 70s – the future of LCA in the ISO 14000 family. Int J Life Cycle Assess 18: 1–4

Finkbeiner M (2014) The international standards as the constitution of life cycle assessment: the ISO 14040 series and its offspring. Chapter 3"Background and Future Prospects in Life Cycle Assessment" (Klöpffer W ed). In: LCA Compendium – The Complete World of Life Cycle Assessment (Klöpffer W, Curran MA, series eds). Springer, Dordrecht, pp 85–106, doi 10.1007/978-94-017-8697-3

4 Experiences with Carbon Footprint of Products in the World: The UK

Tom Cumberlege

The Carbon Trust, SE1 9NT London, UK

4.1 Why Do Businesses Footprint and Label Products?

Businesses undertake the footprinting of products and services for many reasons. Footprinting concerns opportunities for cost reduction, product design optimization, supplier engagement, supply chain risk minimization, etc. Footprinting data provides value as it helps business to understand how to optimize the design and creation of their products or services and helps to highlight how existing operations can be adapted to improve efficiency.

A business will choose to certify and label its products to create value by influencing customers by demonstrating a product's green credentials, enhancing the brand, differentiating a product, meeting the procurement requirements of business/public sector customers, etc.

Within the UK, it is increasingly clear that different customers are influenced to different extents by such labels (at this point in time). As businesses can provide products for a range of different customers; it is therefore important to bear in mind the extent to which customers for different products can be influenced when deciding which products to prioritize for labeling.

Customers can be categorized as consumers (B2C), businesses (B2B), retailers (B2R), and public sector/government (B2G). In general, products sold B2B and B2G see the largest immediate benefits from labeling; sales through certain B2R channels also benefit, as do B2C sales of "green products" (i.e., those with a demonstrable environmental advantage over competitors). The lowest immediate benefits are observed for B2C products with no demonstrable environmental advantage over competitors).

4.2 The Development of Product Carbon Footprinting in the UK

Businesses that see compelling reasons for footprinting can relatively easily quantify and communicate the footprint of their products using the right methodology, tools, and guidance. Without a clear scheme structure that makes footprinting and certification a simple process, many businesses can be discouraged by the complexity and costs that can arise.

There is a range of factors that can make a life cycle assessment of a product or service a complex exercise for businesses such as the need to understand all the environmental impacts (not just carbon) across the life cycle and the need for specialist knowledge and management of large volumes of data.

4.3 PAS 2050

The PAS 2050 (BSI 2008) was established to accelerate the adoption of a common approach to carbon footprinting among businesses. The objective of the specification was to provide a clear methodology that businesses could adopt to increase the consistent approach to quantifying the GHG emissions of goods and services. It is important to note that this was the first single-criteria (carbon) assessment, whereas previous life cycle assessment studies evaluated multicriteria impacts.

PAS 2050 for assessing product life cycle GHG emissions was published by the British Standards Institute in 2008. It was the first published specification on product carbon footprinting in the world and was co-sponsored by the Carbon Trust and DEFRA. Although it was established in the UK, the specification was designed to establish an international approach to product carbon footprinting. Over half of the comments received during the consultation period came from international stakeholders.

The Carbon Trust provided the technical author for the PAS 2050 (BSI 2008) and developed in parallel the "Code of Good Practice on GHG emissions and reductions claims" (the Code); this is a guide to making claims about GHG emissions. As a mission-led organization, the Carbon Trust's objective for creating the guidance was to help businesses make clear, robust consistent claims about their product footprint results.

4.4 *The Carbon Label*

In parallel to the development of the PAS 2050, the Carbon Trust developed the carbon reduction label in 2007. The rationale for the label was to provide independent certification of a business' product or service to the PAS 2050. Once awarded the label, a business could demonstrate the accuracy of their footprint assessment and communicate credible, comparable data.

Since 2007, over 28,000 individual stock keeping units (SKU) have had the accuracy of their carbon footprint independently certified by the Carbon Trust for 178 organizations around the world, in 27 different countries and on every continent.

When it was launched, the Carbon Trust carbon label communicated the footprint results alongside the functional unit of the product. A functional unit is the quantified performance of a product system for use as a reference unit (according to PAS 2050:2008). The purpose was to communicate the carbon footprint of the primary function of that product or service. Additional contextual information was often provided to indicate how the consumer could compare the footprint figure against similar products. An example of this is shown in Fig. 2.4 where the footprint of fresh orange juice 360 g CO_2 per 250 ml serving is compared against the lower footprint of 260 g CO_2 per 250 ml for concentrated orange juice. The fresh orange juice has the higher footprint due to the additional energy required to chill the product both in transportation, retail, and use phase.

The label also provided information to consumers on how they could play their role to reduce the impact of the product still further. For example, the label indicates

Fig. 2.4 The carbon labels for fresh and concentrated orange juice

the carbon footprint for washing with biological detergent. The footprint of the non-biological detergent is also shown along with the potential impact that the consumer could make to reduce the emissions by washing their clothes at 30° rather than 40°.

The PAS 2050 and the Carbon Trust carbon label were launched when public interest in climate change was at its highest in 2007 and 2008. Since this time, public interest has been falling as both the economic crisis and international terrorism has occupied an increasing proportion of news coverage. In addition, many companies assumed that consumers (B2C) would drive the most demand for carbon labeling. However, over time, B2B and B2G customers have driven the volume of certification and labeling services. There are a number of reasons for this trend:

- Consumers signal (via surveys) that they are concerned about environmental issues; however, their spending habits indicate that these intentions always come after price and quality considerations.
- Consumers lack time in the retail environment to evaluate on pack labeling and make informed decisions.
- Businesses and governments are more strategic and have time and resources to evaluate purchasing choices and influence their supply chain partners via their purchasing policies.

Additionally, over this time period, the Carbon Trust has been increasing the proportion of supply chain footprinting and engagement activity to businesses seeking to reduce their upstream carbon emissions.

4.5 Evolution of the Carbon Label Scheme

While the PAS 2050 and the subsequent release of the Greenhouse Gas Protocol Product Standard provides a clear methodology and rules for footprinting, it has not yet been widely adopted by businesses. Collecting and calculating product footprints can still remain too high a cost for business despite the benefits of improved reputation, identified cost savings, and risk mitigation. Larger corporates with resources for sustainability active with few products and services have been more likely to adopt footprinting compared to smaller businesses or corporates with a large range of SKU (stock keeping units).

There are two factors that influence the cost of footprinting: understanding and accuracy. In order to have valuable footprint data that can be used to make decisions with supply chain partners, data needs to be of sufficient granularity to inform understanding of where the greatest impacts lie. Second, the quality of data needs to be of sufficient accuracy to provide confidence to business and consumer decisions. When the granularity (understanding) of the footprint is increased and the data quality (accuracy) is also increased, then the costs associated with footprinting process also increase. By slightly reducing the accuracy and granularity of the footprint to a level that still informs business decisions and provides meaningful information to consumers, the costs can be reduced significantly (Fig. 2.5).

The revision to the PAS 2050 (BSI 2011) and launch of the Greenhouse Gas Protocol Product Standard in 2011 (WRI and WBCSD 2011) allowed greater flexibility in this approach. To mirror a more business friendly approach to quantifying footprints that are "good enough" to provide confidence in decision-making and inform consumers, the Carbon Trust carbon label has also evolved to provide greater flexibility.

Feedback from consumers suggest that they welcome carbon labels as they have provided them with additional information. The benefit for a business that labels its products or services is that they have proven the steps they have taken to understand the indirect impacts of their activities across the life cycle of their products.

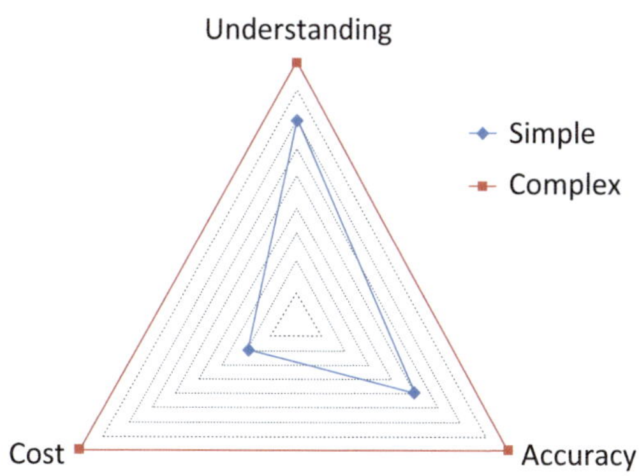

Fig. 2.5 How data granularity (understanding) and quality (accuracy) influence the cost of footprinting

Fig. 2.6 The Carbon Trust carbon labels

However, the footprint number did not necessarily *mean* anything to consumers. Consumers do not know whether a footprint result is high or low as they lack a broader context to judge these impacts. Not enough businesses are calculating product footprints to improve the contextual understanding of consumers, and governments have also not identified improving public understanding as a priority.

Consumers also have very little time to review the detailed product footprint information provided on the pack and will generally just refer to on pack symbols that signify a product's characteristics. Examples of this can be seen in both traffic light system for nutritional information or Fair trade labels.

As a result of the need to both reduce the cost and complexity for businesses and the need to improve the simplicity for consumers, the Carbon Trust's Carbon label has evolved into the formats shown in Fig. 2.6.

- The CO_2 *Measured* label can improve a business's reputation by clearly communicating its achievements in accurately measuring its carbon footprints and disclosing the results.
- The *Reducing* CO_2 label offers a simple and effective way for a business to communicate that it has measured and certified the carbon footprint of its products and services. The Reducing CO_2 label also allows it to communicate the carbon footprint measurement, and it has committed to reduce it.

  Disclosure of the carbon measurement number alongside these labels is no longer a mandatory requirement.
- The *Carbon Neutral* label is awarded to organizations measuring the life cycle impact of their product or service and purchasing verifiable offsets for these emissions in line with the PAS 2060[2] carbon neutral methodology.

[2] PAS 2060: Specification for the demonstration of carbon neutrality.

Two further improvements have been made to the Carbon label scheme to allow certifications of:

1. Single footprinting of *average SKUs*. This certification is provided to similar but not identical SKUs, e.g., crisps in a small or large bag where the average footprint could be communicated by weight or by serving size.
2. Carbon footprint *model certification*. A certified carbon footprint model allows businesses to efficiently deliver accurate and reliable product carbon footprints at scale. Carbon Trust Certification will assess a company's methods and processes for producing carbon footprints.

4.6 International Footprinting Schemes

The Carbon Trust is involved in the development of product footprinting and labeling schemes in Hong Kong, Mexico, Brazil, and Taiwan. It has also provided in-depth advice to China concerning its national scheme, as well as Malaysia.

Hong Kong Working with CMA (the Chinese Manufacturers' Association of Hong Kong) means:

- To develop and pilot a carbon product footprinting and labeling scheme for Hong Kong
- Featuring in priority order: a recognition pathway to the Carbon Trust carbon label to assist exporters and information creation which helps companies reduce costs and carbon
- The driving of behaviors toward lower carbon-impact solutions when designing, making, and purchasing products
- A domestic label for business-to-business and business-to-consumer communications

Work is underway to explore suitability to expand this scheme into China.

Mexico Working with SEMARNAT (Secretaría del Medio Ambiente y Recursos Naturales, the equivalent of the Environment Department of Mexico), means:

- To develop and pilot a carbon and optional water sustainability product footprinting and labeling scheme for Mexico
- Featuring in priority order: a domestic label to communicate the environmental performance of products both business to business and business to consumer; the driving of behaviors toward lower carbon/water impact solutions when designing, making, and purchasing products; information creation which helps companies improve the energy, resource, and water efficiency of products; and an option for recognition with the Carbon Trust carbon label.

Brazil Working with MDIC (Ministério do Desenvolvimento, Indústria e Comércio Exterior, the equivalent of the Environment Department of Brazil) means:

- To develop a carbon footprinting and optional water sustainability and certification and labeling scheme for products in Brazil
- Piloting the approach first with the aluminum, glass, chemical, cement, steel, professional services, and poultry sectors and then across other sectors
- Featuring in priority order: information creation which helps companies improve the energy, resource, and water efficiency of products; communication of the environmental performance of products especially business to business and also via a domestic label; the driving of behaviors toward lower carbon/water impact solutions when designing, making, and purchasing products; and an option for a recognition pathway to the Carbon Trust carbon label.

Taiwan Working with ITRI (the Industrial Technology Research Institute of Taiwan) means to develop and pilot a mutual recognition scheme between the Taiwan product carbon footprint label and the Carbon Trust carbon label.

4.7 Achieving Critical Mass Through Sharing and Integration

Understanding the carbon and water impact of products and services is extremely useful to companies and their customers but is currently undertaken by too few. What is needed is scale. Scale will greatly help meet economic and environmental needs, but scale needs a low-cost, simple, efficient approach and to be well recognized and understood by both producers and consumers.

The schemes above to varying extents incorporate the Carbon Trust's 8 years' experience working with companies and scheme bodies, to make product footprinting and certification simpler and cheaper. These schemes are designed to significantly reduce costs while maximizing value for participating companies, as the only route to achieving high-volume footprinting, which in turn provides companies with the data needed to make the huge energy and resource efficiency savings which are only achievable when supply chains work together to more efficiently deliver the needs of their ultimate customer.

As part of its mission, Carbon Trust is committed to bringing down the cost and complexity of footprinting and certification by working with partners internationally to share modular elements of schemes. Such integration means sharing any of the following to a greater or lesser extent: scheme organization, sales, scheme methodology, product category rules (PCRs), tools, calculators, data, verification methods, and labels, which simultaneously reduce costs by sharing infrastructure, improve comparability by aligning how results are calculated, and enhance value derived from labels.

The establishment of such a common framework also enables the use of alternative labels for different markets which are better recognized by consumers or the establishment of a consolidated label. This allows a viscous circle to be broken – ensuring the volume of labels breaks through the tipping point where consumers recognize and understand these labels, which provide enhanced value to companies using them and which further increase volumes.

These schemes are based upon international standards GHG Protocol Product Standard (WRI and WBCSD 2011a) and ISO14067:2013 (ISO 2013a), with significant enhancements layered upon these to maximize value and minimize cost and complexity.

4.8 Conclusion

Businesses undertake product carbon footprinting to help them understand their value chain impacts. Footprint data can inform supplier engagement strategies, product design decisions, and sustainability reporting. It is only when the business benefits of footprinting are added to the business benefits of certification and labeling and many businesses readily see financial benefits of undertaking footprinting and labeling.

Justifying the cost of labeling has always been easier for high-revenue products and always will be, but as all surveys see a continuous rise in the business benefit of labeling each year, across all products, sold to all customer groups, the number of products for whom this approach has immediate returns is continually rising.

Over time B2C will come to be a critical driver for certification and labeling; however, over recent years, the two main forces driving global demand for certification and labeling have been (1) public sector regulations and (2) private sector business demand. Both have affected the growth of product certification and labeling around the world.

As discussed in this section, this trend indicates that it is mainly business transactions, not consumer demand, that are driving the uptake of product footprinting and certification.

Public sector regulatory demands for certification processes and labeling are expected to advance within the sustainable purchasing and procurement domain, whereas private sector demand is expected to escalate strongly in the coming years.

5 Experiences with Carbon Footprint of Products in the World: France–The French Initiative on Consumer Product Environmental Footprinting and Communication

Sylvain Chevassus

French Ministry for Environnent, Energy and Sea (MEEM), La Défense cedex, 92055 Paris, France

5.1 *Context and Goals of the Initiative*

Highlighted by the National Environment Round Table ("Grenelle de l'environnement"), household consumption of goods and services represents a major challenge in reducing our impact on the environment, in terms of combating the greenhouse effect and moving toward a more energy and resource-efficient economy. While the goal is ambitious (including an environmental component in consumer purchasing choices and providing the entire production and distribution chain with new indicators should in turn intensify their efforts to better ecodesign products), the means to achieve this are also important. This applies both as regards the regulation laws and the partnership work initiated within the ADEME-AFNOR stakeholder platform, which is dedicated to the LCA-based methodological developments.

In a context where some private front-runners had opened the way with carbon labels (retailers, e.g., Casino) and where consumer expectations on sustainable consumption were growing, two environmental framework laws were passed in 2009–2010 that contain articles introducing a general right for the consumer to be informed about the environmental impacts of products and organizing a national pilot of 1 year. In the meantime, since 2008, the ADEME-AFNOR stakeholder platform has been developing a general environmental footprinting methodology (BPX 30-323) and product category rules (PCRs) – 28 PCRs to date. ADEME is also constructing a public generic product life cycle database, as well as calculators. These tools aim to facilitate a general implementation.

5.2 *Governance of the Initiative*

The Ministry for Environnent, Energy and Sea (MEEM) is responsible for the general policy and for the regulatory context.

At the technical level, in order to implement environmental information of products, ADEME and AFNOR have set up a general stakeholder platform, including more than 600 organizations and experts. This platform supervises several sector working groups (detergents, cosmetics, furniture, textile, food products, toys, etc.), a methodological working group (in charge of the horizontal footprinting methodology), and a communication format working group. The platform validates the documents produced by the different working groups.

5.3 General Platform (ADEME-AFNOR Stakeholder Platform)

The objective of this initiative is to progressively allow the consumer to use the information concerning the environmental impacts of a product throughout its life cycle as a choice criterion when deciding on a purchase. This environmental communication should also allow comparison of products belonging to the same category and, when relevant, between product categories. As a consequence, ADEME was asked to lead the elaboration of methodologies to assess the environmental impacts of mass market products and to develop a generic database that quantifies the environmental impacts of products to make the assessments possible.

The environmental information has to respect the following principles:

- Packaging and product system to be studied
- Life cycle approach, from cradle to grave
- Multicriteria approach

5.3.1 General Methodology

This platform is adopted in 2009 the first version of the general methodology and principles for the environmental communication on mass market products (BP X 30-323). The key points of this document are:

- Selected approach is life cycle assessment (ISO 14040:2006 and ISO 14044:2006).
- Carbon footprint of products is required whatever the category.
- Environmental indicators are the same within a category (defined by the same functional unit).
- The number of indicators per category has to be limited (maximum five for evaluation, three for communication).
- The communication format has to be harmonized in order to facilitate comparison.
- ADEME has to develop a public secondary database in order to simplify the assessments.

This methodology is consistent with ISO 14040:2006 and ISO 14044:2006 standards. It contains requirements about key issues in LCA such as end-of-life and recycling aspects, allocation rules, cutoff criteria, and exclusions.

The BPX 30-323 has been revised twice since then, notably in order to take into account EU developments. As a result, the coherence with the PEF is improved on many aspects, even though on some issues – notably end-of-life calculation – the BPX 30-323 keeps its specificities. A new revision can be considered when the PEF itself is revised.

5.3.2 Sector-Specific Working Groups

Sector working groups were initiated in order to develop simplified (but more sector-specific) methodologies per product category for the environmental impact assessment of products. These working groups involve representatives from industry, retailers, consultants in life cycle assessment, nongovernmental organizations (environment and consumers), administrations, and technical experts.

The objective is that a consumer should have the same kind of indicators on products that he/she could want to compare. These working groups produce the product category rules, on the basis of LCAs and sector expertise.

About 11 sector working groups have produced 28 PCRs to date (shoes, wood furniture, shampoos, sofas, toilet paper, TV, bedding, disposable nappies, food, drinks, pet food products, detergents, printed paper, bicycles, shower gel, textiles, milk, coffee, mobile phones, hotel night, etc.). The PCRs must be consistent with the general methodology while specifying sector approaches, indicators, and data requirements (primary, secondary, etc.).

5.3.3 Communication Format

The working group on communication format focuses on the different possibilities for the communication of environmental impacts to the consumers (how and where to communicate). This group first elaborated a comparison table in which the advantages and disadvantages of each possibility are provided. In 2015, it produced recommendations (yet to be validated at the political level) for a harmonized logo.

5.4 Database Development

As soon as the national initiative on consumer product environmental footprinting and communication was launched, ADEME was mandated to set up a national LCI/LCIA database.

Base IMPACTS® is now available online, in French and English, at: www.base-impacts.ademe.fr.

In terms of format, ADEME's database is based on the JRC's ILCD dataset format. Each process dataset is imported as an aggregated life cycle inventory (LCI) dataset and characterized through the methods recommended by the JRC, allowing the LCIA indicators to be released to the public. In terms of content, two modes have been set up to feed the database in terms of LCI datasets: purchase (mode 1) and development (mode 2). A contribution mode (mode 3) remains to be set up.

Mode 1 relies on the adaptation of existing LCI process datasets through framework contracts with PE International, Cycleco, ecoinvent, and Quantis. Fourteen subsequent contracts have already been signed, mainly with PE (and Cycleco for textile), covering intermediate systems such as electricity, transportation, plastics, wood, steel, end of life, etc. The main sectoral modeling rules beneath these datasets are described in the metadata and in the documentation available on the database website.

For sectors lacking data, ADEME has set up collaborative projects to develop LCI datasets. A unit process version of each dataset is also released (available at ADEME but not directly through the database). Choices made to model the systems are gathered and detailed through public and thorough methodological reports. Two projects are now over on agriculture (Agri-BALYSE®) and pulp and paper production with COPACEL; one is ongoing on food processing (ACYVIA®), and new ones remain to be launched with industrial partners: on chemical production, wastewater treatment plant (with the aim to allocate the impacts per chemicals), plastic or textile recycling, WEEE treatment, etc. To come along with this work on process datasets, around 1.500 aquatic ecotoxicity characterization factors have been developed following the ® model through a partnership between ADEME and Cycleco and will be soon implemented into the Base IMPACTS®.

Two environmental footprint calculators are also proposed on the Base IMPACTS® website (for shoes and TV monitors) as well as a link with ecodesign with the Bilan Produit® tool.

5.5 *National Pilot*

As requested by law, the Ministry organized a national pilot on consumer product environmental footprinting and communication in 2011–2012. The trial covered the quantification of environmental impacts and the communication of the environmental footprint to the consumer. Two hundred thirty companies are applied; 168 of them, of all sizes, were selected. All important sectors were represented, with about one-third from the food and drink area. Several foreign companies were part of the selection (Chile, Colombia, Sweden) as well as French branches of multinationals. This experiment allowed to test several issues (calculation methodologies, data, communication, consumer reaction, costs, SMEs, imported products, etc.). A governmental report (MEEM) was sent to the parliament, which concludes that the experiment was a success but several things need to be consolidated (technical tools and data, implementation costs, verifiability, etc.). The report also calls for strengthening cooperation with the EU and international levels.

5.6 Outlook

France is active at the national, EU, and international levels. The French private and public sectors are actively participating in the EU-PEF pilot. France is also involved in several UN and UNEP activities (10 YFP, UNEP/SETAC Life Cycle Initiative, international dialogue on LCA and databases. etc.) and has participated in discussions at WTO and OECD.

While seeking consistency with the EU level, France continues its national activities. Methodological and database development is ongoing. A stakeholder agreement has been found on a harmonized logo. The aim is to continue to consolidate the technical basis while potentially starting to implement on a wider scale, starting with proactive sectors – depending on political decisions and in coherence with EU developments.

6 Experiences with Carbon Footprint of Products in the World: France, Casino

Corinne Mercadie

EMC Distribution Groupe Casino, 94400 Vitry sur Seine, France

6.1 Background and Development

Due to the approach to improve the environmental profile of Casino products, the Casino Group worked on an innovative indicator, taking into account the product life cycle and resulting environmental impacts in terms of emissions of greenhouse gases. Casino's point of view was in particular: creating a methodology allowing the measure of performance for all food products. The retailer wished to detect and valorize the improvements made by the supplier.

Casino contracted BIO Intelligence Service, specialized in life cycle assessment, to develop an evaluation method adapted to the way a retailer works (quite "far" from the product food production). From there, the Carbon Index, specific to Casino, was born.

This environmental labeling project started in 2006 by an experiment of 13 products, in combination with some suppliers, BIO Intelligence Service, and the support of ADEME and NGOs.

Casino collected the primary data needed for this project from their suppliers. The secondary data were provided by the environmental expert consultant.

But it was not possible to calculate the Carbon Index of all the products, because some recipe ingredients had never been analyzed at this time, e.g., peanut, pistachio, spinach, and goat milk. Therefore, Casino charged a consultant to do a simplified life cycle assessment of these ingredients when possible.

In 2010, more than 700 products displayed this information on the packagings, creating competition between the products. It is the consumer to decide on the environment impact of his/her own food consumption, relating to the Carbon Index of the products chosen.

The only nonfood product evaluations were done about dustbin bags. Volume, resistance, and closing systems explained the Carbon Index difference for consumers.

6.2 *Definition of the Casino Carbon Index of Products*

The Carbon Index is an estimation of the amount of greenhouse gas emitted during the main stages of a product life cycle. It is expressed in gram CO_2 equivalents per 100 g of the end product.

This is the sum of five indicators corresponding to the five basic steps of the product life cycle:

1. Agricultural steps (crops, livestock, etc.): This indicator takes into account the greenhouse gas emissions from the agricultural production of raw materials. BIO Intelligence Service was responsible of this calculation based on the exact recipe of the product and additional specific data collected from the Casino suppliers (performance, loss during processing, etc.) and used robust emission factors from public databases.
2. Transport (raw materials, packagings, and end products): This indicator corresponds to greenhouse gas emissions due to the transport of raw materials from the fields and farms to Casino suppliers, as well as transport of packagings (primary, secondary, and palletizing) from the packaging manufacturers to Casino suppliers and finally transport of the end products to Casino warehouses. The calculation of this indicator includes the distances and the various possible ways of transport.
3. Packaging materials (manufacture, end of life): This indicator corresponds to greenhouse gas emissions from the production of packagings (weight of the individual components, materials used) and their end of life, e.g., the processing of the resulting waste.
4. Manufacturing (processing and product packagings): This indicator reflects the greenhouse gas emissions linked to the energy consumption of the processing site. It is based on the energy data consumption (electricity, gas, fuel oil) of the supplier concerned, taking into account the country of the manufacturing site.
5. Distribution (logistics and retail parts): This indicator calculates the greenhouse gas emissions caused by the retail and work around the products (storage, warehouses, office, transportation from warehouses to shops and from shops to consumers). It is based on the results of the Casino carbon footprint and varies

depending on product net weight and its conservation mode (ambient, fresh, frozen).

The use phase was excluded because consumers can use the same product differently (cooking, dilution, other ingredients added, etc.). For example, an egg can be hard-boiled, soft-boiled, poached, fried, and shirred and can be in the composition of biscuits, pie, etc. How to consider the use phase of such diversity?

6.3 Specifics of the Carbon Index

The specification of calculation is an essential feature of the Carbon Index.

Casino provided the suppliers a computer application called "supplier kit." It was developed by BIO Intelligence Service and calculates the three indicators "transport," "packaging," and "manufacture" from simple data (distances in kilometers, weight, etc.), specific for each product and supplier.

6.4 Optimization of the Carbon Index

In 2010, a process began to reduce the values of the Carbon Index.

Indeed, Casino wanted the clients to understand the impacts of the different products for a more sustainable consumption. But Casino also wanted to reduce the environmental impacts of products.

At the distribution level, Casino continued to work on the efficiency of its logistics system to reduce the impacts.

At the three steps related to suppliers (packaging, transportation, manufacturing), it was possible to provide several ways for achieving optimization simulations through the supplier kit.

Thus, since the establishment of the process, several products have seen their Carbon Index reduced through various means:

- Using recycled materials in packagings
- Deleting cardboard sheets on pallets
- Changing suppliers of raw materials
- Reducing the supply circuit delivery of raw materials or supplies

6.5 From Carbon Index to the Environmental Index

In 2011, the French government launched a project that worked on multi-environmental indicators, the "Environmental Index." It considers three environmental indicators: GHG, water consumption, and eutrophication.

The parameters of the project are quite similar to those of the Carbon Index, but changes are due to the fact that Casino had to comply with the French methodology ADEME-AFNOR BPX30-323. This document considers, for example, that the transport by the consumers does not need to be included in the evaluation of the retailer part. However, the use phase of the product has to be included, contrary to the initial Carbon Index.

Casino created a web tool to collect the primary data and to calculate the Environmental Index. This web tool is used since 2012 for all calculations. The manufacturers input directly the data, and Casino checks the coherence before validating them, allowing then the automatic calculation of the index.

In 2015, Casino keeps collecting manufacturer data in the web tool to continue the evaluations of products and tries, in parallel, to provide relevant information to the French work group on product environmental footprinting.

The functional unit chosen is "100 g of end product" for solid products and "100 ml of end product" for liquid ones.

All stages of the life cycle recommended in the document BPX30-323 are taken into account in the methodology of this project:

- Raw material production
- Packaging production
- Production and packaging stages
- Transport of raw materials, packagings, and end products
- Distribution of products
- Flows related to transport and energy infrastructures
- The use of the products on customer premise
- End of life cycle of end products and packagings

The specificity of the Casino Environmental Index is that the indicators are aggregated and that the rating has a reality message for the consumer.

Environmental information based solely on several environmental criteria expressed in absolute values can dampen customers or even cancel out the potential benefits of an environmental labeling initiative on products. Indeed, leaving the customer choose between the different environmental impact categories makes it very complicated: *is it better for me to opt for climate warming? Water consumption? Eutrophication?*

In order to facilitate interpretation of the results and help customers in their decision-making, LCA-based weighting-aggregation methods can be used: the results obtained for each environmental impact category are "converted" into scores which are then aggregated into a single score, thereby facilitating the comparison of several points.

Aggregation thus provides customers with simple, easy-to-understand information which can also be used as a decision-making tool to prioritize and grade the environmental stakes both by ecodesign manufacturers and policy-makers.

The Environmental Index defined by Casino represents the environmental impact of 100 g of product compared to the environmental impact of the total daily consumption of food of a French person, accounting for three impact indicators (greenhouse gas emissions, water consumption, and eutrophication) aggregated using the PRIOR® method. The Environmental Index is expressed in percentage per 100 g of product. An Environmental Index of 30 % means that the impact of 100 g of product represents 30 % of the environmental impact of the total daily food consumption of a French person. If the sum of the products consumed over 1 day shows a score greater than 100 %, it means that the customer's consumption of food products has more impact on the environment than an average French person.

The Environmental Index is aimed at providing the same level of information as the nutrition label featured on food products. The nutrition label, given per 100 g of product (or, in some cases, per portion) provides information on the nutritional value of a food giving rise to fully informed food choices.

The analogy with the nutrition label is relevant: to help customers opt for healthy/sustainable foods. It took several years for customers to master and exploit the nutrition label. It has now become a source of must-have information for customers. The same trend is to be expected for environmental information.

The Environmental Index is aimed at providing clear and objective information so that customers can make sustainable purchase choices briefed with the corresponding environmental information.

7 Experiences with Carbon Footprint of Products in the World: Japan

Akira Kataoka

Japan Environmental Association for Industry (JEMAI), 101-0044 Tokyo, Japan

7.1 Overview

In 2008, the Japanese government made a cabinet decision to approve an action plan for achieving a low-carbon society. The plan was formulated as a practical roadmap to shift toward a low-carbon society, as recommended by the Prime Minister Yasuo Fukuda's speech at the G8 Hokkaido Toyako Summit. In the same year, the Ministry of Economy, Trade, and Industry (METI) conducted the preliminary feasibility study before launching the CFP project in 2009 as one of the measures against global warming. The Japan Environmental Management Association for Industry (JEMAI) joined the project and provided its knowledge and experience garnered through its type 3 environmental labeling program (JEMAI 2015a).

In March 2012, the CFP project was completed as a pilot project and transferred to the hands of JEMAI in April. It was renamed as the CFP communication program (JEMAI 2015b) with some changes to improve the cost-effectiveness of the program. Since then, this labeling program continues to grow without any financial assistance from the government.

The program aims to visualize "carbon hotspots" in a product's life cycle as well as promote the communication between companies and consumers to accelerate the move toward a low-carbon society.

LCA is used in the CFP communication program to calculate the amount of GHG emissions associated with products, and the program is conform with ISO14040:2006 (ISO 2006a), ISO 14044:2006 (ISO 2006b), and ISO/TS 14067:2013 (ISO 2013a). There is no legal framework for the program, which means all related activities are undertaken on a voluntary basis.

7.2 The Japan Environmental Management Association for Industry (JEMAI)

JEMAI is the program holder of the CFP communication program. It is a public corporation organized by the membership of approximately 700 organizations. Established in 1962 when industrial pollution became one of the major concerns in Japan, JEMAI's activities since the outset have expanded and now include environmental assessment, technological development, surveys regarding pollution, and other global environmental issues (JEMAI 2015c).

7.3 Databases

Three databases (the basic secondary database, the distance data between countries and regions, and the heating value database) and one library (the available data library) as secondary data are disclosed to improve the convenience for calculation and the practical use of the CFP communication program.

- *Basic secondary database*: It is made up of data verified with the Basic Data Verification Criteria by the Unit Data Review Panel which consists of third-party intellectuals. The database is the result of the CFP pilot project (FY2009–FY2011) which developed the tentative database of GHG emission factors for the CFP pilot project.
- *Available secondary database*: It is a library of unit data made available for the CFP program and complements the basic secondary database.
- *Distance data between countries and regions*: This is for calculating marine, land, and air transportations between countries and regions.
- *Heating value database*: This data was utilized for the creation of the basic database and is available for estimating similar fuel classifications.

7.4 Product Category Rules

The CFP communication program uses the CFP-product category rules (CFP-PCRs), which define the basic rules of the CFP communication and CFP quantification for the respective product categories. They aim to provide interested parties with information on the conditions under which to conduct the CFP quantifications and to enhance better understanding of the communication contents.

So far, approximately 100 PCRs have been created covering ten industrial sectors: construction, food, houseware, industrial product, IT equipment, material, office equipment, printing stuff, service, and textile.

7.5 CFP Logo

A decision was made to create the CFP logo at the first meeting of the Study Group for Development and Promotion of the CFP Project held in June 2008. Proposals for the logo design were invited from the public, and the one using a kitchen scale was selected from 515 proposals (Fig. 2.7). The idea behind the design is that CO_2 (GHG) is not visible but scalable. Products carrying the CFP logo were placed on the market in October 2009.

7.6 Procedure for Using the CFP Logo

7.6.1 Application

The permission to use the CFP logo is granted through the three-step application procedure: (1) selection of an existing CFP-PCR or development of a CFP-PCR for a new product category, (2) CFP calculation and verification, and (3) application for the registration and publication of the CFP-PCR.

7.6.2 Certification and Verification Processes

A CFP quantification and a CFP declaration draft are examined from the following three basic perspectives: conformity to relevant rules, conformity to an applicable CFP-PCR, and ensuring of traceability of data.

The CFP program offers two methods of verification: (1) product-by-product verification and (2) system certification. Both methods are third party verified and equally valid.

1. *Product-by-product verification*: Upon receiving an application for a CFP verification from a company wishing to make a CFP declaration, JEMAI selects a

Fig. 2.7 Official logo of Japanese CFP shown on the package of the product

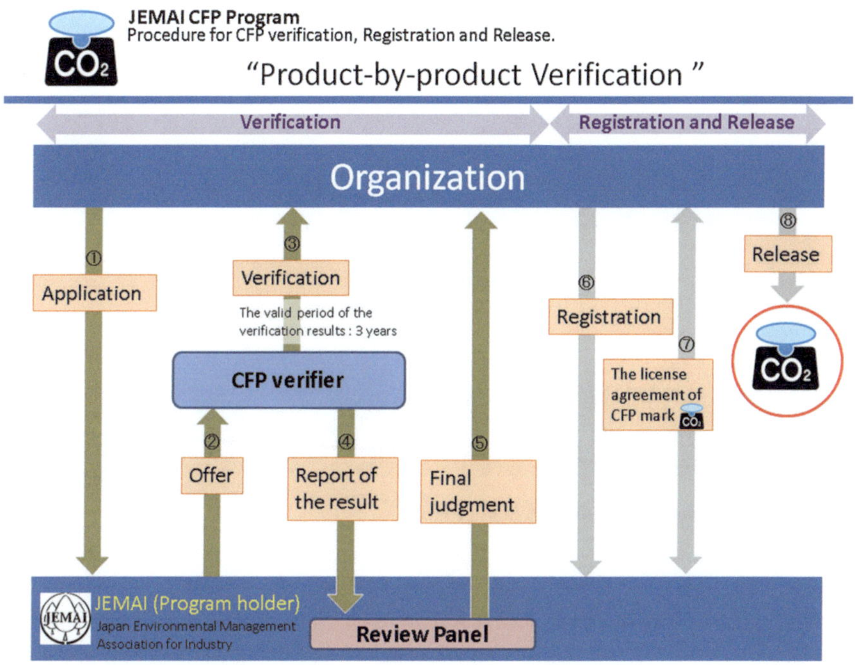

Fig. 2.8 Product-by-product verification method in Japan

CFP verifier from among the licensed reviewers. A selected verifier conducts a CFP verification for the product in the application and makes an approval/disapproval decision. The review panel then validates the result of the verification submitted by the verifier and makes a final judgment on whether to approve. Figure 2.8 illustrates the rules of the product-by-product verification method.

2. *System certification*: The objective of the CFP system certification is to validate that the quantification, verification, and disclosure of a CFP system, which has been internally set up within the applicant company, meet the requirements and that the result of the CFP quantification and declaration made by the applicant company is reliable.

   A CFP system certification body registered with JEMAI conducts an audit and certifies the aforementioned CFP system that is in accordance with the

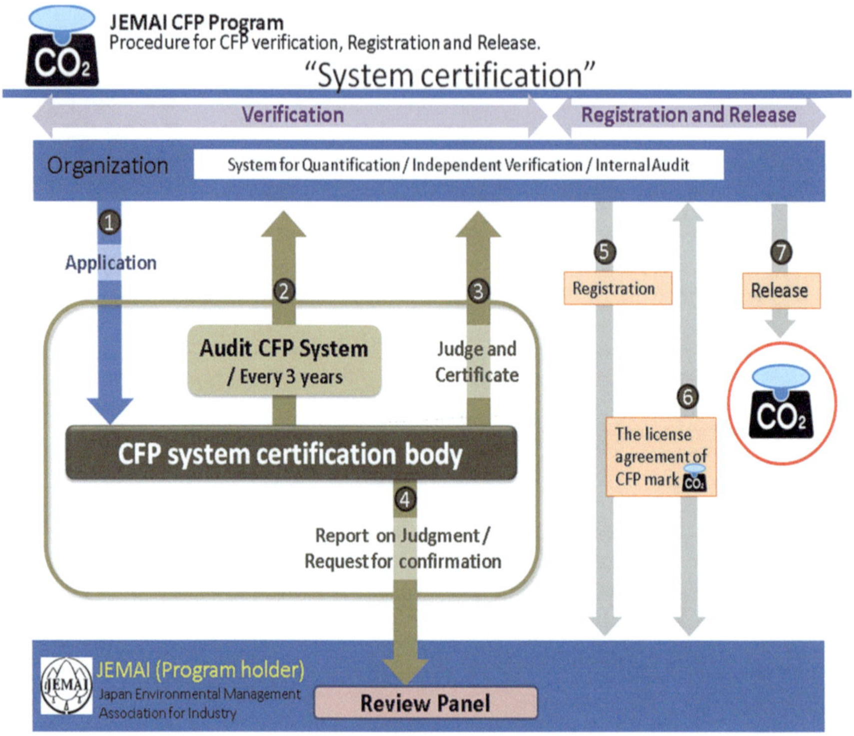

Fig. 2.9 CFP system certification method in Japan

requirements prescribed by JEMAI. The review panel then checks the result of the audit. Once the CFP system is certified, the applicant company may apply for the registration and publish the CFP declaration by conducting an independent verification within the system.

Figure 2.9 illustrates the rules of the CFP system certification method.

7.7 *State of the Art and Future Plans*

By May 2015, 189 companies have registered 1,067 products in the CFP communication program using 105 PCRs, and 5,253 GHG emission factors have been registered on the CFP secondary database. The numbers indicate the growing recognition of the CFP communication program.

Since 2008, a carbon offset program utilizing the CFP communication program has been undertaken by METI and other related ministries with the aim to increase the efforts for reducing Japan's CO_2 emissions. METI implemented a campaign titled the *Donguri* (acorn) Reward System in collaboration with private sector companies. The campaign uses the acorn logo on the products registered in the campaign to signify that by using carbon credits, those products offset 100 % of the carbon emission calculated by the CFP communication program. Sixty-four companies joined the campaign in the fiscal year 2014.

Currently, JEMAI is in the process of combining the CFP communication program and the EcoLeaf (type III environmental labeling program), and the process is expected to be completed in 2018. Three factors need to be taken into consideration for this change: an improvement of interoperability with overseas programs, a consideration for the full-scale adoption of multiple environmental aspects, and the eventual launch of the unified program compatible with international movements.

8 Experiences with Carbon Footprint of Products in the World: Republic of Korea

Eunah Hong

Korea Environmental Industry & Technology Institute (KEITI), 122-706 Seoul, Republic of Korea

8.1 Background

For the Republic of Korea, the average temperature has risen by 1.5 °C over the past 100 years, an increase far higher than the average global temperature rise, bringing about serious impacts on primary sectors of Korean industries. Agriculture, for instance, locations suitable for crop cultivation change and crop damage by blights and pests also increases.

The Korean government started responding to climate change since 2008 by declaring "Low Carbon, Green Growth" as the new national vision to lead Korea's future development for the next 60 years. The vision for the green growth represents three purposes: (1) maximize synergy of environment and economy, (2) enhance national prestige to international level, and (3) improve quality of life and green revolution. In align with the vision, the government has implemented practical strategies and promotion plans including climate change adaptation, green technology, and energy independence (Fig. 2.10).

The CFP labeling is based on Article 57 "Framework Act on Low Carbon, Green Growth" and Article 18 "Support for Environmental Technology and Environmental Industry Act." The labeling was selected as the main policy for green revolution of life, the ninth national strategy among ten low-carbon green growth strategies.

For sustainable development, GHG reduction is needed not only in the production sector but in the consumption area. In that context, CFP labeling is a very efficient system to lead green production of companies and green consumption.

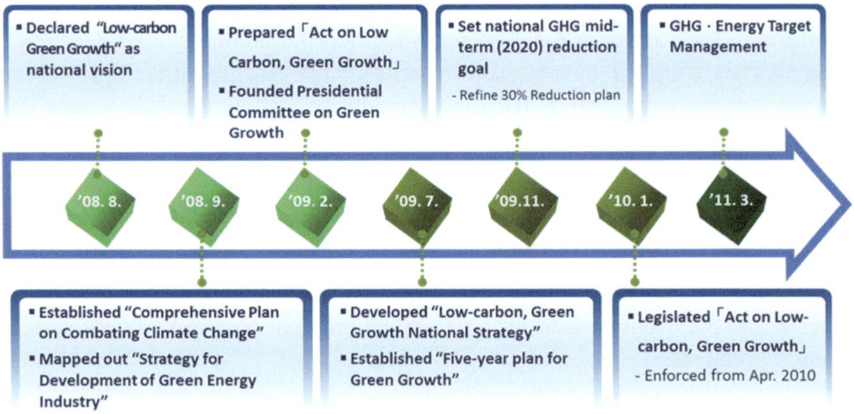

Fig. 2.10 Progress of the Korean government's response to climate change

8.2 *History*

Korea has prepared the foundation of CFP labeling since 1998 by establishing the life cycle inventory database for carbon emission calculation. The CFP labeling system, by the Ministry of Environment, was launched in February 2009 after pilot certification in 2008. Korea Environmental Industry and Technology Institute (KEITI), which is an affiliate organization of the Ministry of Environment, has operated the CFP labeling system (Korean Guideline 2009).

In 2011, KEITI implemented the second phase named "low-carbon product certification." The certification is issued when products reduce GHG emissions by improving process or fuel efficiency.

In order to save time and costs of companies, KEITI introduced the "certification of product category verification system" in 2012. Under this certification, a company can grant its internal verifiers with the responsibility and authority to autonomously perform document reviews and site audits required to certify carbon emissions.

In 2014, KEITI launched the third phase "carbon neutral product certification." The certification is issued when products offsets GHG emissions to zero by external reduction activities.

The system aims to sensitize consumers to GHG emissions generated for the use of products and services, thereby encouraging consumers to choose low-carbon products. CFP labeling plays a key role in establishing a culture of low-carbon and green consumption.

8.3 Legal Framework

- Article 18 (Certification of Eco-label) of the Support for Environmental Technology and Environmental Industry Act
- Article 57 (Diffusion of Culture in Production and Consumption for Green Growth) of the Framework Act on Low Carbon Green Growth
- Regulations on Certification of Carbon Footprint Labeling (Ministry Notification No. 2014–150)

CFP labeling is a certification system based on the law, but it is not legally binding, and all the activities are undertaken on a voluntary basis.

8.4 Assessment Standards of Carbon Footprint of Products

CFP is an index for a specific environmental impact, i.e., the greenhouse effect, and it quantifies the GHG emissions of a product over its entire life cycle by using LCA. In that sense, the methodological framework for carbon footprint assessment is based on environmental labeling and declaration – ISO 14025:2006 (ISO 2006c) and ISO 14020 series, in addition to ISO 14040:2006 (ISO 2006a)/14044:2006 (ISO 2006b).

CFP-product category rules (PCRs) are the basic system for CFP quantification for the respective product categories in accordance with ISO/TS 14067:2013 (ISO 2013a) (carbon footprint of products).

8.4.1 Implementing Agencies

The Ministry of Environment governs the overall management of carbon footprint labeling.

KEITI is responsible for the development and revision of the guidelines for carbon footprint of products, certification of carbon footprint label, and follow-up management (KEITI 2015).

The Korean Environment Preservation Association (KEPA) provides education programs to train certification inspectors for carbon footprint labeling.

8.4.2 Labels

The carbon footprint label is marked on a product to specify the CO_2 equivalent of greenhouse gas emissions generated in the entire life cycle of the relevant products and services, from production, transportation, distribution, and usage to end of product life.

Fig. 2.11 Phases of the carbon footprint labeling

Carbon footprint labeling comprises three phases: certification of carbon emissions (phase I), certification of low-carbon products (phase II), and certification of carbon neutral products (phase III) (Fig. 2.11).

8.4.3 Databases

The Korean life cycle inventory database (LCI DB) is developed to improve the convenience for calculation and the practical use of the environmental product declaration system and the CFP labeling system.

The Korean LCI DB development shall be in accordance with ISO 14040 series. The Ministry of Environment and the Ministry of Trade, Industry, and Energy developed the LCI DB jointly since 1998.

The management of LCI DB supply system and the standardized guideline for national LCI DB were developed in 2002. The plan of LCI DB administration and dissemination was established in 2004.

The Korean LCI DB consists of total 417 modules as of the end of 2014, and those are open to the public (Table 2.2).

KEIT has developed the "TOTAL" software for the benefit of Korean companies that desire to acquire Korea's environmental product declaration system. TOTAL (tool for type III and LCA) is compatible with a range of LCI DB at home and abroad and has the same format as the environmental product declaration system performance report forms.

8.4.4 PCRs

CFP-product category rules (PCRs) are the basic rules for quantifications of the CFP labeling system. CFP-PCRs specify the calculation method for the amount of greenhouse gas (GHG) emissions generated during production, transportation, distribution, use, and end-of-life products.

Table 2.2 Current LCI DB status

Life cycle stage	Data category	No. of unit
Pre-manufacturing stage	Construction material	26
	Rubber	8
	Metal	51
	Basic component	30
	Basic chemical	92
	Water resource	11
	Energy	23
	Pulp and paper	9
	Plastic	37
	Others	20
Manufacturing stage	Metal processing	13
	Component processing	0
	Plastic processing	24
	Others	0
Transportation stage	Land transport	20
	Air transport	0
	Sea transport	22
	Others	0
End-of-life stage	Landfill	3
	Incineration	10
	Recycling	16
	Others	2
Total		*417*

Korea's CFP-PCRs are three guidelines: general products, energy-using products, and use scenarios for each energy-using product. There are 50 PCRs in total for Korea's CFP labeling:

- The guidelines for general products (guidelines set I) for products whose usage do not consume energy
- The guidelines for energy-using products (guidelines set II) for products the usage of which consume energy
- The guidelines for usage scenarios for each energy-using product (guidelines set III) for energy-using products, each of which is under a different usage scenario

8.4.5 Guidelines for Certification of Low-Carbon Products

The guideline for certification of low-carbon products is composed based on the criteria for carbon emission level and carbon reduction rates. To be certified as a low-carbon product, the product shall meet both the criteria for certain carbon emission level and carbon reduction rates. Until the end of 2017, however, certification can be granted to products which satisfy only one of the two criteria.

8.4.6 Guidelines for Certification of Carbon Neutral Products

The guideline for certification of carbon neutral products includes the following requirements: the calculation of carbon emissions, securing and cancelation of carbon offsetting units, and follow-up of the management for certification.

The certification of carbon neutral products is given to products with the certification of low-carbon products, if the product offsets GHG emissions to zero through offset activities including CER (certified emission reduction) purchase or forestation fund-raising.

The formula to calculate the amount of carbon offset per product unit as below:

$$G_{product}(\text{the amount of carbon offset per product unit}) = G_{unit}(\text{the amount of carbon emissions per unit}) \times M_{product}(\text{the yield})$$

- The "amount of carbon emissions per unit" means the amount of carbon emissions of the product certified as low-carbon products and applying for the certification of carbon neutral products
- The "yield" means the entire output for 3 years from the date the product was certified as low-carbon products

The carbon emission from the certified product shall be offsetted through the standard offset activities: offset unit utilization and forestation funding.

- "Offset unit utilization" is to purchase CERs issued by external reduction investment and offset the carbon emission.
- "Forestation funding" is to provide financing for forestation and offset the carbon emission.

8.4.7 Guidelines for Certification of Product Category Verification System

The guidelines set requirements for certification of verification system that can calculate and manage the carbon emissions in each product category.

The certification of product category verification system is a process under which applicants producing multiple products with the same function are given the responsibility and authority to calculate and verify carbon emissions, following the assessment of whether the applicant has established an appropriate system such as the capacity and organizational structure in accordance with the guidelines to calculate and verify the carbon emissions of relevant products.

The guidelines cover factors necessary for an applicant company for the certification of product category verification in order to calculate carbon emissions of products and include systemic documentation with regard to the organizational structure, human resources, data collection and calculation process, verification review procedure, and operational and management requirements.

As for methods to calculate carbon emissions and definitions of terms for those products which are not covered in these guidelines, the guidelines for carbon footprint of products shall be applied.

The guidelines require certified companies to establish, document, implement, maintain, and improve a product category verification system on an ongoing basis.

Certified companies are given the responsibility and authority for the product category verification system and the efficient performance of carbon footprint labeling on their own products in accordance with the requirement of these guidelines. By doing so, they can also establish a foundation for systemic management of GHGs and environmental pollutants generated from business sites and products.

8.4.8 Current Status and Future Plan

As the number of carbon footprint-labeled products continues to increase, carbon footprint labeling has emerged as the core certification system in efforts to address climate change. The number of carbon footprint-labeled products has reached 1,667 at the end of 2014, thereby ranking the Republic of Korea at the second place in the world with regard to the number of certified products.

As of 2014, a total of 1,667 products are awarded the carbon footprint label, recording an annual average increase of 33 % since it was first introduced in 2009. Certification of carbon emission (phase I), certification of low-carbon product (phase II), and certification of carbon neutral products (phase III) have been granted to 1,390, 264, and 13 products, respectively (Fig. 2.12). The proportions of major product categories (as of 2014), non-durable goods such as detergents and food, account for 38 % followed by energy-using products such as automobiles and computers (39 %), production goods (20 %), services (1 %), and durable goods (2 %).

Asia Carbon Footprint Network (ACFN) began as an endeavor to share information and reinforce cooperation among operating agencies of carbon footprint labeling in the Asia region. Launched in October 2013 (ACFN 2013), ACFN is a voluntary consultative body designed to enhance cooperation for carbon footprint labeling among 14 agencies across nine countries in Asia including China, Thailand, and the Philippines. KEITI and the Economic and Social Commission for Asia and the Pacific (UNESCAP) serve as a secretariat for ACFN.

Asia Carbon Footprint Seminars and ACFN workshops aim to transfer know-how and experiences of carbon footprint labeling from more advanced countries to the countries that wish to introduce and improve the systems. The efficacy and utilization methods of carbon footprint labeling in Korea are disseminated in countries without equivalent systems in order to expand the international carbon footprint labeling regime.

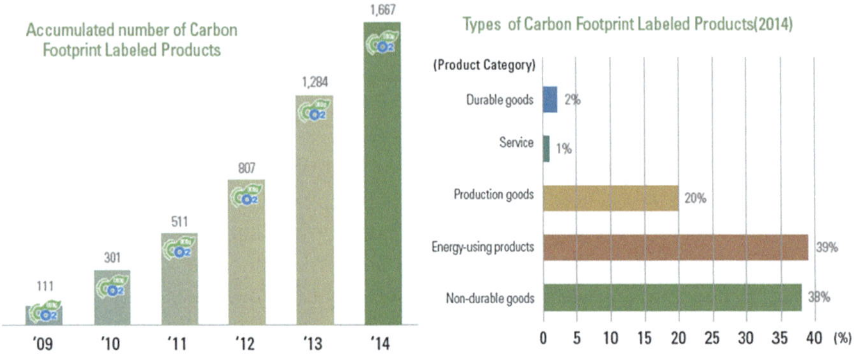

Fig. 2.12 Trends in the number of CFP and types of CFP (2014)

Future directions will aim to boost the satisfaction of certified companies by reinforcing incentives for carbon footprint labeling and expanding support for the certification of SMEs and high-potential enterprises.

Further plans include proactive responses to other certification systems currently under debate in the international community, such as carbon neutral certification and water footprint.

ACFN membership will be expanded through the establishment of an online and offline cooperative platform in order to vitalize the network.

9 Experiences with Carbon Footprint of Products in the World: Thailand

Pongvipa Lohsomboon

Thailand Greenhouse Gas Management Organization (TGO), 10210 Bangkok, Thailand

Thumrongrut Mungcharoen

National Science & Technology Development Agency and Faculty of Engineering, Kasetsart University, 10900 Bangkok, Thailand

9.1 Introduction

The Intergovernmental Panel on Climate Change (IPCC) report released in February 2007 (IPCC 2007a), based on the work of some 2,500 scientists in more than 130 countries. It concluded that humans have caused all or most of the current global warming problems which could lead to large-scale food and water shortages and have catastrophic effects on wildlife (IPCC 2007b). In response, international, regional, national, and local initiatives are being developed and implemented to reduce GHG concentrations in the atmosphere. Calculation of carbon embedded in a product or CFP is gaining popularity. CFP describes the sum of GHG emissions accumulated during the whole life cycle of a product (good or service) in a specified application (Quack 2010). Thailand has recognized the significance of climate change and global warming by becoming a member of the United Nations Framework Convention on Climate Change (UNFCCC) on 28 December 1994 and later ratifying the Kyoto Protocol on 28 August 2002. It was placed in "Non-Annex I" category, a non-binding treaty. Due to the report by the United Nations Development Program, Thailand's CO_2 emissions rose very quickly at an average of 12.8 % a year during 1990–2004. The country's per capita emissions now rank the 22nd among the world's top 30 carbon emitters (Hossain and Selvanathan 2011). The LCA methodology and the LCI database are the two key elements for determining the CFP. Thailand started the LCI database in 2005 and the National LCI database project in 2007. After having obtained sufficient LCI datasets with quite a number of local LCA experts, Thailand launched the pilot project in February 2009. It is a joint collaboration between the Thailand Greenhouse Gas Management Public Organization (TGO) and the National Metal and Materials Technology Center (MTEC) under the National Science and Technology Development Agency. TGO acts as a label issuing authority, while MTEC acts as a technical supporter with the help of Thai LCA experts (Mungcharoen 2009, 2010a). The CFP labeling scheme has been well accepted and should increase the competitiveness of Thai industries in the world market.

9.2 *Background*

9.2.1 National CFP Pilot Project

The Thai CFP project initiated in February 2009 was accomplished due to several LCA experts and the good collaboration among partner organizations, namely, TGO, MTEC, FTI, and several universities. The 76-page "technical guidelines on carbon footprint of products" for Thailand (in Thai) was completed and published in December 2009 (Mungcharoen 2010b). It is available for free download from the website of TGO (TGO 2015). Among 25 pilot companies, the first sets of CFP labels (Fig. 2.13) were awarded to 23 products from 16 companies on 25 December 2009. The companies (and their products) include Thainamthip Limited (325 cc Coca-Cola can); Thai Ceramic Co., Ltd. (ceramic wall tile); Asia Fiber Public Co., Ltd. (Nylon 6 Textured Yarn); CPF Co., Ltd. (Teriyaki Chicken (Thailand); Public Co., Ltd. (pineapple juice from concentrate 200 l); Bangsue Chia Meng Ricemill Co., Ltd. (jasmine rice 5 kg); Carpets International (Thailand) Public Co., Ltd. (Axminster Carpet); Thai Union Manufacturing Co., Ltd. (Kaeng Khiaw Waan Tuna "Sealect" brand 185 g can); Otani Tire Co., Ltd. (Tractor tile No.F-17 12.4-24 (R1) 40.66 kg); Eastern Polypack Co., Ltd. (food box with cover 34 g); Thai Airways International Public Co., Ltd. (chicken curry "kiew-wan" and chicken

Fig. 2.13 Carbon footprint label of Thailand

curry "massaman," steamed Thai Hom Mali Rice); SIG Combibloc Ltd. (aseptic carton 125 mm, 200 mm, 250 mm); and Betagro Public Co., Ltd. (broiler starter feed, broiler grower feed, broiler finisher feed).

The certified products are eligible to use the CFP label for 2 years.

Thai Airways International Public Company Limited (THAI), one of the 16 existing companies, is the pioneer as the first airline in the world to receive the CFP label on its signature dishes. From the beginning of 2010, THAI in-flight menu features CFP information for passengers, starting with two Thai signature dishes, chicken massaman curry with steamed Thai Hom Mali Rice (1.36 kg CO_2e per 250 g serving), and green curry "kiew-wan" with steamed Thai Hom Mali Rice (1.39 kg CO_2e per 250 g serving).

The first sets of CFP labels were awarded to 23 products from 16 companies on 25 December 2009. Moreover, approximately 20 local CFP experts and the "technical guidelines on carbon footprint of products" (TGO and MTEC 2009) for Thailand were obtained from this CFP pilot project.

9.2.2 Criteria for CFP Label's Registration

The guideline has several specifications. It can be applied to all products and be used for full carbon footprint or "business-to-consumer (B2C)" assessment. This assessment is also known as the cradle-to-grave assessment considering all life cycle stages. The guideline is also applicable to "business to business (B2B)," also known as the cradle-to-gate assessment as it only covers raw material acquisition, manufacturing, and distribution up to the factory gate. The partial carbon footprint can, however, be publicized only on websites, brochures, reports, etc. but cannot be displayed on the product packaging. The companies are authorized to use the label during the 2-year certification period.

9.2.3 The Use of GHG Emission Factors for CFP Calculation

The six major greenhouse gases as identified in the Kyoto Protocol, e.g., CO_2, CH_4, N_2O, hydrofluorocarbons (HFCs), perfluorocarbons (PFCs), and sulfur hexafluorides (SF_6), are included in the CFP assessment. The GHG emissions will be evaluated and presented in terms of mass of carbon dioxide equivalent (CO_2e), and the CFP calculations shall use the equivalency factors for global warming potential over 100 years (GWP_{100}) for each greenhouse gases mentioned in the latest version of IPCC report.

The four phases of LCA as defined by ISO14040/44 that include goal and scope, inventory analysis, impact assessment, and interpretation must be followed to perform the CFP assessment. The national guideline on CFP introduces the most common methodology approach to calculate the CFP. It converts the primary and secondary data of inputs/outputs since raw material acquisition, manufacture, use, waste management, and final waste disposal in all stages to GHG emissions by multiplying their loadings with the respective coefficients or "CO_2 emission factors." Full lists of CO_2 emission factors can be cited from the TGO's carbon footprint webpage. The factors are in reference to various sources including:

1. National LCI database (http://www.thailcidatabase.net)
2. Peer-reviewed journals, technical report, or theses in the context of Thailand
3. Inventory databases in LCA software
4. Publications from international organizations, e.g., UNEP, FAO, etc.

These factors are updated every 6 months based on new information, particularly as more and more verified datasets from the national LCI data become available.

9.2.4 Product Category Rule (PCR) Development

As specified in ISO 14025, product category rules (PCRs) are essential for the concept of environmental and climate declarations. Hence, in order to enable the transparency and credibility of the CFP calculation, the PCR approach is therefore applied to the CFP labeling in Thailand. This enables the manufacturers to perform the CFP assessment following the related PCRs in which they want to get the CFP label. If there are products without PCR in Thailand, the assessment conductor/ consultant must submit a draft of product specifications following the template format provided on the TGO website or contact TGO for further information.

There are two types of PCRs developed in different ways. Initially, PCRs were developed for each product by the company (or their consultant) who applied for the carbon footprint label using a standard format developed by TGO. After these company, PCRs were put before the National Technical Committee for consultation, if any objections from other companies manufacturing the same product arise the rules that would have to be reconsidered by the National Technical Committee. Without so doing, the company PCRs would remain applicable. The next PCR or the national PCR is developed by TGO with support from the National Technical Committee in consultation with various stakeholders such as industries, government and nongovernment organizations, and universities. The typical procedures for the national PCR development are shown in Fig. 2.14.

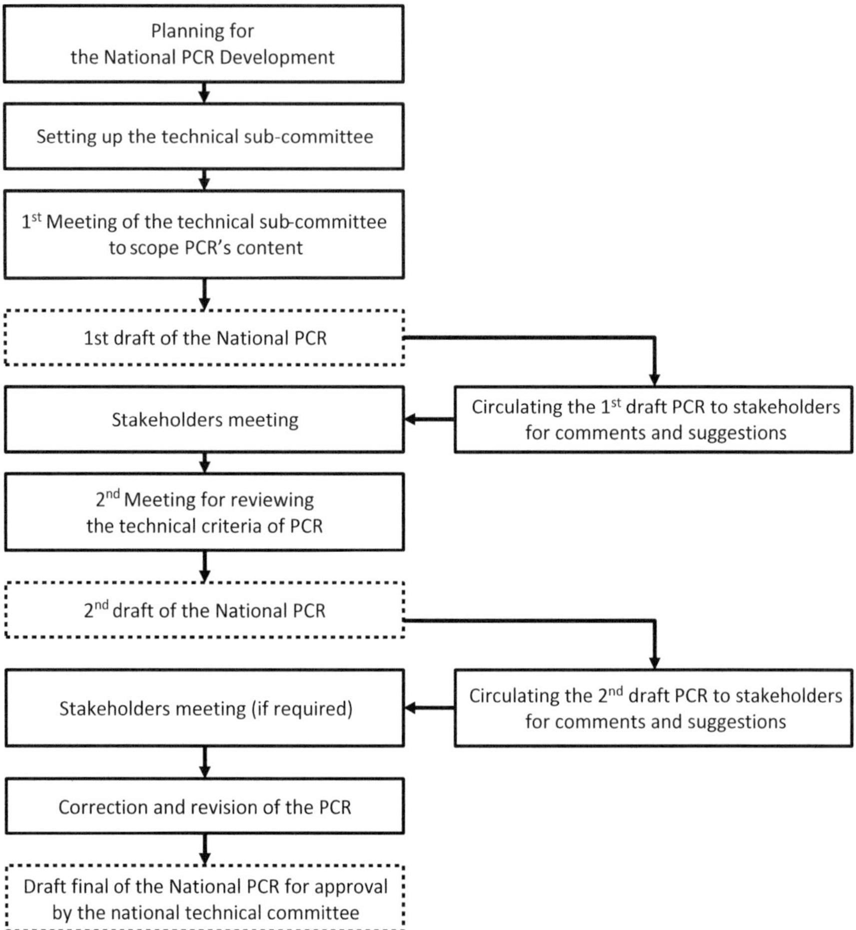

Fig. 2.14 Procedures for the national PCR development

The activity starts from planning for the national PCR development in the focused sector. After the technical subcommittee is set up for drafting national PCR, the final PCR will be consulted with various stakeholders before reporting to the National Technical Committee for approval. The contents provided in the Thai PCR are as follows:

1. Scope of the PCR
2. Description of product (includes product definition and components)
3. Referenced standards and PCRs
4. Terms and definitions
5. Scope of the assessment with calculation unit and life cycle stages to be covered
6. General requirements applied to all stages, i.e., life cycle flow diagram, range of data collection, primary data collection method and requirements, secondary

data collection method and requirements, period of data collection, allocation method, cutoff criteria, scenarios, and others
7. Data presentation requirements
8. Appendices for additional information in calculation of CFP

As of May 2015, there are 161 approved PCRs. The detailed information of PCRs is available on the webpage of TGO.

9.2.5 Verification and Certification Procedures

After completion of carbon footprint calculation, the company has to contact a registered verifier to verify and approve the results. The affirmation statement is then issued by the registered verifier, which has to be submitted to TGO's Carbon Business Office (CBO). CBO acts as the secretary to present to the "working group on promotion of carbon footprint of products and organizations in Thailand" (CF-WG) for approval and issuance of the license. The CF-WG, chaired by TGO Executive Director, consists of ten members including CF experts, representatives of FTI, Marketing Association, Department of Alternative Energy Development and Efficiency (Ministry of Energy), Pollution Control Department (Ministry of Natural Resources and Environment), and Thai Industrial Standards Institute (Ministry of Industry).

It is imperative to note that the verifiers registered with TGO have to pass through very strict quality control criteria. They should first undergo rigorous training in LCA and CF and should also have conducted at least three product carbon footprint case studies. It must also be ensured that there is no conflict of interest between the company and the registered verifier. The roles and responsibilities of CF-WG to the CFP are to process the application for CFP, to verify and approve the use of carbon footprint label, to approve and register the carbon footprint consultants, and to provide advice on the principles, criteria, and procedures for CFP.

9.2.6 Current Situation

With the support and effort from the collaborative organizations such as MTEC, FTI, the National Food Institute, the Thailand Textile Institute, the Thailand Environment Institute, and the universities in promotion of CFP label program in Thailand, through several activities and projects brought about increased recognition of CFP label from manufacturers, especially from the food industries. 1,570 products from 363 companies have received the CFP label as of 31 May 2015, denoting that there is a rapid growth in CFP certification.

The CFP-labeled products cover a wide range such as chicken meat, chicken seasoning, jasmine rice, canned food, fruit juice, beverage, animal feed, wall tile, instant rice vermicelli, textiles, food packaging, beverage packaging, etc.

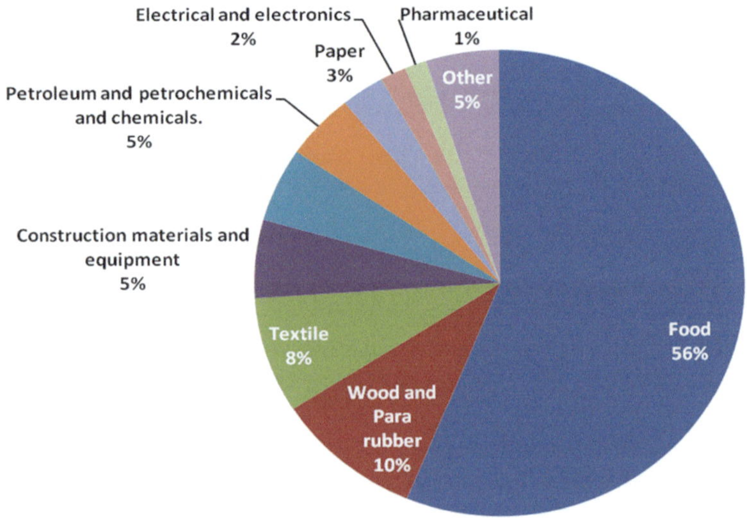

Fig. 2.15 Classifications of CFP-labeled products by types of industries

The classification of CFP-labeled products by types of industries is shown in Fig. 2.15. Food industry has the highest share in the total CFP-labeled products as it contributes about 44 %, followed by the products from wood and Para rubber industry, textile industry, construction industry, and petroleum and petrochemical and chemical industry with about 10 %, 8 % 5 %, and 5 % shares, respectively.

There are several other local products that have received the CFP label such as minced and preserved pork (Moo Yor), honey, silk, Thai bean cake, macadamia nuts, coffee roast, Saa paper, mulberry paper umbrella, etc. These products are known as OTOP that stands for "One Tambon One Product," which is a project promoted by the Royal Thai Government since 2001 in order to boost the rural economic development by using the principles about development of local products to global with the local self-reliance creativity and resources. TGO in collaboration with the Department of Environmental Quality Promotion (DEQP) and the Ministry of Natural Resources and Environment has been working to promote CFP label to the OTOP products as the values of the exported "OTOP" products account for 10,000 million Thai Baht with a total inflow of money being more than 100,000 million THB. The implementation of CFP label to the local communities is a measure to raise the environmental standard of the "OTOP" products, to satisfy the demands for low-carbon products and to increase the opportunity in exporting products to other countries in the future.

For CFP technical supporting system, the GHG emission factors (EFs), generated from the Thai life cycle inventory database by MTEC, have been supplied to TGO every 6 months (Mungcharoen and Olarnrithinun 2014). More than 600 EFs are publicly available on the TGO website to be used for CF calculation of both products and organizations. Regarding to the local CFP experts, there are 74 qualified CFP consultants listed and 39 qualified CFP verifiers certified by TGO. The

CFP verifiers have to attend the CFP training course, organized by TGO at least once every 2 years to refresh or update their knowledge in CFP. Moreover, the national CFP technical committee which is now in its sixth year has met regularly every 2 months to provide technical recommendations including the update the CFP national guidelines and the PCRs.

Regarding the regional collaboration and networking, TGO and MTEC are founding and active members of the Asia Carbon Footprint Network (ACFN) established in 2013 (ACFN 2013).

9.2.7 Outlook

Since the initiative on CFP pilot project in 2009, there are several organizations and companies in Thailand that have expressed their interests on calculation and reduction of CFP toward "low-carbon economy" trends. The product's carbon footprint reduction (CFR) or global warming reduction label, initiated by TGO in 2014, is the label that demonstrated an achievement in reduction of the product's carbon footprint as required by the TGO's Carbon Labeling Program. An assessment of CFR shall be based on the concept of product's life cycle; raw material acquisition, transportation and distribution, production, usage, and end-of-life disposal to account and compare between the product's carbon footprint for the base year and the present year. This determines and evaluates the reduced carbon footprint of the product against the TGO's requirement. Products eligible for awarding CFR registration are:

1. The products that confirm the CFR requirement, such that it achieves present year reduction compared to base year's carbon footprint (which is not less than 2 %)
2. The products that confirm the CFR requirement, such that it achieves carbon footprint reduction that can lower or equal the benchmark of category set by TGO

As of April 2015, 65 products from 13 companies are registered for CFR, including ceramic tiles, wall and floor tiles, dish cleaner, textiles, cement, rice bags, cooking oil, and others which can reduce greenhouse gas emissions 116,472 t CO_2e.

Additional to the CFP and CFR labels, TGO also promotes CoolMode label for textile products, carbon offset label, and carbon neutral label (TGO 2015). On the broader aspect, MTEC has promoted the applications of LCA-LCI database to support the national green growth policy (Mungcharoen and Olarnrithinun 2014) and most recent green product trends, including water footprint, environmental footprint, and upcycle carbon footprint.

10 Conclusion

Atsushi Inaba
Department of Environmental and Energy Chemistry, Kogakuin University, 163-8677 Tokyo, Japan

In this chapter, the experiences with CFP in the UK, France, Japan, Korea, and Thailand were introduced, and the results and outcomes were described.

CFP has attracted attention because the consumers can see it displayed on the packages directly in the store, and most likely, they compare the labeled products with each other. However, in the UK and Japan, the CFP marks are displayed without any numerical values. Also the reduction mark, introduced in Korea and Thailand, which shows the reduction ratio of GHG emissions of the targeted product compared to the former product, does not indicate an absolute figure. So it can be assumed that producers hesitate to make the CFP transparent.

CFP is a type 3 label showing the calculation result of the GHG emission using LCA. However, it can strongly be assumed that consumers cannot judge the quality of the CFP number shown on the package, if it is a positive or a negative indicator. So the argument is that, in order to recommend the environmental conscious product to consumers, type 1 label is more convenient than type 3 label, because type 1 discloses the criteria to be harmless for the environment.

The focus of *Chap. 2* lies on CFP, i.e., on "product," but recently, a CFP of "organization" has been conducted in many enterprises. And also, "environmental footprint" became popular, which assesses various environmental categories, not only global warming. There is a big social movement toward "organization" from "Product" and from "CFP" toward "environmental footprint."

The multi-issues methods of "environmental footprint" are classified into two groups, "products" and "organization."

For example, there is the activity of "The Sustainability Consortium (TSC)" based in the USA, which was launched in 2009 under the leadership of Walmart, the supermarket chain (TSC 2015). They are developing the methods to assess the environmental impacts, together with the social impacts, of the products mainly sold in the supermarket.

The Environment Directorate-General of the European Commission published "The Guidance of PEF" (product environmental footprint) (European Commission 2013a) and "The Guidance of OEF" (organization environmental footprint) (European Commission 2013b) in 2012, and then their pilot projects were launched in 2013 (European Commission 2013c) and (European Commission 2013d).

Also, ISO/TS 14072:2014 is the technical specification to show the methodologies of LCA for organizations, comparing to ISO 14044 (2006), the standard for LCA of products. To support ISO/TS 14072, UNEP/SETAC Life Cycle Initiative launched the working group "Organizational LCA (O-LCA)" (Martínez-Blanco et al. 2015a, b, c) in 2013 and published its guidance (UNEP/SETAC 2015).

These activities are introduced in detail in Chap. 8 of this book (Martínez-Blanco et al. 2016).

In this movement, CFP is the origin of the communication tool to disclose environmental information directly to consumers. The communication rules are now discussed in order to publish ISO 14026/AWI (ISO 2015), which is the technical specification of "communication of footprint," in 2017. The experiences with CFP are most profitable to consider the methodologies for the communication between consumers and producers. It will be helpful to establish and realize "sustainable consumption and production" in the near future.

Acknowledgment We want to acknowledge the contribution for GHG Protocol and Scope 3 from Ms. Cynthia Cummis of WRI and Mr. Masayuki Kanzaki of JEMAI for type 3 label. They deserve a lot of thanks as they acted like coauthors in providing valuable information.

References

ACFN (2013) Asia Carbon Footprint Network. http://www.acfnetwork.net. Accessed 11 July 2015

BSI (2008) PAS 2050:2008, specification for the assessment of the life cycle greenhouse gas emissions of goods and services. http://www.hspg.com/assets/policies-certificates/certificates/pas2050-2008-specification.pdf#search='PAS2050+2008'. Accessed 30 May 2015

BSI (2011) PAS 2050:2011, specification for the assessment of the life cycle greenhouse gas emissions of goods and services. http://shop.bsigroup.com/en/forms/PASs/PAS-2050/. Accessed 30 May 2015

Carbon Trust (2008) CTC745 "Code of Good Practice for GHG Claims", Carbon Trust. www.carbontrust.com. Accessed 23 June 2015

EPD (2015) The International EPD system. www.environdec.com. Accessed 30 May 2015

European Commission (2013a) Product Environmental Footprint (PEF) guide. http://ec.europa.eu/environment/archives/eussd/pdf/footprint/PEF%20methodology%20final%20draft.pdf#search='PEF+Guide'. Accessed 11 July 2015

European Commission (2013b) Organisation environmental footprint guide. http://ec.europa.eu/environment/archives/eussd/pdf/footprint/PEF%20methodology%20final%20draft.pdf#search='PEF+Guide'. Accessed 11 July 2015

European Commission (2013c) Guidance for the implementation of the EU Organisation Environmental Footprint (OEF) during the Environmental Footprint (EF) pilot phase. http://ec.europa.eu/environment/eussd/smgp/pdf/Guidance_organisations.pdf#search='Guidance+for+the+implementation+of+the+EU+OEF'. Accessed 11 July 2015

GEDnet (2015) http://gednet.org. Accessed 30 May 2015

Hossain M, Selvanathan EA (2011) Climate change and growth in Asia. Edward Elgar Publishing, Cheltenham, UK, 288 pp. ISBN 978 1 84844 245 0

IPCC (2007a) Intergovernmental Panel on Climate Change Assessment Report 4 (AR4), Working Group I Summary for Policymakers (SPM), February 2007. https://www.ipcc.ch/publications_and_data/ar4/wg1/en/spm.html. Accessed 11 July 2015

IPCC (2007b) Intergovernmental Panel on Climate Change Assessment Report 4 (AR4), Working Group II Summary for Policymakers(SPM), April, 2007. https://www.ipcc.ch/publications_and_data/ar4/wg2/en/spm.html. Accessed 11 July 2015

IPCC (2014) Intergovernmental Panel on Climate Change (IPCC) Assessment Report 5, 2014. https://www.ipcc.ch/report/ar5/. Accessed 11 July 2015

ISO (1999a) ISO 14024: environmental labels and declarations – principles and procedures. Type I environmental labelling. International Organization for Standardization, Geneva
ISO (1999b) ISO 14021: environmental labels and declarations – self-declared environmental claims. Type II environmental labelling. International Organization for Standardization, Geneva
ISO (2000) ISO 14020: environmental labels and declarations – general principles. International Organization for Standardization, Geneva
ISO (2006a) ISO 14040: environmental management – life cycle assessment – principles and framework. International Organization for Standardization, Geneva
ISO (2006b) ISO 14044: environmental management – life cycle assessment – requirements and guidelines. International Organization for Standardization, Geneva
ISO (2006c) ISO 14025: environmental labels and declarations – principles and procedures. Type III environmental declarations. International Organization for Standardization, Geneva
ISO (2006d) ISO 14064-1: greenhouse gases – part 1: specification with guidance at the organization level for quantification and reporting of greenhouse gas emissions and removals. International Organization for Standardization, Geneva
ISO (2013) ISO/TS 14067: greenhouse gases – carbon footprint of products – requirements and guidelines for quantification and communication. International Organization for Standardization, Geneva
ISO (2015) ISO/AWI 14026: environmental labels and declarations – communication of footprint information. International Organization for Standardization, Geneva
ISO/DTS 14027: environmental labels and declarations – Product Category Rule (PCR) development. Type III environmental declarations. International Organization for Standardization, Geneva, Switzerland
JEMAI (2015a) Eco-leaf environmental label. http://www.ecoleaf-jemai.jp/eng/label.html. Accessed 30 May 2015
JEMAI (2015b) CFP Communication Program. http://www.cfp-japan.jp/english/. Accessed 30 May 2015
JEMAI (2015c) Japan Environmental Management Association for Industry (JEMAI). http://www.jemai.or.jp/english/. Accessed 30 May 2015
KEITI (2015) Carbon footprint labeling website (in English): http://www.edp.or.kr. Accessed 11 July 2015
Korean Guideline (2009) Guidelines for carbon footprint of products. http://www.edp.or.kr/information/data.asp?bbs_code=6 (in Korean). Accessed 11 July 2015
Martínez-Blanco J, Inaba A, Finkbeiner M (2015a) Activities of guidance of organizational LCA by UNEP/SETAC WG. Jpn J LCA 11(2):97–103
Martínez-Blanco J, Inaba A, Finkbeiner M (2015b) Scoping organizational LCA – challenges and solutions. Int J Life Cycle Assess 20(6):829–841
Martínez-Blanco J, Lehmann A, Chang Y-J, Finkbeiner M (2015c) Social organizational LCA (SOLCA) – a new approach for implementing social LCA. Int J Life Cycle Assess 20 (11):1586–1599
Martínez-Blanco J, Inaba A, Finkbeiner M (2016) Life cycle assessment of organizations. Chapter 8 "Special Types of Life Cycle Assessment" (Finkbeiner M ed). In: LCA compendium – the complete world of life cycle assessment (Klöpffer W, Curran MA, series eds). Springer, Dordrecht, pp 333–394
MEEM (2014) Consumer product environmental display: evaluation report to the Parliament. http://www.developpement-durable.gouv.fr/Bilan-au-Parlement-de-l.html
METI (2009) TS Q 0010. http://www.cms-cfp-japan.jp/english/specifications/pdf. Accessed 11 July 2015
Mungcharoen T (2009) LCA Activities in Thailand and Initiatives in ASEAN. http://icgsi.eng.cmu.ac.th/program.html. Accessed 11 July 2015
Mungcharoen T (2010a) Progress of LCI Database Development in Thailand. http://www.sntt.or.jp/DB2010/arc/8AM03ThumrongrutMUNGCHAROEN.pdf. Accessed 11 July 2015

Mungcharoen T, Olarnrithinun S (2014) Life cycle inventory database and its applications to support public policy. In: Proceedings EcoBalance 2014, Tsukuba, Japan, 28–30 Oct 2014
Quack D (2010) Requirements on consumer information about product carbon footprint. ANEC Final report, Oko-Institut e.V., Germany
TGO (2015) http://thaicarbonlabel.tgo.or.th/. Accessed 5 June 2015
TSC (2015) The sustainability consortium. http://www.sustainabilityconsortium.org/. Accessed 11 July 2015
UN (2002) Report of the world summit on sustainable development. Johannesburg, South Africa, 26 Aug – 4 Sep 2002, United Nations, http://www.unmillenniumproject.org/documents/131302_wssd_report_reissued.pdf
UNEP (2003) Marrakesh process. http://esa.un.org/marrakechprocess/. Accessed 11 July 2015
UNEP/SETAC (2015) Organizational life cycle assessment – guidance for organizations to conduct LCA considering their value chain (under publication). UNEP/SETAC Life Cycle Initiative, United Nations Environment Programme and Society for Environmental Toxicology and Chemistry, Paris, France. http://www.lifecycleinitiative.org/wp-content/uploads/2015/04/o-lca_24.4.15-web.pdf. Accessed 11 July 2015
WRI and WBCSD (2011) Product Life Cycle Accounting and Reporting Standard/standards/product-standard. http://www.ghgprotocol.org/files/ghgp/public/Product-Life-Cycle-Accounting-Reporting-Standard_041613.pdf. Accessed 11 July 2015

Chapter 3
Water Footprinting in Life Cycle Assessment: How to Count the Drops and Assess the Impacts?

Markus Berger, Stephan Pfister, and Masaharu Motoshita

Abstract Freshwater scarcity is a relevant problem for more than one billion people around the globe. Therefore, the analysis of water consumption along the supply chain of products is of increasing relevance in current sustainability discussions. This chapter aims at providing insight into the scientific development and practical application of water footprinting. In a comprehensive literature review, more than 30 water footprint methods, tools, databases, and standards have been identified and discussed. The scopes of different water footprint approaches vary regarding the types of water use accounted for, the distinction of water courses, and the consideration of temporal and regional aspects such as water scarcity and sensitivity of population or ecosystems for impact assessment. In order to illustrate the application of water footprinting, several case studies representing different levels of complexity (crops to cars) and scientific standards (liter to disability-adjusted life year, DALY) are presented. Subsequently, key methodological challenges are identified ranging from the adequate resolution of inventory flows to the consideration of water quality aspects. As the most advanced methods require the highest resolution inventory data, the trade-off between precision and applicability is a key challenge, which needs to be addressed in future database and method developments. Such future developments are the subject of the closing section, which, e.g., provides an outlook on the consensus impact assessment model being currently developed by the UNEP/SETAC Life Cycle Initiative. Moreover, the increasing relevance of water footprinting in decision-making and communication strategies is discussed along with opportunities and limitations of water footprinting.

M. Berger (✉)
Chair of Sustainable Engineering – Institute of Environmental Technology, Technische Universität Berlin, Office Z1, Strasse des 17. Juni 135, 10623 Berlin, Germany
e-mail: markus.berger@tu-berlin.de

S. Pfister
Institute of Environmental Engineering (IfU), ETH Zurich, HPZ E 33, John-von-Neumann-Weg 9, CH-8093 Zurich, Switzerland

M. Motoshita
Research Institute of Science for Safety and Sustainability, National Institute of Advanced Industrial Science and Technology (AIST), 16-1 Onogawa, Tsukuba, Ibaragi 305-8569, Japan

M. Finkbeiner (ed.), *Special Types of Life Cycle Assessment*, LCA Compendium – The Complete World of Life Cycle Assessment,
DOI 10.1007/978-94-017-7610-3_3

Keywords LCA • Life cycle assessment • Life cycle impact assessment • Life cycle management • Special types of life cycle assessment • Sustainability • UNEP/SETAC life cycle initiative • Water consumption • Water demand • Water footprint • Water footprint network • Water resource • Water scarcity • Water use

Acronyms

ADP	Abiotic depletion indicator
AoP	Areas of protection
AWS	Alliance for water stewardship
CTA	Ratio of water use to consumption availability
CV	Coefficient of variation
DALY	Disability-adjusted life year
DTA	Demand-to-availability
EIA	Environmental impact assessment
EWR	Environmental water requirement
EWS	European water stewardship
PDF	Potentially disappeared fraction
PEF	Product environmental footprint
WAVE	Water accounting and vulnerability evaluation
WBCSD	World business council on sustainable development
WF	Water footprint
WFN	WaterStat database
WSI	Water stress index
WTA	Ratio of water use to water availability
WULCA	Water use in LCA

1 Introduction

1.1 Yet Another Footprint...

Water is a viable resource on our planet as it is crucial to sustain life and cannot be replaced by any other substance. However, freshwater is scarce in some regions, countries, and even continents. Therefore, the use of freshwater in agriculture and industry can lead to severe problems for both humans and ecosystems.

Yet, when assessing the environmental performance of products, attention is usually drawn on the energy consumed along a product's lifespan or the emission of greenhouse gases. In contrast, the consumption of water has often been neglected even though its environmental impact can be substantial, especially with regard to agricultural products that are grown in water-scarce regions.

This severe deficiency is currently addressed by many stakeholders including research, industry, and politics. Next to developing methods which enable the quantification of a product's "water footprint," there is also a need to achieve consensus on how such water footprints are to be determined in order to ensure consistency and comparability.

In the introduction, this chapter provides an overview of global freshwater scarcity and resulting consequences, discusses the development of water footprinting, introduces involved stakeholders, and defines a water footprint-related terminology used throughout this chapter. Subsequently, the various approaches for water footprinting, which range from simple accounting tools to complex impact assessment models, are discussed along with their possibilities, data demands, and limitations. Since water footprinting is still a rather young discipline, representative case studies showing the potential and difficulties on the volumetric and impact assessment levels are introduced. Finally, urgent remaining challenges are presented along with an outlook on future method, database, and application developments.

1.2 Status of Water Resource and Demand: Global Picture and Regional Aspects

1,400,000,000 km^3 – that is the total amount of water available on our planet. It covers two thirds of the Earth's surface. However, only 3 % of this volume is freshwater, of which 69 % is locked up in glaciers and polar ice caps (Gleick 1996). The remaining 13 million km^3 of usable freshwater sustains life on our planet but is distributed very unevenly around the globe. While some regions in Columbia, Indonesia, or New Zealand abound in water (>3000 mm annual precipitation), other places, such as the Atacama desert, the Sahel zone, or Saudi Arabia are extremely dry with less than 100 mm precipitation per year.

During the past century, water use grew twice as fast as the world's population (UN and FAO 2007). The main reason for this was agricultural irrigation which is responsible for about 70 % of global water withdrawal and 85 % of global water consumption (Shiklomanov 2003). The share of water withdrawal by the main consumption clusters "agriculture," "industry," and "domestic" is shown in Fig. 3.1.

Today, 1.2 billion people live in such water-scarce regions and another 1.6 billion people suffer from economic water shortage. This means they do not have access to safe drinking water due to missing opportunities to withdraw, purify, or transport water from aquifers and rivers (UN and FAO 2007).

Figure 3.2 shows areas facing physical and economic water scarcity. As a consequence of climate change, population growth, and changing consumption patterns in emerging nations, water scarcity is expected to increase significantly in many parts of the world (Alcamo and Henrichs 2002).

1.3 Consequences Resulting from Water Scarcity

In addition to the physical depletion of freshwater resources, water scarcity threatens the basic living conditions of all species on earth. According to

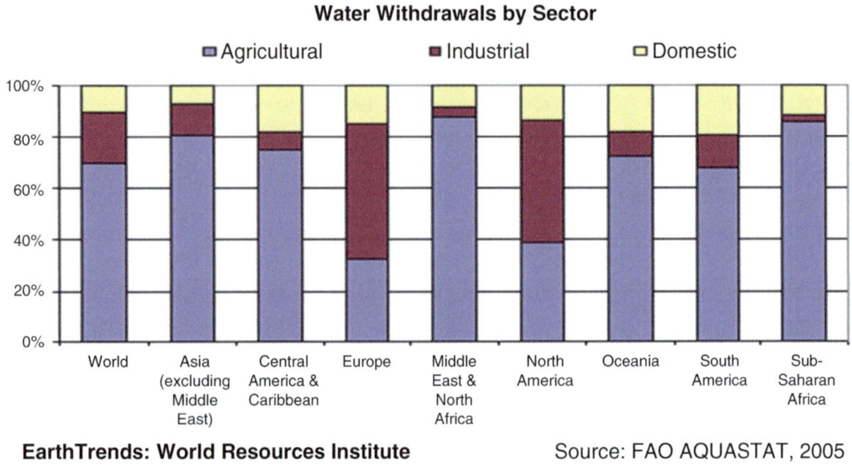

Fig. 3.1 Water withdrawal by sector (FAO 2009)

Vörösmarty et al. (2010), almost 80 % of the human population and ecosystems in 65 % of continental discharge in the world are already threatened concerning water security.

For human beings, the shortage in fundamental supply for domestic use (drinking, bathing, cleaning, etc.) results in no accesses to safe water and subsequent health effects by infectious diseases. In the report by Prüss-Üstün from World Health Organization (Prüss-Üstün et al. 2008), it is estimated that 6.3 % of total death and 9.1 % of total health damage (indicated by disability-adjusted life years (DALYs) accounting the years of premature deaths and life lived with disability) are attributed to unsafe water, inadequate sanitation, or insufficient hygiene. As a second impact pathway, water scarcity restricts agricultural production and causes insufficient food supply. As malnutrition due to the shortage of food supply is related to not only agricultural water scarcity but also other economic/social factors, it is not easy to quantify the effects of water scarcity. However, comparing the minimum requirements of agricultural water (Falkenmark and Rockström 2004) with actual use (FAO 2015), shows that around 36 % of countries (15 % in population) are estimated to be short in agricultural water use to avoid malnutrition. In addition, it is obvious that competing water demand will become more severe in the future because of increasing demand with population and economic growth.

While the pathways from water scarcity to the impacts on ecosystem are complex, the growth of terrestrial plants and habitats of water-dependent species are obviously influenced by water shortage. Especially, many of water-dependent species are already in danger of extinction (Wetlands International 2010). While extinction of water-dependent species is arisen from several factors including degradation in water quality, water scarcity can be obviously one of the major factors affecting biodiversity of water-dependent ecosystems as drastically illustrated in the example of the Aral Sea.

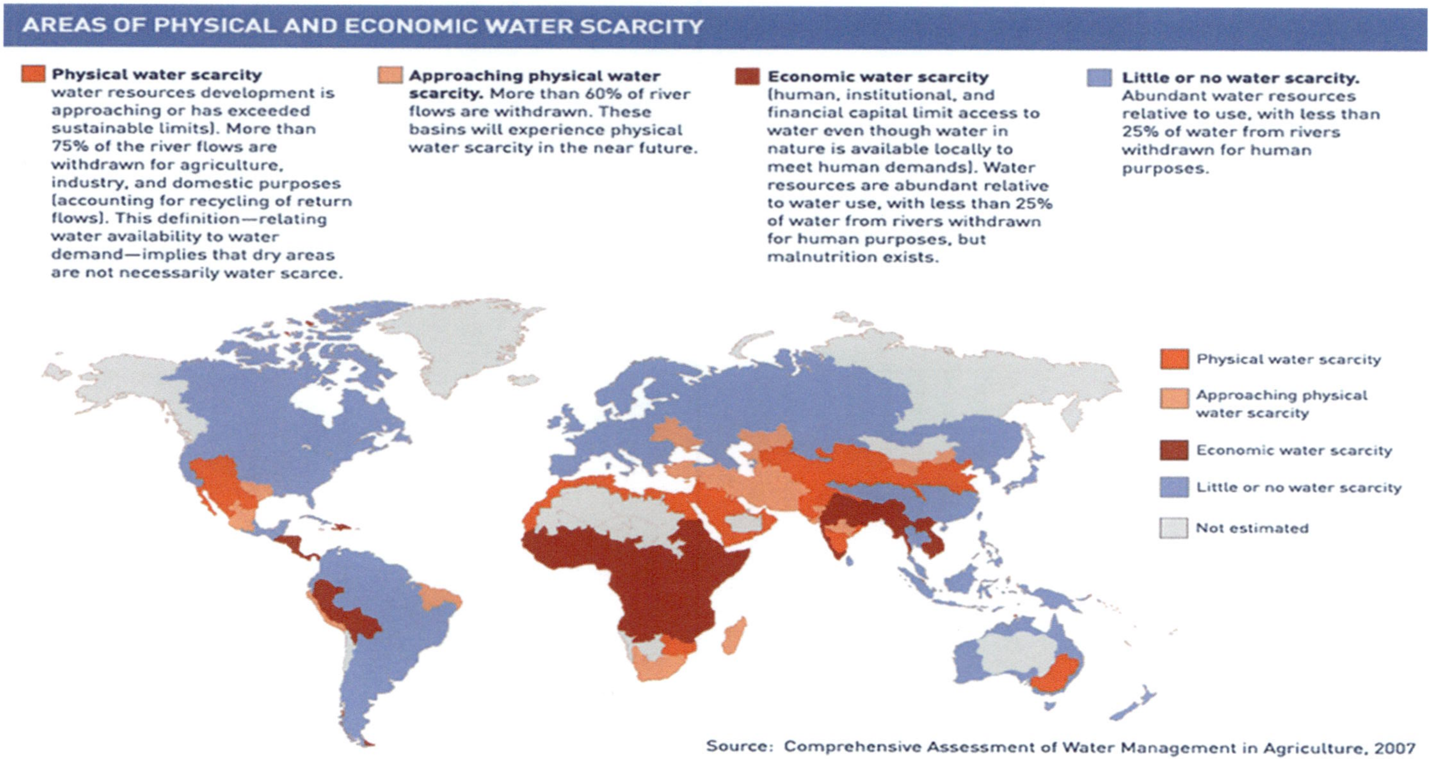

Fig. 3.2 Areas of physical and economic water scarcity (FAO 2009)

1.4 Assessment of Water Use Across Product Life Cycles

Considering global freshwater scarcity and the resulting consequences, the analysis of water use across the supply chains of goods and products seems urgent. In general, water use can be assessed on the volumetric level of virtual water, within life cycle assessment, and by means of a water footprint analysis. While methodological details are explained in Sect. 2, this section provides an overview on the development of the three approaches.

In the 1990s of the last century, the virtual water concept was developed which accounts for the consumption of ground and surface water (blue water), the evapo (transpi)ration of rainwater (green water), and the pollution of freshwater (gray water) (Allan 1998). A decade later, the water footprint was introduced as a tool which expresses the virtual water content of products, organizations, people, and nations (Hoekstra and Hung 2002) in a spatially and temporally explicit way. By revealing surprisingly high volumes, like 140 L per cup of coffee (Chapagain and Hoekstra 2007) or 2700 L per cotton T-shirt (Chapagain et al. 2006), the consumer's attention has been drawn on the amounts of water consumed or polluted during the production of daily goods (Berger et al. 2014). Even the WF (water footprint) of nations and global virtual water imports and exports have been analyzed based on WF estimates of products, consumption patterns, and trade statistics (Hoekstra and Mekonnen 2012b; Suweis et al. 2012).

Life cycle assessment (LCA) is a commonly accepted and widely applied environmental management tool measuring the various environmental consequences caused throughout products' life cycles (Schnoor 2009). However, when assessing the environmental performance of a product by means of LCA, attention is usually drawn on the pollution of freshwater resources by means of emission-oriented impact categories such as eutrophication, acidification, or human- and ecotoxicity (Guinee et al. 2002). In contrast, the use of freshwater along a product's lifespan was usually neglected in the past. As discussed in Berger and Finkbeiner (2010), this can be explained by the historic background of LCA. This method was developed in industrial countries that usually do not suffer from water scarcity and was traditionally used to assess industrial products, which require rather low amounts of water in their production. However, there are also specific methodological challenges making the impact assessment of water use difficult. For instance, freshwater scarcity varies around the globe, different types of water, like ground and surface water, fulfill different ecological functions, and different water qualities enable different uses.

In contrast to the volumetric virtual water and water footprint approaches, LCA aims at assessing regional impacts in addition to the volumes of water used along a product's life cycle. This different interpretation of the water footprint as a volumetric- or impact-oriented indicator has led to strong dispute in the scientific community. Some scholars highlight the need of additional interpretation as 1 m^3 of rainwater consumption in Brazil does not compare to 1 m^3 of groundwater consumed in Egypt (Pfister and Hellweg 2009; Ridoutt and Huang 2012). In contrast, other authors argue that global freshwater appropriation is more important, as impacts are hard to predict and water is a global resource subject to virtual trade via products (Gerbens-Leenes et al. 2009; Hoekstra and Mekonnen 2012a). However, as problems occur due to a regional and not a global shortage of water (Ridoutt and Huang 2012), consensus is increasing that the water footprint should measure impacts in addition to volumes.

This idea is also reflected in the recently published international standard on water footprinting (ISO 14046 2014) which defines the water footprint as "metric (s) that quantifies the potential environmental impacts related to water." In contrast to the water footprint definition of Hoekstra and Hung, a volumetric water consumption should be regarded as water inventory but not as water footprint.

Similar to the impact category climate change and the carbon footprint (ISO 14064 2006), consequences of water use can be assessed as an impact category within LCA (ISO 14044 2006) or as a stand-alone water footprint (ISO 14046 2014). For this reason, water is addressed twice in the LCA compendium. While a certain methodological overlap is unavoidable, this chapter focuses on the individual water footprint analysis as a special form of LCA. In contrast, the chapter "water use in LCA" in the volume "life cycle impact assessment" (Pfister 2015) discusses the integration of water as a new impact category in LCA. In general, the differences can be found in the scope of the two approaches. An impact category "water use" in LCA usually focuses on consequences resulting from the depletion of freshwater resources. Freshwater quality alterations can also be addressed but care should be taken to avoid double counting with existing categories such as eutrophication or toxicity if mutually exclusive impact pathways are described (Berger and Finkbeiner 2013). When the water footprint is accomplished as a stand-alone analysis, the scope should be broader to address as many water-related impacts as possible.

1.5 Stakeholders

Water footprinting is basically driven from all relevant stakeholders. There is no stakeholder group actually opposing to it.

There are numerous initiatives on the government level putting water high on the priority list. As an example, the Chinese government decided to make the securing of drinking water a top priority and significantly raised fines for water polluters. The security of drinking water and purification of relevant rivers and lakes, in combination with major pollution and emissions control and urban waste treatment efforts, were highlighted in the country's 11th Five-Year Plan (2006–2010) of Environmental Protection (China Daily 2007).

As far as companies are concerned, there are both individual commitments, e.g., Nestlé defined water as one of their three key areas for their company's business strategy and essential to creating a better and healthier world in the twenty-first century (Nestle 2009), and cooperative commitments, e.g., the Food and Beverage Industry Partnership on water involving several multinationals (Nestlé, The Coca-Cola Company, and Pepsico International, etc.).

A forum having a long history in research concerning water is the UNESCO-IHE Institute for Water Education (UNESCO 2009) which has been founded in 2003 and continues the work of the IHE that began already in 1957. The overall aim of the institute is to promote global education and knowledge for integrated water resource management and to assist developing countries and countries in transition in meeting their water-related capacity building requirements. Next to several education programs, the activities of the institute comprise project work and research in the areas of wastewater treatment, water resource management, water footprinting, etc. The work in terms of water footprinting has mainly been put forward by Arjen Hoekstra who founded the Water Footprint Network (Water Footprint Network 2009) in cooperation with lots of stakeholders from industry, academia, and other organizations in October 2008.

The Water Footprint Network aims at promoting sustainable, fair, and efficient use of freshwater worldwide by advancing the water footprint methodology introduced by Arjen Hoekstra, increasing the water footprint awareness, and encouraging forms of water governance. Activities undertaken in the water footprint network mainly comprise the development of standards and tools (methods, guidelines, and criteria) for water footprinting, water footprint impact assessment, and reduction and offsetting of negative consequences of water footprints. Moreover, the Water Footprint Network assists companies, organizations, and governmental institutions in implementing water footprint accounting and in developing sustainable water policies. Besides, the network has accomplished water footprint studies for a wide range of products itself and supports communication and exchange of knowledge as well as education concerning water footprinting.

Being a founding member of the Water Footprint Network, the World Business Council on Sustainable Development (WBCSD) is active in water footprinting as well. More than 60 companies cooperate in the council project on water and sustainable development which intends to "get water higher on everyone's business agenda by providing frameworks and tools to support water management plans, as well as sharing best practice across sectors" (WBCSD 2013). Besides publishing successful experiences in water management, a tool to support business water footprinting (WBCSD 2013) has been developed (see Sect. 2.4).

With a particular focus on Life Cycle Assessment (LCA), the UNEP/SETAC Life Cycle Initiative task force 2 (UNEP/SETAC 2009) works on methodological approaches promoting water accounting in LCA. The task force has established a framework (Bayart et al. 2010) providing recommendations concerning the inventory modeling of water use and the development of impact assessment methods for water consumption. Furthermore, a comprehensive criteria-based method comparison has been established (Kounina et al. 2013) and significant aspects and modeling choices of impact assessment methods have been analyzed (Boulay et al. 2015b, c). Currently, the working group develops consensus impact assessment models to harmonize the diverse assessment approaches available today (Boulay 2015a).

Another organization involved in water footprinting is the International Organization for Standardization (ISO). Recently, the ISO subcommittee for life cycle assessment (ISO/TC 207/SC 5) (ISO 2009), has published an international standard presenting principles, requirements, and guidance for water footprinting (ISO 14046 2014). The new water footprint standard completes the series of standards concerning LCA, eco-efficiency, and carbon footprint which are provided by ISO.

The CEO Water Mandate (UN 2009), which has been established by the UN Global Compact in 2007, is a relevant comprehensive and visible cross-sectoral, public-private partnership on water. It represents both a call-to-action and a strategic framework for responsible water management by business. Being voluntary in nature, it is built around six core areas of responsibility with which its endorsers must commit to and show improvement: direct operations, supply chain and watershed management, collective action, public policy, community engagement, as well as transparency. The mandate developed a transparency framework to provide endorsers with a review of innovative practice and common approaches for reporting on water management and efficiency. With membership limited to UN Global Compact members, the mandate features endorsers with sector- and geographic diversity, including companies such as Coca-Cola, Dow Chemical, Levi Strauss, Nestlé, PepsiCo, Royal Dutch Shell and Unilever (Morrison et al. 2009).

The Alliance for Water Stewardship (AWS) is an initiative which has developed a global freshwater stewardship certification program. This voluntary water stewardship program offers independent certification rewarding responsible water management with recognition and competitive advantage. The certification scheme is designed to be applicable both to water "users" (businesses) and water "providers" (utilities). Originally introduced by The Nature Conservancy, the Water Stewardship Initiative, and the Pacific Institute, AWS is expanding to include participation from various stakeholders, including NGOs, water utilities, and businesses (Morrison et al. 2009).

From NGOs, e.g., the World Wildlife Fund (WWF) is very active in water footprinting. WWF is a founding member of the Water Footprint Network. In addition, there are numerous NGOs specifically addressing the water issue, e.g., The Blue Planet Project (http://theblueplanetproject.blogspot.com/), International Water Association (http://www.iawq.org.uk), International Water Resources Association (http://www.iwra.siu.edu), IRC International Water and Sanitation Centre (http://www.irc.nl), WaterAid (http://www.wateraid.org.uk), WaterLife Foundation (http://www.waterlife.org), or the World Water Council (http://www.worldwatercouncil.org).

1.6 Terminology

In order to enable a consistent terminology throughout this chapter, the terms water use and water consumption, which are often used synonymously, need to be defined. Adopting the terminology proposed by Owens (2001), water use describes the total withdrawal of freshwater which can be differentiated into consumptive, degradative, and borrowing water use. Consumptive water use (or water consumption) denotes the fraction of total water use which is not returned to the originating drainage basin due to evapo(transpi)ration, product integration, or discharge into other basins and the sea. Degradative water use is the part of withdrawal returned to the basin after quality degradation (e.g., wastewater discharge). In contrast, borrowing water use expresses withdrawal and discharge with low or no quality degradation (e.g., turbined water in hydropower).

Other important terms are water scarcity and water stress but they are used inconsistently throughout the literature. However, they are generally used to indicate the potential impact of using or consuming water in impact assessment as they indicate the vulnerability of the water resource or of access to it. While in the ISO standard (ISO 14046 2014) water scarcity refers to impacts based on water consumption and water stress includes also quality issues, other publications assess water stress as a function of competition (i.e., use or consumption-to-availability ratios) and therefore these terms are used synonymously.

Some technical terms are described in the ISO standard on water footprinting (ISO 14046 2014), but there is no complete definition in the standard.

2 Water Footprint Methods

By means of a comprehensive literature review a broad set of methods enabling the accounting and impact assessment of water use has been identified. They can be categorized as stand-alone and LCA-based methods (Fig. 3.3). Moreover, databases and tools which facilitate water footprinting as well as the new ISO standard have been included in the review.

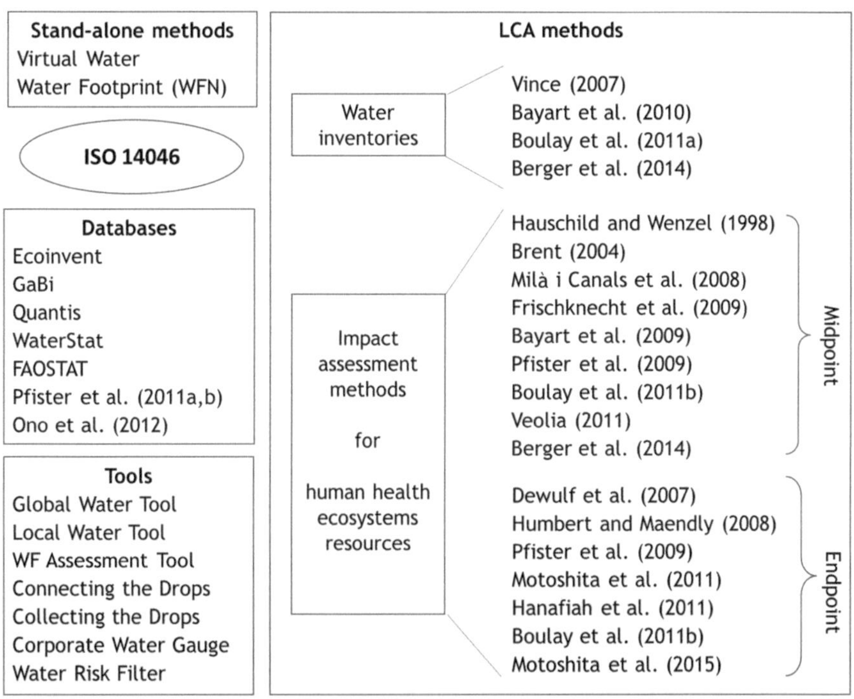

Fig. 3.3 Water footprint methods, databases, and tools identified and classified in the literature review

2.1 *Virtual Water and Water Footprint According to WaterStat Database*

Stand-alone methods like Virtual Water (Allan 1998) and the Water Footprint as defined by the WaterStat Database (WFN) (Hoekstra et al. 2011) enable the analysis of water use throughout products' or organizations' supply chains. Results are usually presented on a volumetric level and potential regional consequences are discussed on a qualitative level.

Based on the concept introduced by Allan (1998), several authors divide water into three categories: green, blue, and gray water. The green water consumption describes the evapotranspiration of rainwater during plant growth, which is especially relevant for agricultural products. Blue water consumption is the volume of ground and surface water that evaporates during production. Thus, it comprises the amount of water that is not returned into the environmental compartment from which it has been withdrawn initially. As the water that is returned to the environment (e.g., effluent of wastewater treatment plants) can be of lower quality, the gray water describes the total amount of water that is polluted by that effluent. Hence, gray water equals the volume of water required to dilute the used water until it reaches commonly agreed quality standards.

The water footprint according to Hoekstra (Hoekstra and Hung 2002) was introduced in 2002 and relies on the virtual water concept, but additionally includes spatial and temporal information (Water Footprint Network 2009). Accordingly, the quantitative water footprint of a product is the same value as its virtual water content. Furthermore, water footprints were calculated for individuals, organizations, or nations by multiplying all products and materials consumed with their respective virtual water content and by adding the direct water consumption of the person, organization, or nation (Water Footprint Network 2009).

2.2 *Methods to Assess Water Use in Life Cycle Assessment*

Methods analyzing water use in LCA can be categorized into accounting and impact assessment approaches. Accounting methods remain on the volumetric level and provide the basis for any subsequent impact assessment.

2.2.1 Accounting Methods

LCI schemes developed by Vince (2007), Bayart et al. (2010), and Boulay et al. (2011a) propose a detailed accounting of water use which considers volumetric, geographical, watercourse, and quality information in order to satisfy inventory requirements of modern impact assessment methods. The accounting scheme of

Berger et al. (2014) additionally considers effects of atmospheric moisture recycling within basins.

Impact Assessment Methods

On the midpoint level, the basic and common concept of indicators for assessing the potential environmental impacts of water use or consumption is to express the physical resource availability compared to the demand by taking the ratio of water use or consumption to water availability (WTA or CTA). This formulation represents how severely water resources are used or consumed compared to available resource amounts. Many of currently developed methods apply this concept to their indicators for assessing potential impacts in general without referring to a specific user (Pfister et al. 2009; Boulay et al. 2011b; Berger et al. 2014).

On the other hand, some of these methods integrate some unique aspects into their modeling. The potential impacts on aquatic ecosystems are explicitly considered by deducting environmental water requirement (EWR) from the available amount of water resources in the model of Milà i Canals et al. (2008).

In the context of physical scarcity of water resources, hydrological processes is another critical point. For instance, even if water is consumed by evaporation in an area, some part of them may reproduce the amounts of available water in the same or other areas through natural return flow by precipitation. These hydrological processes are modeled in the indicators by Berger et al. (2014). Specifically regarding return flow to groundwater, Milà i Canals et al. (2008) also consider the effects of replenishment in their indicator for assessing the impacts of groundwater depletion. However, the abiotic depletion indicator (ADP) (Guinee et al. 2002) is applied to model the impacts in their indicator for groundwater depletion for the sake of consistent assessment with other abiotic resources. Thus, the amount of resources is squared in the numerator of their indicator for groundwater depletion and the ratio of demand to squared availability is normalized with that of antimony in the same was as other abiotic resources. In this context, the indicator for groundwater depletion proposed by Milà i Canals et al. (2008) has a conceptual difference in the form of the assessment indicator.

As illustrated in Kounina et al. (2013), water consumption will result in impacts on endpoint level related to three areas of protection, "human health," "ecosystem quality" and "resources." The cause-effect chain on human health is relatively clear compared to that on ecosystem quality and resources. Therefore, several models provide indicators to assess potential damages on human health resulting from water consumption. Malnutrition damage due to agricultural water scarcity (Motoshita et al. 2008, 2014; Pfister et al. 2009; Boulay et al. 2011b) and infectious diseases damage due to domestic water scarcity (Motoshita et al. 2011, 2014; Boulay et al. 2011b) have been modeled as a consequence of water consumption at global scale. The impact pathways of water consumption to ecosystem quality are more complicated and the targets and approaches for assessing potential damage are very diverse. From the aspect of terrestrial ecosystem, the potentially disappeared fraction (PDF) of terrestrial plants is estimated as a proxy of plant growth prevention resulting from water consumption at a global scale (Pfister et al. 2009). Detailed pathway of water consumption to terrestrial species loss are described by modeling the relationship between the risks of species loss and the changes of groundwater table level due to water consumption in the context of the Netherlands (Van Zelm et al. 2011). However, there are limitations to expand this kind of detailed analysis to global scale due to the lack of knowledge on region-specific situations of the pathway and data availability of parameters for the analysis. Regarding the aquatic ecosystem, Hanafiah et al. (2011) focus on fish species and model the impacts on them resulting from loss of river discharge as a consequence of water consumption at global scale. Verones et al. (2013) assess the impacts on species in wetlands (birds, water-dependent mammals, reptiles and amphibians) resulting from wetland area change because of water consumption at global scale. From the viewpoint of costal wetland species, the impacts of water consumption on both terrestrial and aquatic species (plants, fish, algae and a crustacean) is estimated as a result of salinity increase in coastal wetlands at global scale (Amores et al. 2013). On the other hand, water use (including nonconsumptive use from reservoirs for hydropower) may also affect aquatic ecosystem and cause fish species loss per unit power production. This cause-effect chain is estimated for dams in Switzerland and the United States (Humbert and Maendly 2008). Potential damages on resources have been attempted to describe from different aspects. Dewulf et al. (2007) regards the loss of accumulated exergy of water resources as potential damage on resources by water consumption. On the other hand, energy consumption for desalinated water production is accounted as potential damage of water consumption to compensate the lack of water resources based on the concept of "backup technology" (Pfister et al. 2009). While the former represents the lost energy "for other water users" that may be theoretically utilized, the latter targets on the energy that will be consumed "by other water users" and deprived from "other users including non-water users." Even though both approaches assess potential damage on resources in terms of energy, the target users of potential damage are different.

All of the above methods are related to the assessment of physical freshwater scarcity. For an assessment of quality degradation, impact assessment methods on pollution of freshwater (like eutrophication, aquatic acidification, chemical toxicity, etc.) are applicable (e.g., Guinee et al. 2002). Impact assessment methods which focus on the pollution of freshwater are not included in this review.

An early summary of the water footprint approaches available along with a discussion of individual strengths and weaknesses has been published in a review paper of Berger and Finkbeiner (2010). Many of the methods developed after the publication of this review paper are summarized in the work of Kounina and colleagues (2013). Regarding midpoint level and human health damage on endpoint level, the consistency and difference in current models are tested and some significant parameters relevant to the differences among models are revealed in the method comparison analysis of Boulay et al. (2015c). These will be helpful for deeper understandings on the differences in current models.

2.3 Databases for Inventory Analysis

In addition to water footprint methods as such, several databases have been identified which provide water use and consumption data for various products and materials. Databases can be divided into typical life cycle inventory (LCI) databases like GaBi (Thinkstep 2016) and ecoinvent (Ecoinvent centre 2015), sector and country-specific databases (FAO 2013; Pfister et al. 2011b, c; Ono et al. 2012), and distinct water footprint databases like the Quantis Water Database (Quantis 2013) or the WaterStat database (WFN 2013b).

2.4 Tools

Moreover, several tools like the Global Water Tool (WBCSD 2013), the Local Water Tool (GEMI 2013c), the WF Assessment Tool (WFN 2013a), Collecting the Drops (GEMI 2013a), Connecting the Drops (GEMI 2013b), the Corporate Water Gauge (CSO 2013), and the Water Risk Filter (WWF 2013) have been identified which facilitate the accounting of a company's (direct) water use and assess environmental, operational, legal, and reputational risks.

2.5 Water Footprint ISO Standard

The international community recently finalized an international standard on water footprinting (ISO 14046 2014). Aiming at "providing transparency, consistency, and credibility for assessing water footprint and reporting water footprint results of products, processes or organizations," the standard includes "principles, requirements, and guidelines" on water footprinting. After defining a consistent terminology and describing underlying principles, the methodological framework is presented and guidance on reporting and critical review is provided.

In line with the LCA structure (ISO 14044 2006), the framework of a water footprint analysis comprises the goal and scope definition, the water footprint inventory analysis, the water footprint impact assessment, and the interpretation of results.

The standard clearly states that the water footprint assessment is an impact-based measure. Contrary to the definition of Hoekstra and colleagues (2011), a water footprint inventory can be reported but shall not be termed "water footprint." It is stated that a water footprint assessment can be used as both a stand-alone analysis and part of an LCA containing additional environmental information. The water footprint assessment should be a comprehensive analysis comprising water availability and water pollution aspects. If only single aspects of such a comprehensive analysis are taken into account, this should be reflected in the name of the study. For example, a "water availability footprint" considers only the volume of water consumed and the resulting impacts. In contrast, a "water eutrophication footprint" assesses the impacts of eutrophication caused by water pollution but neglects the consumed volume.

Rather than proposing a specific inventory and impact assessment method, the standard defines criteria which have to be fulfilled in an ISO compliant water footprint study. For instance, elementary flows should include information concerning quantities, type of watercourse, water quality, types of water use, geographical location, time, and emissions. In impact assessment, water availability footprints should be determined by means of characterization models assessing "the contribution of the product, process, or organization to pressure on water availability." In a similar way, water footprints addressing water degradation should be determined by characterization models describing "the contribution of the product, process, or organization to impacts related to water degradation." The preferred water footprint profile contains several impact categories measuring water availability and degradation footprints.

In order to illustrate the methodological guidance provided in ISO 14046, a technical report is currently under development (ISO TR 14073 2014), which contains various practical case studies from different areas of application

2.6 *Comparison of Water Footprint Methods, Databases, and Tools*

In order to enable a comparison of the water footprint methods, databases, and tools identified in the (updated) literature review, a set of criteria has been developed for assessing their scope and applicability. These criteria comprise:

- The type of water analyzed
- The type of usage considered
- The inventory data required/provided
- Areas of protection (AoP) addressed
- Availability of characterization factors
- ISO 14044 compliance regarding comparative assertions disclosed to the public

As shown in Table 3.1, most methods focus on consumptive blue water use. Green and gray water is mainly considered by stand-alone methods in order to address rainwater evapotranspiration of agricultural products and degradative freshwater use, respectively. The analysis revealed that the inventory requirements of water footprint methods differ significantly. In general, scientifically advanced methods show a need for higher resolution inventory data. In addition to volumes and regional information, methods like Veolia (2011) or Boulay et al. (2011b) require information on watercourses and water qualities. In some cases, even temporal information is required to acknowledge varying water scarcity throughout

Table 3.1 Scope and characteristics of the water footprint methods, databases, and tools identified in the literature review

		Method	Water analyzed			Type of usage considered		Inventory data required/provided	
			Green	Blue	Gray	Water use	Water consumption	Volumes	Geography
Stand-alone		Virtual water	x	x	x	x (gray)	x	x	
		Water footprint (WFN)	x	x	x	x (gray)	x	x	x
Life cycle assessment methods	LCI	Vince (2007)		x		x	x	x	x
		Bayart et al. (2010)		x		x	x	x	x
		Boulay et al. (2011a)		x		x	x	x	x
		Berger et al. (2014)		x			x	x	x
	LCIA (midpoint)	Hauschild and Wenzel (1998)		x			x	x	x
		Brent et al. (2004)		x		x		x	x
		Milà i Canals et al. (2008)	x	x		x	x	x	x
		Frischknecht et al. (2009)		x			x	x	x
		Bayart et al. (2010)		x			x	x	x
		Pfister et al. (2009)		x			x	x	x
		Boulay et al. (2011b)		x		x	x	x	x
		Veolia (2011)		x		x	x	x	x
		Berger et al. (2014)		x			x	x	x
	LCIA (endpoint)	Bösch et al. (2007)		x		x[a]	x	x	
		Humbert and Maendly (2008)		x[a]		x		x	
		Pfister et al. (2009)		x			x	x	x
		Motoshita et al. (2011)		x			x	x	x
		Hanafiah et al. (2011)		x					
		Boulay et al. (2011b)		x		x	x	x	x
		Motoshita et al. (2014)		x			x	x	x

			Areas of protection addressed in impact assessment				Availability of characterization factors	ISO 14044 compliance[b]
Watercourse	Quality withdrawal	Quality discharge	Human health	Ecosystems	Resources	Unspecified		
		x			–		–	–
		x		x			For the main basins	Yes
x	x	x						
x	x	x			–		–	–
x	x	x						
					x		To be calculated	Yes
						x	For South Africa	No, weighting
x				x	x		For main basins	Yes
					x		For basins/ countries	No, weighting
x	x	x	x				For 7 countries	Yes
						x	For basins/ countries	Yes
x	x	x	x				For basins/ countries	Yes
x	x	x				x	To be calculated	Yes
					x		For basins/ countries	Yes
					x		Fixed exergy content	Yes
				x			To be calculated	Yes
			x	x	x		For basins/ countries	Yes[c]
			x				For countries	Yes
				x			For 214 river basins	Yes
x	x	x	x				For basins/ countries	Yes
			x				For basins/ countries	Yes

(continued)

Table 3.1 (continued)

	Method	Water analyzed			Type of usage considered		Inventory data required/provided	
		Green	Blue	Gray	Water use	Water consumption	Volumes	Geography
Databases	Ecoinvent		x		x	x	x	
	GaBi		x		x	x	x	
	Quantis		x		x	x	x	x
	WaterStat	x	x	x	x (gray)	x	x	x
	FAOSTAT		x			x	x	x
	Pfister et al. (2011a)		x			x	x	x
	Pfister et al. (2011b)		x			x	x	x
	Ono et al. (2012)	x	x		x	x	x	
Tools	Global Water Tool	x	x		x	x	x	x
	Local Water Tool	x	x		x	x	x	x
	WF Assessment Tool	x	x	x	x	x	x	x
	Collecting the Drops		x		x	x	x	x
	Connecting the Drops		x	x	x	x	x	x
	Corporate Water Gauge		x		x	x	x	x
	Water Risk Filter		x	x	x	x	x	x

Updated from Berger and Finkbeiner (2010)
[a]Barrage water only
[b]For comparative assertions disclosed to the public
[c]Unless aggregated eco-indicator 99 result is used

			Areas of protection addressed in impact assessment					
Watercourse	Quality withdrawal	Quality discharge	Human health	Ecosystems	Resources	Unspecified	Availability of characterization factors	ISO 14044 compliance[b]
x								
x								
x	x	x						
		x						
					–		–	–
x								
x								
x								
x		x			-		-	-
x	x	x						
x								
x	x	x						

the year (Hoekstra et al. 2012; Pfister and Baumann 2012). Even though this increased level of precision is appreciated from a scientific point of view, such inventory requirements are hard to fulfill – especially if complex background systems are involved. Hence, the trade-off between "precision" and "applicability" needs to be addressed in future studies and in the new international standard. Considerable differences have been detected concerning the availability of characterization factors. While some methods comprise characterization models but no factors (e.g., Hauschild and Wenzel 1998), other methods provide characterization factors on both drainage basin and country levels (e.g., Pfister and Hellweg 2009; Berger et al. 2014). As some of the impact assessment models contain a weighting step (e.g., Frischknecht et al. 2009), they cannot be used in water footprint studies comprising comparative assertions disclosed to the public.

The level of detail provided in LCI databases differs significantly. While LCI databases like GaBi and ecoinvent only provide information on the volumes and watercourses used, additional regional, quality, and even temporal information can be found in distinct water footprint databases. A similar variation concerning inventory requirements has been identified in the water footprint tools.

A detailed follow-up characterization of methods has been accomplished by the water use in LCA (WULCA) working group of the UNEP/SETAC Life Cycle Initiative (Kounina et al. 2013). This work is based on the review scheme of the International Reference Life Cycle Data System (JRC-IES 2011). In a recent work, WULCA analyzed the influence of key methodological choices on the resulting characterization factors in a broad scope of impact assessment methods (Boulay et al. 2015b, c).

The review presented in this chapter, which updates the publication of Berger and Finkbeiner (2010), clearly shows that there is not only one "water footprint." Next to stand-alone methods, databases, and tools, most methods have been developed in an LCA context. Impact assessment models range from rather simple scarcity indicators up to comprehensive endpoint models which describe complex cause-effect chains. In addition to differences concerning the addressed areas of protection and the availability of characterization factors, water footprint methods differ significantly regarding their inventory data requirements. While for some impact assess models the volume und the regional information are sufficient to enable applicability, other methods require additional quality or even temporal information. In order to support this theoretical comparison and to test applicability, some of the water footprint methods have been applied in industrial case studies, whose results are presented in the following section.

3 Lessons Learned: Water Footprint Case Studies

3.1 Volumetric Studies

From the perspective of product water use, there are numerous studies accounting for volumes of water use/consumption related to products. Especially, accounting water use/consumption associated with agricultural products has been focused in many studies because they are one of the biggest users of freshwater. Volumetric values of water use/consumption are easy to imagine and illustrate the significance of water related to products. As a good example, Humbert et al. (2009) showed that just one cup of coffee (100–200 mL) uses 13–42 L of water (more than 100–200 times of its product volume) throughout its life cycle. This type of volumetric information will succeed to draw interests of general public. In addition to agricultural products, some of industrial products have been also analyzed with regard to their volumetric water use/consumption, like a study on diapers by Sauer et al. (1994) which represents an initial volumetric analysis. Virtual water is the concept of accounting volumetric water saved by trade from consumer perspective and studies with many types of products and nations have been conducted as mentioned in the previous section.

Volumetric studies give good insights of water volumes related to product life cycles and may attract attentions of general public by providing imaginable information about environmental loads of products. On the other hand, there is a potential risk for misleading because the relevance of the same volume of water differs from region to region. Figure 3.4 shows the relationship between total water withdrawal and surplus amount of available water resources in countries based on the AQUASTAT database (FAO 2015). Surplus amount of available water resources is calculated by subtracting withdrawal from total available water and indicates the amount of water left for other users. Even in countries with similar total withdrawal, surplus amounts of available water resources vary with several orders of magnitude. This fact supports that a volume of water use/consumption has spatially different meanings. In the beginning of water footprint studies, volumetric analysis was dominant; however, impact assessment is currently recognized as the mandatory step for water footprint assessment. This is also mentioned clearly in the ISO standard of water footprint (ISO 14046 2014).

3.2 Impact-Oriented Studies

3.2.1 Water Footprint Assessment of European Passenger Cars as Part of a Complete Product Life Cycle Assessment (Reproduced from Berger et al. 2012)

Introduction Volkswagen has been analyzing the environmental effects of its cars and components by means of LCA for many years (Volkswagen 2010). However,

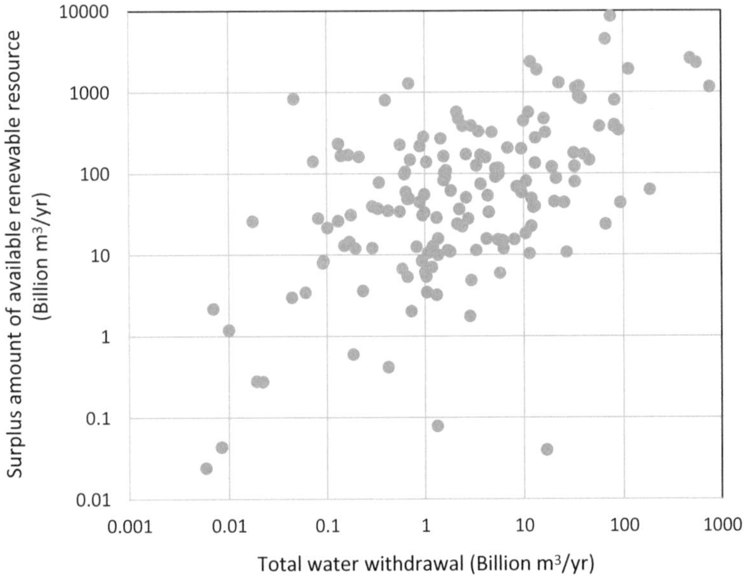

Fig. 3.4 Relationship between total water withdrawal and surplus amount of available water resource in each country

due to lack of data and appropriate impact assessment models, the consumption of freshwater has not been considered until recently. Therefore, the aim of this study was to analyze freshwater consumption along the life cycles of three Volkswagen car models on both inventory and impact assessment levels.

Methodology Water consumption was analyzed along the life cycles of the three cars, comprising the production, use, and end-of-life phases. In order to apply impact assessment methods evaluating the consequences of the water consumption determined from the LCI models, the basic volume is not enough. Regionalized water inventories, which state the location where water consumption occurs, are needed to consider regional water scarcity conditions, the vulnerability of ecosystems, or socioeconomic parameters affecting the sensitivity to water scarcity-induced health damages (Berger and Finkbeiner 2010). Such geographically explicit water inventories are determined in a top-down approach. First, the car's total water consumption is divided into the shares consumed by the life cycle stages production, use, and end-of-life. For further specification, the water consumed in the production phase is assigned to manufacturing steps and to 15 material groups. Now the water consumption caused by the manufacturing steps and material groups is allocated to specific countries based on production mixes, location of suppliers, production sites, etc. Based on these inventories, seven impact assessment methods, which represent different levels of sophistication and model different impact pathways, were applied.

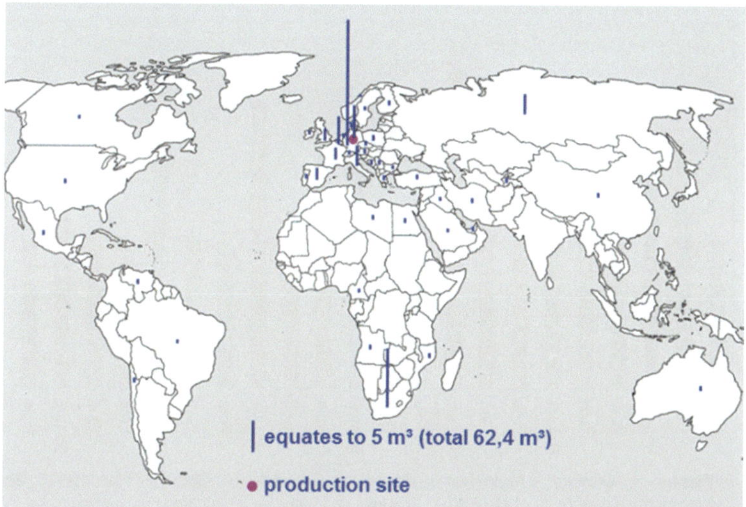

Fig. 3.5 Global water consumption throughout the life cycles of the Golf 1.6 TDI (Berger et al. 2012)

Results and Discussion Water inventory. The water consumption along the life cycles of the three cars amounts to 51.7 m^3 (Polo 1.2 TDI), 62.4 m^3 (Golf 1.6 TDI), and 82.9 m^3 (Passat 2.0 TDI). When assuming fossil fuel consumption, 95 % of the water is consumed in the production phase of all three cars which mainly results from the production of steel and iron, precious metals, as well as polymers.

The top-down regionalization revealed that water consumption takes place in 43 countries worldwide (Fig. 3.5). Less than 10 % are consumed directly at the production sites in Pamplona, Wolfsburg, and Emden resulting mainly from painting and evaporation of cooling water. Hence, more than 90 % of the water consumption along the cars' life cycles is caused by the material and energy production in the background system.

Impact Assessment Based on the regionalized water inventories, the impact assessment models of the ecological scarcity method (Frischknecht et al. 2009; Motoshita et al. 2011; Pfister et al. 2009) were applied in order to evaluate consequences resulting from water consumption in different countries.

Figure 3.6 shows the results obtained by means of the water inventory and impact assessment methods normalized to the Polo. Since absolute results for the Polo differ among the scenarios, Fig. 3.6 only allows for comparing the cars within one impact category.

The results of the ecological scarcity method and the impact category freshwater deprivation depend on two factors: the volume of water consumed and the physical water scarcity at the place of consumption. While the ecological scarcity method uses the WTA ratio as a weighting factor directly, freshwater deprivation uses a

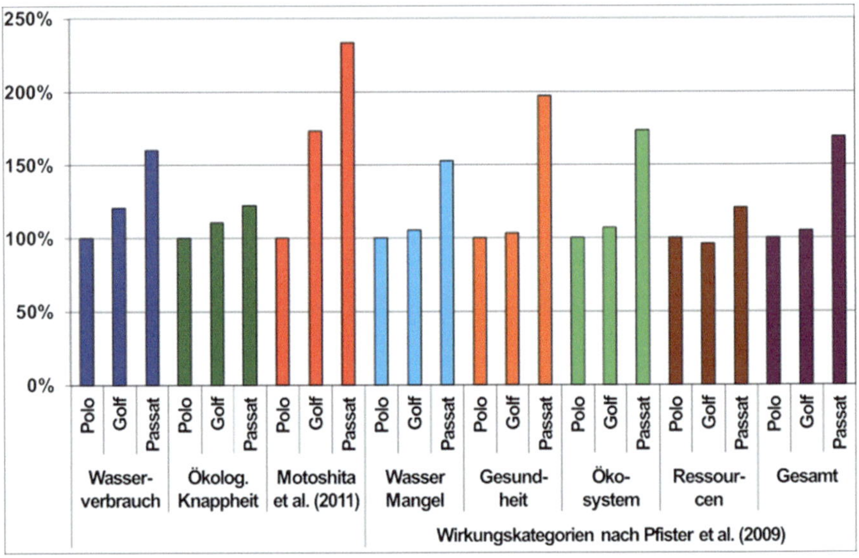

Fig. 3.6 Relative comparison of results on the inventory and impact assessment levels

water stress index (WSI) as a characterization factor which is based on WTA, but additionally considers seasonal variation of water availability (Pfister et al. 2009). Despite different proportions, both methods are dominated by the water consumption in similar countries mainly Germany (due to high volumes) as well as Spain, Belgium, and South Africa (due to high scarcity).

While the method of Pfister et al. (2009) assessing health damages from malnutrition considers physical water scarcity and socioeconomic aspects, the method of Motoshita et al. (2011) measuring health damages from infectious diseases considers only socioeconomic aspects. As physical water scarcity is high and the level of development is rather low, human health impacts measured according to Pfister et al. (2009) are dominated by the water consumed in South Africa resulting from the PGM production. Damages determined in the method of Motoshita et al. (2011) are mainly caused by relatively low amounts of water (78–191 L) consumed in the aluminum production in Mozambique. In contrast, due to high sanitation standards and a high degree of development, the water consumption in countries like Spain or Australia doesn't cause damages to human health, despite high physical water scarcity in these countries.

Ecosystem damage denotes the loss of biodiversity and is influenced by water scarcity and the local sensitivity of vascular plants (Pfister et al. 2009). Again, the water consumption in South Africa dominates the impact assessment result with 56 % (Golf) to 67 % (Passat). Damages caused by the depletion of resources only occur in countries where water withdrawal exceeds the renewability rate (WTA >1). As this is not the case in Central Europe, where most of the water is consumed, large shares of water consumption do not contribute to resource damage. This impact category is dominated by the water consumption in Spain and Ukraine which contribute 55 % (Passat) to 67 % (Polo) to the overall result depending on the car.

Comparison Between Cars By showing the results of the water inventory and impact assessment methods normalized to the Polo.

Figure 3.6 allows for a comparison of the three cars. It can be seen that the increased water consumption of the Golf and Passat are to a similar extent reflected by the ecological scarcity method and the model of Motoshita et al. (2011), showing that these methods lead to similar conclusions as the inventory in this scenario. Yet, in the categories developed by Pfister et al. (2009), the impacts of the Polo and Golf are regarded as rather similar despite different water consumption. This can be explained by two facts. First, similar water consumption is weighted higher at the Polo's production site in Spain than at the Golf's production site in Germany. This compensates the advantages of the lower water consumptions in the material production resulting from the reduced weight of the Polo in comparison to the Golf. Second, some impact categories, especially the one developed by Pfister et al. (2009) measuring damages to human health, are dominated by the water consumption of the PGM production in South Africa. As the PGM contents of the Polo and Golf are comparable, results of these impact categories are similar, too. Since the Passat contains more PGM than the Polo and Golf, the same reasoning can explain the higher impacts in the human health categories. In contrast, the water consumption in South Africa does hardly affect damage to resources since WTA is below 1 in most watersheds, which, according to Pfister et al. (2009), means that no depletion of water resources occurs. For that reason the Passat scores only slightly worse in this impact category due to the larger water consumption of the larger material production.

3.2.2 Water Footprint Assessment of Biofuels (Based on Pfister et al. 2011a)

Introduction Biofuels have become a major driver for crop production in recent years and they compete with food and fiber production. Since crop production is very diverse and depending on climate conditions, the water footprint of bioenergy is very dependent on the crop type and origin used to produce it. This example illustrates the water availability footprint of different biofuel types from specific crops using global average and country-specific data and compares it to cotton as another crop that has an alternative fossil feedstock (polyester) and is not as essential as food production.

Methodology In a first step, the water consumption in crop production is estimated applying a global model on a 5 arcmin resolution (<10 km). The CROPWAT method (FAO 1999) is applied to each model cell with different climate input data and coupled with crop production and yield data (Monfreda et al. 2008) to calculate irrigation water requirements per kilogram of crop harvested. Since these are not necessarily met, we applied a map with shares of irrigated area in each model cell (Siebert et al. 2007) and assumed only this share of irrigation water requirement to be met. This gives a lower and upper estimate, since irrigation might take place also in areas, where it is not reported.

The water stress index (WSI) from Pfister et al. (2009) is applied on each model cell. Using a geographic information system (GIS), the WSI, which is specified on a spatial resolution of >10.000 watersheds, is combined with the water consumption model, which has >2 million model cells.

In order to derive country average values, both the inventory (water consumption) and the water stress weighted consumption (impact) are aggregated on country average (weighted by the production volume of the crop in each model cell within the country).

The biofuel production was modeled based on Zah et al. (2007) and we applied water consumption and impacts for crop production in the specific countries reported as well as for assuming a global crop mix. The variability of the values within a region have been quantified by the coefficient of variation (CV), which is the standard deviation divided by the mean of all values within the region (weighted by production volume in each model cell). In order to compare the different biofuels among each other and with cotton, the net fossil fuel saving by these biotic products are calculated per GJ of fossil fuel saved (including the production process from crop to product).

Results and Discussion Figure 3.7 shows the result of the crop modeling on high spatial resolution and the resulting map for country average crops for the example of wheat. It can be seen that there is a high variability in water consumption but even a much more distinct picture is resulting after applying impact assessment. The aggregation from model cell to country level might not always be intuitive: Some countries like the US have many regions with high water consumption and footprint but the country average is rather low since the main production is in areas of low water footprint.

The different water consumption and related footprint as a function of the cultivation location is also reflected in the results for different biofuel options. Table 3.2 shows that the variability among the fuels is very high, ranging from almost zero (for palm oil biodiesel) to 36 m^3 equivalents of highest water scarcity (for rapeseed) per GJ of fossil fuel replaced. On the other hand, rapeseed biodiesel can be produced with very low water consumption in Switzerland and similar countries. The CV of each crop/region combination indicates that especially for large and climatically diverse countries such as the US or Brazil, the variability within the country is very high and therefore the water footprint is heavily depending on where the feedstock is sourced from. Compared to cotton (as a replacement for polyester) biofuels have rather low water footprints and one might conclude that wearing cotton is the worse option to stop fossil fuel consumption than driving biofuels. However, the quality trade-off in the cotton case needs to

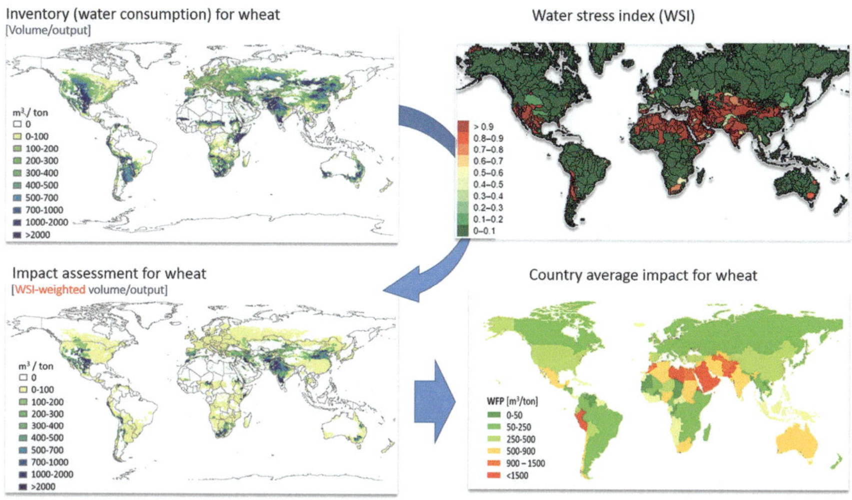

Fig. 3.7 Result for wheat production. Water consumption estimates (*upper left*) are multiplied with the WSI (*upper right*) to arrive at water scarcity footprint result (impacts, *lower left*). Based on the production volume high-resolution results are aggregated to country averages (*lower right*) (Based on Pfister et al. 2011a)

Table 3.2 Inventory (BW) and impacts (WFP) in m^3 per ton crop and m^3 per GJ fossil fuel saving. The variability of WFP within a region is quantified by the production-weighted coefficient of variation (CV) of all model cells

		M^3/ton crop		m^3/GJ fossil fuel saving		Variability
Crop	Origin	BW	WF	BW	WF	CV (WF)
Maize	US	202	42	50	10	2.5
Maize	Global	239	90	58	22	2.0
Sugar cane	Brazil	35	1.1	16	0.5	5.3
Sugar cane	Global	62	34	31	17	1.2
Palm oil	Malaysia	14	0.6	2.4	0.1	N/A
Palm oil	Global	104	3.6	12	0.4	N/A
Soybean	US	794	85	119	13	2.7
Soybean	Global	688	144	98	20	2.1
Rapeseed	Switzerland	109	10	10	0.9	N/A
Rapeseed	Global	648	401	58	36	1.3
Cotton	Global	1698	1314	109	84	0.8

be considered and the lower relevance of clothes, since the demand for fibers is much lower than for fuels.

The reported results have relatively high uncertainty of the modeled water consumption values. On global average, high and low estimates differ by more than a factor 2 and even more in specific cases (further discussed in Pfister and Bayer 2014). Nevertheless, the spatially explicit modeling helps to better capture inventory results but also to better assess the environmental impacts on this level of spatial detail. Therefore, it is suggested to use aggregated results on country or global level based on the impacts rather than applying country average inventory to country-level WSI values. If results of a water footprint study are heavily affected by crop production, it is suggested to apply a higher spatial and also temporal resolution (see Pfister and Bayer 2014) and potentially collect more detailed information on effective irrigation in order to improve the results.

4 Remaining Challenges in Water Footprinting

4.1 Definitions of Elementary Flows for Water Consumption

Proper water footprinting needs a water balance over each of the analyzed processes. This means all the inputs and outputs need to be reported as described in Fig. 3.8 and by Pfister (2015): water flows from (inputs) and to the environment (outputs) and flows of water from product inputs and outputs (water content of products including tap and wastewater flows). For agricultural and forestry processes, the environment (agricultural or forest land) needs to be included for analyzing "green water" use and net soil water use (Δ soil moisture) compared to the reference land cover. Current inventory databases do not account for this issue and it is not considered essential in most cases. "Δ precipitation flows" represents rainwater collection systems (in industry or agriculture). The most important flow for water footprinting is Δ evapotranspiration, which is evapotranspiration caused by the process and is generally the largest share of "blue water" consumption (complemented by product integration and discharge to salt water). While salt water is typically not considered in water footprinting, it can be necessary to balance the water flows over a process, especially if seawater cooling systems are involved. Distinction of blue water consumption into evaporation, product integration, and release to see is also important for the recycling effect for water emissions to air (WAVE, Berger et al. 2014).

Green Water The water consumption from soil moisture (green water) is a controversial issue in water footprinting as it is a natural water supply received through land use (soil moisture provision is a function of the land). Only the difference in consumption of such water compared to the natural situation might be included (Pfister et al. 2009). Therefore these indirect forms of water consumption need to be linked to land use inventories. Nunez et al. (2013) derived regionalized indicators for determining green water consumption of the "reference land use" with global coverage. Net green water consumption is calculated as the difference of green water consumption between the analyzed activity and the reference land use.

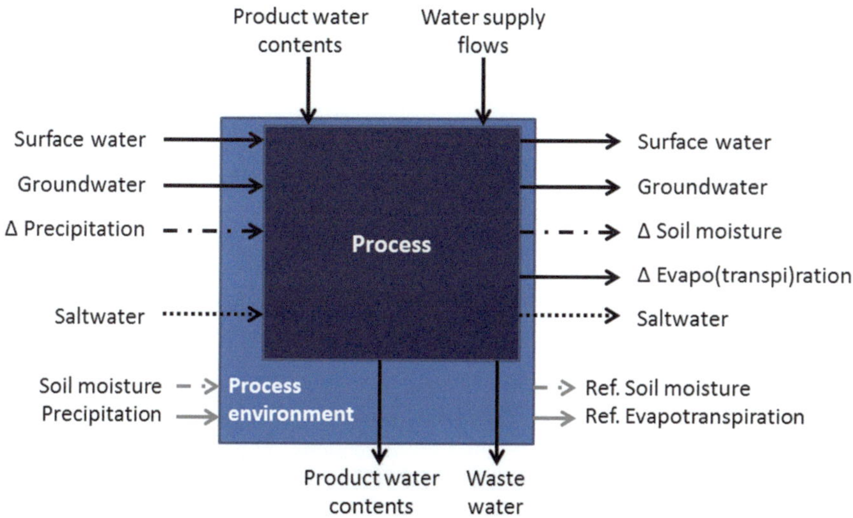

Fig. 3.8 Inventory flows relevant to the assessment of water use impacts. Exchanges with the environment are indicated by horizontal flows, while the vertical flows are technosphere flows (products). The process box shows the system boundaries for the process considered, and the process environment box is relevant for agriculture for the water flows of natural water supply from soil and precipitation. These flows can be considered as "neutral" flows as they present flows of the natural or reference environment (ref. flows) and are therefore considered to occur outside the system (Based on Pfister 2015)

Water Source While most existing impact assessment methods do not distinguish surface and groundwater use, the water balance as depicted in Fig. 3.8 cannot only be balanced for total freshwater consumption in each activity but also for different types of water flows. This can be important, since an activity might consume surface water but have a negative consumption of groundwater. An example is flood irrigation in agriculture, where typically river water is partially consumed as blue water but partially feeding into groundwater an having a negative consumption (i.e., production of groundwater), which can lead to an overall beneficial effect, even if there is a net water consumption (Verones et al. 2012). However, data availability is very low on a global level and therefore might be only considered in foreground systems.

Water Quality As mentioned above, water quality classes might be integrated, differentiating all the flows by pollution level as suggested by Boulay et al. (2011a) to be used with the corresponding impact assessment method (Boulay et al. 2011b). While this is useful to address water quality as a "sum-parameter" from a resource perspective, it is very difficult to include in water accounts and water quality impacts might be better addressed by accounting for all substance flows related to water use and a consequent impact assessment, such as suggested by the structure of existing LCI databases.

Regionalization Spatial distinction of water consumption is essential due to the high variability of related impacts, as outlined above. Nevertheless, standard software tools have only minimal spatial differentiation such as country level if at all. However, a watershed perspective is necessary, since water flows transferred from one watershed to another need to be considered in the assessment. This is necessary because if withdrawal and discharge occur in different watersheds, this causes consumptive use in one watershed and negative consumptive use in the watershed of release (Lin et al. 2012). Coupling regionalization of inventory and impact assessment in LCA and water footprinting tools is therefore an important challenge to address in order to make proper water footprint applicable for practitioners.

Temporal Resolution Temporal aspects are typically not accounted for in existing databases. However, water use-related environmental effects depend on the timing of the use. In terms of feasibility, monthly time steps are applicable for assessing water scarcity impacts (Hoekstra et al. 2012; Pfister and Bayer 2014). To facilitate such assessment, inventories need to include this information too, at least for foreground processes. It is mainly important for agricultural production, where large variability in water use exists among the months as a function of the growth season (Pfister and Bayer 2014).

4.2 Accounting for Degradative Use

While consumptive assessments for a water scarcity footprint has been quite advanced, proper inclusion of water quality issues are more debated. This has mainly two reasons: in LCA the quality changes are basically addressed through emissions to water, while in the WFN approach a very basic dilution-oriented assessment has been used, which only focused on nutrient emissions and resulted volumes of water polluted (gray water) that could be confused with real water volumes quantified for water consumption. In many reports (e.g., Gerbens-Leenes et al. 2009) these numbers are summed up, which has no scientific meaning. To overcome these issues and combine state-of-the-art assessments for water quality and quantity, Ridoutt and Pfister (2013) suggest to combine these impacts based on endpoint LCA methods in order to aggregate the damages in one unit and calculate water consumption equivalents based on the ReCiPe method (Goedkoop et al. 2009).

Boulay et al. (2011a) finally suggest that water quality degradation has an impact beyond those tackled by pollution and consequent effects and suggest division of water flows into quality classes. The main issue is the lacking information on water quality since >100 substance concentrations are needed for the method and potentially this assessment leads to double counting with pollution effects on human health.

The combination of different LCIA methodologies to arrive at an aggregated water footprint including quality and quantity issues needs some more exploration but it seems suitable to provide guidance for practitioners in the near future. In contrast, the quantification of further impacts due to water quality decrease needs additional research and especially data to be implemented in practice.

4.3 Level of Spatial and Temporal Detail

While Sect. 4.1 discusses challenges related to spatially and temporally explicit elementary flows in inventory databases, it should be noted that there are methodological issues as well.

In theory impact assessment models should try to be as temporally explicit as possible and consider water use on the monthly level. However, it is often ignored that such a level of temporal detail requires the consideration of inter-monthly storage capacities which can buffer water-scarce periods throughout the year (Pfister and Baumann 2012). Thus, a consumption-to-availability ratio of a dry month can overestimate scarcity when the water reservoirs created during earlier wet months are ignored. Moreover, the temporal resolution of water scarcity assessments also determines the required spatial resolution. Large basins can have flow times of several months from spring to mouth, which makes a monthly assessment difficult.

5 Outlook: The Future of Water Footprinting

5.1 Consensus Model of WULCA

The WULCA working group is committed to provide harmonized methods and indicators for a generic midpoint indicator for assessing water scarcity as well as an endpoint indicator for addressing impacts on human health by spring 2016. The work is ongoing and the final agreed method will be determined in an expert workshop early 2016. The status of the work has been summarized in Boulay et al. (2015) and is depicted in Fig. 3.9. The main focus is on providing a generic midpoint indicator for assessing water scarcity, which can be used for water scarcity footprint assessment such as the global warming potential for carbon footprint (Inaba et al. 2016).

The preliminary recommended method for the generic midpoint indicator is called AWaRe and denotes the relative Available WAter Remaining per area in a watershed, after the demand of humans and aquatic ecosystems has been met (WULCA 2015). It is first calculated as the water availability minus the demand (humans and aquatic ecosystems) and is relative to the area ($m^3\ m^{-2}\ month^{-1}$),

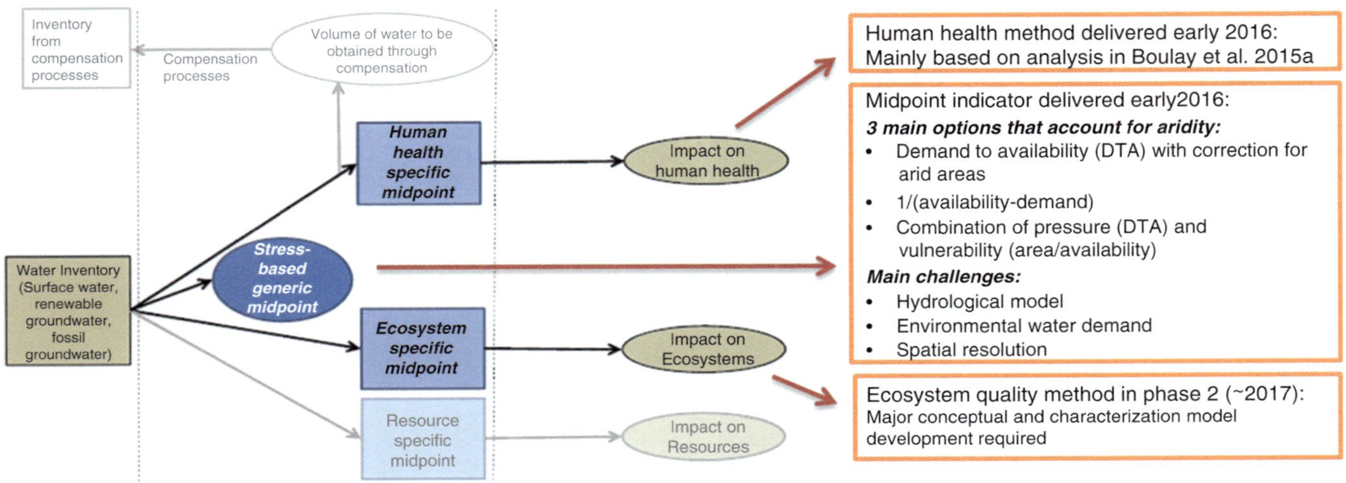

Fig. 3.9 Framework of the harmonization approach. The *left side* indicates the framework of impact pathways (the faded impact pathways are not mature yet). The *right side* boxes indicate the status of the harmonization efforts (Based on Boulay et al. 2015a)

hence representing the area "virtually occupied" to cover the additional water consumption sustainably. In a second step, the value is normalized with the world average result and inverted and hence represents the relative value in comparison with the average m^3 consumed in the world (the world average is calculated as a consumption-weighted average). The indicator is limited to a range from 0.1 to 1000, with a value of 1 corresponding to the world average, and a value of 100, for example, representing a region where there is 100 times less available water remaining per area than the world average. The indicator is calculated at the sub-watershed level and monthly time-step, and then aggregated, if needed, to country and/or annual resolution. This aggregation can be done in different ways to better represent an agricultural use or a domestic/industrial use, based on the time and region of water use. Characterization factors for agricultural and nonagricultural use are therefore provided, as well as default ones if the activity is not known. This method quantifies the potential of water deprivation, to either humans or ecosystems, and serves in calculating a water scarcity footprint as per ISO 14046 (2014). Characterization factor scan be downloaded from the projects homepage: http://www.wulca-waterlca.org/project.html.

For human health impacts, the harmonization effort builds up on the method comparison in Boulay et al. (2015b) and the main question is how to account for the different cause-effect chains (lack of water in food production and domestic use).

Framework of the harmonization approach. The left side indicates the Framework of impact pathways (the faded impact pathways are not mature yet). The right side boxes indicate the status of the harmonization efforts (based on Boulay et al. 2015a)

5.2 More Detailed Databases

As the value chains of products/organizations/services are widely spread to all over the world, both water inventories and impact assessment models need to have detailed spatial resolution at global scale. While most impact assessment models are already developed with different spatial resolutions, like watershed or country level, most of the current inventory databases cover only one specific country or region in general. Some of the databases have been tackling this gap between inventory and impact assessment (Ecoinvent center 2015). Moreover, Lenzen et al. (2013) developed an inventory database at country level which covers the whole world based on multi-region input-output databases. However, there is still a gap with impact assessment models with higher spatial resolutions like watershed or even sub-watershed levels. Boulay et al. (2015a) have already found out that spatial resolution can be a significant factor controlling the uncertainty of assessment results by comparing impact assessment models with different spatial resolution. Thus, inventory database with higher spatial resolution is expected to be established in future.

Additionally, both availability and demand of water vary from time to time. Therefore, temporal resolution is also of importance for both inventory and impact assessment. While several models for impact assessment (Hoekstra et al. 2012; Pfister and Bayer 2014) incorporate temporal resolution at monthly level, inventory databases are generally based on an annual level analysis. For more precise assessment of water footprints, higher temporal resolution is required for further developments of inventory databases.

Regarding more detailed information (e.g., types of water resources, quality of water resources etc.), the necessity is controversial and being discussed. However, the consistency and harmonization between inventory databases and impact assessment models should be carefully considered. While detailed impact assessment models are expected to improve the precision and reliability of results, increasing requirements in data collection will impose considerable burden on practitioners. This trade-off between precision and applicability could be also one of the key issues for the promotion of water footprint and needs to be considered in the development of detailed databases.

5.3 *Use for Decision-Making, Product and Company Labels, and Future Scenario Assessment*

Decision-making processes already consider LCA indicators in some cases, and water footprint has been promoted by different initiatives for product labeling – especially the product environmental footprint (PEF) of the EU commission (EU 2013). However, there is still the need for agreed methods and for a proper communication strategy for such indicators. In principle these indicators should summarize the issue in one number, while all the background information of the study should be made available to the consumer in order to keep transparency (Ridoutt et al. 2015). One main limitation of applying such footprint numbers as product label is the high uncertainty in water footprint assessment, which makes reporting more difficult: i.e., what is a significant difference of two products? However, product labels are very valuable to identify hotspots for private consumers but also businesses. Such labels might show that almonds and cotton clothes are relevant terms of our personal water footprints – while apples are typically not. Such results are more robust than comparing two types of pizza from different producers, where many estimates are required leading to a high overlap of the expected results under uncertainty consideration. Therefore, interpretation of the results needs to be done carefully.

Another risk of product labels selecting a few indicators is the problem shifting from one to another impact category; e.g., from water to land impacts.

Company-based assessments or footprints of whole regions is another application for water footprinting but is following the same procedure, with the same issues. For assessing future scenarios on a regional level or for product and companies, the impact assessment needs to account for the future changes, since most methods will have increased impacts for increased intensity of water consumption. Pfister et al. (2011b) assessed different scenarios for agricultural production and the consequences on the overall water consumption and impact. For this purpose, the water scarcity indicator was recalculated for every scenario, since these include non-marginal changes in the water consumption. As a consequence, product or company water footprints for future states need to account for the fact that depending on the scenario of overall water consumption, the impact assessment factors will vary for most methods.

5.4 *Water Footprint: Cure or Tranquilizer?*

As illustrated in this chapter, the water footprint has developed considerably from a simple volumetric measure to an advanced impact assessment tool which is applicable even in complex industrial case studies. Hence, water footprint results are of increasing robustness and can support stakeholders in industry and politics when analyzing technical or political options. However, two relevant questions remain which are discussed in the following sections:

1. What actions should be taken based on water footprint results?
2. Can the water footprint help to mitigate global water stress?

5.4.1 Actions to Be Taken Based on Water Footprint Results

Considering the water footprint results of various case studies, one might conclude that, e.g., cotton textiles should be avoided (Chapagain et al. 2006), no wheat should be imported from Morocco (Pfister et al. 2011b), and biofuels produced in water-scarce countries (Mekonnen and Hoekstra 2011) should be banned.

Even though this could be preferable from a pure water perspective, there are other aspects which should be considered, too. As shown in Fig. 3.10, water consumption is only one aspect among other environmental interventions, such as global warming or human toxicity. In some cases, e.g., the comparison of biofuels to fossil fuels (Berger et al. 2015), the water and carbon footprints can lead to opposing preferences. Such trade-offs between indicators become even more relevant when including the economic and social dimensions in life cycle sustainability assessment studies (Finkbeiner et al. 2010). Thus, recent developments in water footprinting can help to increase precision and reliability when assessing impacts of water consumption but should not be used as a sole basis for decision-making.

As learned from various case studies, companies are often not primarily interested in absolute water footprint numbers but want to understand where relevant water consumption occurs within the supply chain of their products in terms of volume and impacts. Based on this information, companies can analyze the

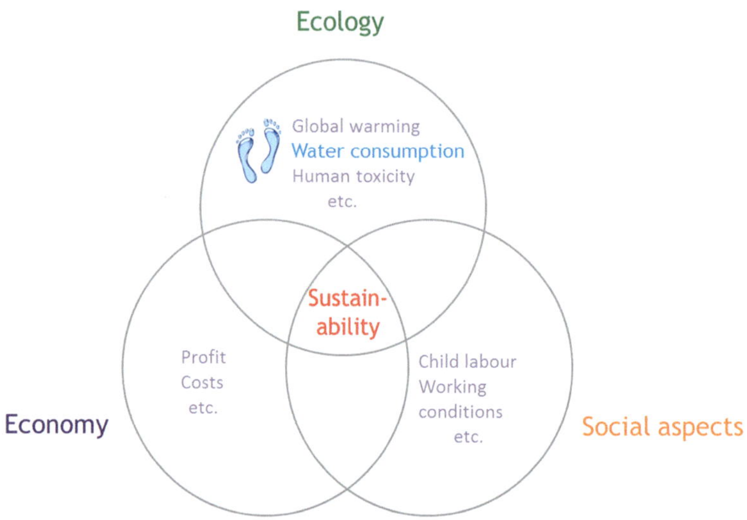

Fig. 3.10 Water consumption as one indicator among others in the context of sustainability

hotspots in greater detail as identified potential impacts do not necessarily mean that there are real damages.

Such a more detailed analysis can be facilitated by, e.g., the water stewardship concept (EWS 2013) which analyzes the water consumption of an organization in greater detail. Taking into account the specific local situation, water stewardship approaches evaluate the environmental, operational, legal, and reputational risks associated with an organization's water consumption. Supporting an "out of the fence approach," water stewardship identifies opportunities and solutions in cooperation with the public, authorities, and other water users within the basin. In addition to reduction and recycling options, solutions to reduce impacts of an organization's water use can also include offsetting measures, such as rainwater collection or drinking water purification projects within the basin.

Similar to the differences between environmental impact assessment (EIA) and LCA, water stewardship analyzes water consumption and the resulting consequences in greater detail – but can hardly be applied throughout a complex supply chain. Therefore, it can be a promising symbiosis to identify potential hotspots by means of the water footprint and analyze real risks and opportunities by means of water stewardship projects.

5.4.2 The Water Footprint: A Means of Mitigating Global Water Stress?

The question whether the water footprint can really help to reduce environmental impacts is difficult to answer. The water footprint can support decision-makers in analyzing potential impacts resulting from water consumption throughout the life cycle of products. This information can be used to identify hotspots, reveal potential for improvement, set reduction targets, and compare alternatives. Especially in combination with water stewardship activities, actions can be taken which reduce impacts on freshwater resources, ecosystems, and human health.

Therefore, the key question is whether this potential is really utilized. So far, water footprints have mainly been determined to inform stakeholders about the volumes and resulting impacts of water consumption occurring along the supply chain of products or business activities. Even though this awareness raising is relevant as such, it is crucial for the water footprint to take the next step: from information to actions that will reduce negative consequences of water consumption. In order to achieve this goal, the water footprint needs to become a management instrument. Similar to LCA, which is successfully implemented as a research and development tool in several companies (Finkbeiner et al. 2001), water footprint results can become one aspect in the complex decision-making process. In addition to numerous conventional parameters like costs, design, or quality, impacts resulting from a product's water consumption will then need to be considered in every decision.

It is hoped that the methodological developments and case studies presented in this paper will support this transition of the water footprint from a mere awareness to a decision tool which will help to mitigate global water stress.

References

Alcamo J, Henrichs T (2002) Critical regions: a model-based estimation of world water resources sensitive to global changes. Aquat Sci 64:352–362

Allan JA (1998) Virtual water: a strategic resource, global solutions to regional deficits. Ground Water 36(4):545–546

Amores MJ, Verones F, Raptis C, Juraske R, Pfister S, Stoessel F, Antón A, Castells F, Hellweg S (2013) Biodiversity impacts from salinity increase in a coastal wetland. Environ Sci Technol 47(12):6384–6392

Bayart JB, Bulle C, Koehler A, Margni M, Pfister S, Vince F, Deschenes L (2010) A framework for assessing off-stream freshwater use in LCA. Int J Life Cycle Assess 15(5):439–453

Berger M, Finkbeiner M (2010) Water footprinting – how to address water use in life cycle assessment? Sustainability 2(4):919–944

Berger M, Finkbeiner M (2013) Methodological challenges in volumetric and impact oriented water footprints. J Ind Ecol 17(1):79–89

Berger M, Warsen J, Krinke S, Bach V, Finkbeiner M (2012) Water footprint of European cars: potential impacts of water consumption along automobile life cycles. Environ Sci Technol 46 (7):4091–4099

Berger M, van der Ent R, Eisner S, Bach V, Finkbeiner M (2014) Water accounting and vulnerability evaluation (WAVE) – considering atmospheric evaporation recycling and the risk of freshwater depletion in water footprinting. Environ Sci Technol 48(8):4521–4528. doi:10.1021/es404994t

Berger M, Pfister S, Bach V, Finkbeiner M (2015) Saving the planet's climate or water resources? The trade-off between carbon and water footprints of European biofuels. Sustainability 7 (6):6665–6683

Boulay A-M, Bouchard C, Bulle C, Deschenes L, Margni M (2011a) Categorizing water for LCA inventory. Int J Life Cycle Assess 16(7):639–651

Boulay A-M, Bulle C, Bayart J-B, Deschenes L, Margni M (2011b) Regional characterization of freshwater use in LCA: modelling direct impacts on human health. Environ Sci Technol 45 (20):8948–8957

Boulay A-M, Bare J, De Camillis C, Döll P, Gassert F, Gerten D, Humbert S, Inaba A, Itsubo N, Lemoine Y, Margni M, Motoshita M, Núñez M, Pastor A, Ridoutt B, Schencker U, Shirakawa N, Vionnet S, Worbe S, Yoshikawa S, Pfister S (2015a) Consensus building on the development of a stress-based indicator for LCA-based impact assessment of water consumption: outcome of the expert workshops. Int J Life Cycle Assess 20(5):577–583

Boulay AM, Bayart JB, Bulle C, Franceschini H, Motoshita M, Muñoz I, Pfister S, Margni M (2015b) Analysis of water use impact assessment methods (part B): applicability for water footprinting and decision making with a laundry case study. Int J Life Cycle Assess 20 (6):865–879

Boulay AM, Motoshita M, Bulle C, Munoz I, Franceschini H, Motoshita M, Margni M (2015c) Analysis of water use impact assessment methods (part A): evaluation of modeling choices based on a quantitative comparison of scarcity and human health indicators. Int J Life Cycle Assess 20(1):139–160

Brent AC (2004) A life cycle impact assessment procedure with resource groups as areas of protection. Int J Life Cycle Assess 9(3):172–179

Chapagain AK, Hoekstra AY (2007) The water footprint of coffee and tea consumption in the Netherlands. Ecol Econ 1(64):109–118

Chapagain AK, Hoekstra AY, Savenije HHG, Gautam R (2006) The water footprint of cotton consumption: an assessment of the impact of worldwide consumption of cotton products on the water resources in the cotton producing countries. Ecol Econ 60(1):186–203

China Daily (2007) 27 November 2007

CSO (2013) The corporate water gauge. Available at http://www.sustainableorganizations.org. Center for Sustainable Organizations. 30 July 2013

Dewulf J, Bosch ME, De Meester B, van der Vorst G, van Langenhove H, Hellweg S, Huijbregts MAJ (2007) Cumulative Exergy Extraction from the Natural Environment (CEENE): a comprehensive life cycle impact assessment method for resource accounting. Environ Sci Technol 41:8477–8483

Ecoinvent centre (2015) Ecoinvent 3 LCI database. Available at http://www.ecoinvent.ch/. 30 July 2013

EU (2013) Commission recommendation of 9 April 2013 on the use of common methods to measure and communicate the life cycle environmental performance of products and organisations (2013/179/EU). Official Journal of the European Union 56

EWS (2013) European water stewardship. Available at http://www.ewp.eu/activities/ews/Global. 18 September 2013

Falkenmark M, Rockström J (2004) Balancing water for humans and nature. The new approach in ecohydrology. Earthscan, London

FAO (1999) Cropwat for Windows 4.3. Rome

FAO (2009) http://www.fao.org/nr/water/issues/scarcity.html. Accessed 16 July 2009

FAO (2013) FAOSTAT. Available at http://faostat.fao.org/. Food and Agricultural Organization of the United Nations. 30 July 2013

FAO (2015) Food and Agricultural Organization of the United Nations. AQUASTAT. Available at http://www.fao.org/nr/water/aquastat/main/index.stm. 3 Mar 2015

Finkbeiner M, Ruhland K, Cetiner H, Stark B (2001) Life cycle engineering and design for environment of the Mercedes-Benz C-Class. Paper presented at the 2001 Environmental Sustainability Conference & Exhibition, 12–14 November 2001, Graz

Finkbeiner M, Schau EM, Lehmann A, Traverso M (2010) Towards life cycle sustainability assessment. Sustainability 2(10):3309–33223322. doi:10.3390/su2103309

Frischknecht R, Steiner R, Jungbluth N (2009) The ecological scarcity method – eco-factors 2006 – a method for impact assessment in LCA. Federal Office for the Environment, Bern
GEMI (2013a) Collecting the drops. Available at http://www.gemi.org/waterplanner/. Global Environmental Management Initiative. 18 Sept 2013
GEMI (2013b) Connecting the drops. Available at http://www.gemi.org/water/overview.htm. Global Environmental Management Initiative. 18 Sept 2013
GEMI (2013c) The local water tool. Available at http://www.gemi.org/localwatertool/. Global Environmental Management Initiative. 18 Sept 2013
Gerbens-Leenes W, Hoekstra AY, Van der Meer TH (2009) The water footprint of bioenergy. Proc Natl Acad Sci U S A 106(25):10219–10223
Gleick PH (1996) Water resources. In: Schneider SH (ed) Encyclopedia of climate and weather, vol 2. Oxford University Press, New York, pp 817–823
Goedkoop M, Heijungs R, Huijbregts M, Schryver AD, Struijs J, van Zelm R (2009) ReCiPe 2008 – a life cycle impact assessment method which comprises harmonised category indicators at the midpoint and the endpoint level. Available from http://lcia.wik.is
Guinee JB, Gorree M, Heijungs R, Huppes G, Kleijn R, de Koning A, van Oers L, Sleeswijk AW, Suh S, Udo de Haes HA, de Bruijn H, van Duin R, Huijbregts MAJ (2002) Life cycle assessment – an operational guide to the ISO standards, vol 7, Eco-efficiency in Industry and Science. Kluwer Academic Publishers, Dordrecht
Hanafiah MM, Xenopoulos MA, Pfister S, Leuven RSEW, Huijbregts MAJ (2011) Characterization factors for water consumption and greenhouse gas emissions based on freshwater fish species extinction. Environ Sci Technol 45(12):5272–5278
Hauschild M, Wenzel H (1998) Environmental assessment of products, vol 2. Chapman & Hall, Thomson Science, New York
Hoekstra AY, Hung PQ (2002) Virtual water trade: a quantification of virtual water flows between nations in relation to international crop trade. In: Value of water research report series 11. UNESCO-IHE, Delft
Hoekstra AY, Mekonnen MM (2012a) Reply to Ridoutt and Huang: from water footprint assessment to policy. Proc Natl Acad Sci U S A 109(22):E1425
Hoekstra AY, Mekonnen MM (2012b) The water footprint of humanity. Proc Natl Acad Sci U S A 109(9):3232–3237
Hoekstra AY, Chapagain AK, Aldaya MM, Mekonnen MM (2011) The water footprint assessment manual – setting the global standard. Earthscan, London
Hoekstra AY, Mekonnen MM, Chapagain AK, Mathews RE, Richter BD (2012) Global monthly water scarcity: blue water footprints versus blue water availability. PLoS One 7(2):e32688
Humbert S, Maendly R (2008) Characterization factors for damage to aquatic biodiversity caused by water use especially from dams used for hydropower. Paper presented at the 35 LCA forum Zurich – Water in LCA, 5 June 2008, Zurich
Humbert S, Loerincik Y, Rossi V, Margni M, Jolliet O (2009) Life cycle assessment of spray dried soluble coffee and comparison with alternatives (drip filter and capsule espresso). J Clean Prod 17(15):1351–1358
Inaba A, Chevassus S, Cumberlege T, Hong E, Kataoka A, Pongvipa L, Mercadie C, Mungcharoen T, Radunsky K (2016) Carbon footprint of products. Chapter 3. In: Finkbeiner M (ed) Special types of life cycle assessment. LCA compendium – the complete world of life cycle assessment. Springer, Dordrecht, pp 11–72
ISO 14044 (2006) Environmental management – life cycle assessment – requirements and guidelines. International Organization for Standardization, Geneva
ISO 14046 (2014) Water footprint – principles, requirements and guidelines. International Organization for Standardization, Geneva
ISO 14064 (2006) ISO 14064-1 greenhouse gases – part 1 – specification for the quantification, monitoring and reporting of entity emissions and removals; ISO 14064-2 greenhouse gases – part 2 – specification for the quantification, monitoring and reporting of project emissions and removals; ISO 14064-3 greenhouse gases – part 3 – specification and guidance for validation, verification and certification. International Organization for Standardization, Geneva

ISO (2009) Available online: http://www.iso.org/iso/standards_development/technical_committees/list_of_iso_technical_committees/iso_technical_committee.htm?commid=54854. Accessed on 23 June 2009

ISO TR 14073 (2014) Water footprint – illustrative examples on how to apply ISO 14046. International Organization for Standardization, Geneva

JRC-IES (2011) ILCD handbook general guide for life cycle assessment—recommendations for life cycle impact assessment in the European context—based on existing environmental impact assessment models and factors. Joint Research Centre – Institute for Environment and Sustainability, Luxembourg

Kounina A, Margni M, Bayart J-B, Boulay A-M, Berger M, Bulle C, Frischknecht R, Koehler A, Canals LM, Motoshita M, Núñez M, Peters G, Pfister S, Ridoutt B, Zelm R, Verones F, Humbert S (2013) Review of methods addressing freshwater use in life cycle inventory and impact assessment. Int J Life Cycle Assess 18(3):707–721

Lenzen M, Moran D, Bhaduri A, Kanemoto K, Bekchanov M, Geschke A, Foran B (2013) International trade of scarce water. Ecol Econ 94:78–85

Lin C, Suh S, Pfister S (2012) Does south-to-north water transfer reduce the environmental impact of water consumption in China? J Ind Ecol 16:647–654

Mekonnen MM, Hoekstra AY (2011) The green, blue and grey water footprint of crops and derived crop products. Hydrol Earth Syst Sci 15(5):1577–1600

Milà i Canals L, Chenoweth J, Chapagain A, Orr S, Anton A, Clift R (2008) Assessing freshwater use in LCA: part I – inventory modelling and characterisation factors for the main impact pathways. Int J Life Cycle Assess 14(1):28–42

Monfreda C, Ramankutty N, Foley JA (2008) Farmingtheplanet:2. Geographic distribution of crop areas, yields, physiological types, and net primary production in the year 2000. Glob Biogeochem Cycles 22(1):1–19

Morrison J, Morikawa M, Murphy M, Schulte P (2009) Water scarcity & climate change: growing risks for businesses & investors. Pacific Institute, Oakland

Motoshita M, Itsubo N, Inaba A (2008) Development of impact assessment method on health damages of undernourishment related to agricultural water scarcity. Paper presented at the Proceedings of EcoBalance 2008 – The 8th International Conference on EcoBalance, 10–12 December 2008, Tokyo

Motoshita M, Itsubo N, Inaba A (2011) Development of impact factors on damage to health by infectious diseases caused by domestic water scarcity. Int J Life Cycle Assess 16(1):65–73

Motoshita M, Ono Y, Pfister S, Boulay A-M, Berger M, Nansai K, Tahara K, Itsubo N, Inaba A (2014) Consistent characterisation factors at midpoint and endpoint relevant to agricultural water scarcity arising from freshwater consumption. Int J Life Cycle Assess. doi:10.1007/s11367-014-0811-5, open access

Nestle (2009) Available online: http://www.nestle.com/. Accessed on 15 July 2009

Núñez M, Pfister S, Roux P, Antón A (2013) Estimating water consumption of potential natural vegetation on global dry lands: building an LCA framework for green water flows. Environ Sci Technol 47(21):12258–12265

Ono Y, Kim YD, Horiguchi K, Itsubo N (2012) The development of water inventory database considering water consumption. Paper presented at the proceedings of EcoBalance 2012 – the 10th international conference on EcoBalance, 21–23 November 2012, Yokohama

Owens JW (2001) Water resources in life cycle impact assessment – considerations in choosing category indicators. J Ind Ecol 5(2):37–54

Pfister S (2015) Water use. Chapter 12 "Life Cycle Impact Assessment" (Hauschild MZ, Huijbregts MAJ eds). In: LCA compendium – the complete world of life cycle assessment (Klöpffer W, Curran MA, series eds). Springer, Dordrecht, pp 223–245

Pfister S, Baumann J (2012) Monthly characterization factors for water consumption and application to temporally explicit cereals inventory. Paper presented at the 8th international conference on LCA in the Agri-Food Sector, Rennes, 2–4 Oct 2012

Pfister S, Bayer P (2014) Monthly water stress: spatially and temporally explicit consumptive water footprint of global crop production. J Clean Prod 73(1):52–62
Pfister S, Hellweg S (2009) The water "shoesize" vs. footprint of bioenergy. Proc Natl Acad Sci U S A 106(35):E93–E94
Pfister S, Koehler A, Hellweg S (2009) Assessing the environmental impacts of freshwater consumption in LCA. Environ Sci Technol 43:4098–4104
Pfister S, Bayer P, Koehler A, Hellweg S (2011a) Environmental impacts of water use in global crop production: hotspots and trade-offs with land use. Environ Sci Technol 45(13):5761–5768
Pfister S, Bayer P, Koehler A, Hellweg S (2011b) Projected water consumption in future global agriculture: scenarios and related impacts. Sci Total Environ 409:4206–4216
Pfister S, Saner D, Koehler A (2011c) The environmental relevance of freshwater consumption in global power production. Int J Life Cycle Assess 16(6):580–591
Prüss-Üstün A, Bos R, Gore F, Bartram J (2008) Safer water, better health: costs, benefits and sustainability of interventions to protect and promote health. World Health Organization, Geneva
Quantis (2013) The Quantis water database. Available at http://www.quantis-intl.com/waterdatabase.php. 30 July 2013
Ridoutt BG, Huang J (2012) Environmental relevance—the key to understanding water footprints. Proc Natl Acad Sci U S A 109(22):E1424
Ridoutt B, Pfister S (2013) A new water footprint calculation method integrating consumptive and degradative water use into a single stand-alone weighted indicator. Int J Life Cycle Assess 18 (1):204–207
Ridoutt B, Fantke P, Pfister S, Bare J, Boulay A-M, Cherubini F, Frischknecht R, Hauschild M, Hellweg S, Henderson A, Jolliet O, Levasseur A, Margni M, McKone T, Michelsen O, Milà i Canals L, Page G, Pant R, Raugei M, Sala S, Saouter E, Verones F, Wiedmann T (2015) Making sense of the minefield of footprint indicators. Environ Sci Technol 49(5):2601–2603
Sauer BJ, Hilderbrandt CC, Franklin WE, Hunt RG (1994) Resource and environmental profile analysis of children s diaper systems. Environ Toxicol Chem 13:1003–1009
Schnoor JL (2009) LCA and environmental intelligence? Environ Sci Technol 43:2997
Shiklomanov IA (2003) World water resources at the beginning of the 21st century. Cambridge University Press, Cambridge
Siebert S, Doll P, Feick S, Hoogeveen J, Frenken K (2007) Global map of irrigation areas, version 4.0.1. Johann Wolfgang Goethe University, Frankfurt
Suweis S, Rinaldo A, Maritana A, D'Odorico P (2012) Water-controlled wealth of nations. Proc Natl Acad Sci U S A 110(11):4230–4233
Thinkstep (2016) GaBi LCI database. Available at http://www.gabi-software.com. 8 March 2016
UN (2009) Available online: http://www.unglobalcompact.org/Issues/Environment/CEO_Water_Mandate/. Accessed on 16 July 2009
UN, FAO (2007) Coping with water scarcity – challenge of the twenty first century
UNEP-SETAC (2009) Available online: http://lcinitiative.unep.fr/. Accessed on 23 June 2009
UNESCO (2009) Available online: http://www.unesco-ihe.org/. Accessed on 23 June 2009
Van Zelm R, Schipper AM, Rombouts M, Snepvangers J, Huijbregts MAJ (2011) Implementing groundwater extraction in life cycle impact assessment: characterization factors based on plant species richness. Environ Sci Technol 45:629–635
Veolia (2011) The water impact index and the first carbon-water analysis of a major metropolitan water cycle. Available at http://www.veoliawaterna.com/north-america-water/ressources/documents/1/10975,Water_Impact_Index-White_Paper.pdf. Accessed 3 July 2013
Verones F, Bartl K, Pfister S, Jiménez Vílchez R, Hellweg S (2012) Modeling the local biodiversity impacts of agricultural water use: case study of a wetland in the coastal arid area of Peru. Environ Sci Technol 46:4966–4974
Verones F, Pfister S, Hellweg S (2013) Quantifying area changes of internationally important wetlands due to water consumption in LCA. Environ Sci Technol 47(17):9799–9807. doi:10.1021/es400266v

Vince F (2007) Proposition for the LCI framework (unpublished)
Volkswagen AG (2010) The life cycle of a car – environmental commendations document progress. Group Research, Environment Affairs Product http://www.volkswagen.com/vwcms/master_public/virtualmaster/en2/unternehmen/environmental_commendations.html. Accessed 10 Mar 2011
Vörösmarty CJ, McIntyre PB, Gessner MO, Dudgeo D, Prusevich A, Green P, Glidden S, Bunn SE, Sullivan CA, Liermann CR, Davies PM (2010) Global threats to human water security and river biodiversity. Nature 467:555–561
Water Footprint Network (2009) Available online: http://www.waterfootprint.org. Accessed on 23 June 2009
WBCSD (2013) The WBCSD global water tool. Available at http://www.wbcsd.org/web/watertool.htm. World Business Council for Sustainable Development. Accessed 30 June 2013
Wetlands International (2010) Biodiversity loss and the global water crisis. Wetlands International, Wageningen
WFN (2013a) The water footprint assessment tool. Available at http://www.waterfootprint.org/tool/home/. Water Footprint Network. 18 Sept 2013
WFN (2013b) The WaterStat database. Available at http://www.waterfootprint.org/?page=files/WaterStat. Water Footprint Network. Accessed 30 July 2013
WULCA (2015) WULCA – consensual method development. Available at http://www.wulca-waterlca.org/project.htmlWorld. Accessed 26 Oct 2015
WWF (2013) The water risk filter. Available at http://waterriskfilter.panda.org/. World Wide Fund for Nature. 18 Sept 2013
Zah R, Böni H, Gauch M, Hischier R, Lehmann M, Wäger P (2007) Life-cycle analysis of energy products: ecological evaluation of biofuels (German). Empa, Technology and Society Lab, St. Gallen

Chapter 4
Eco-efficiency Assessment

Peter Saling

Abstract Quantifying the sustainability benefits of materials, products like chemicals or consumer goods represent an important aspect for the further development of more sustainable solutions in the future. Generating a realistic and validated estimate of innovative potentials by using a quantitative method is essential for the development of new products and processes. The Eco-efficiency Assessment is therefore a key element in industrial sectors to make progress in planning new products, processes, or applications by taking sustainability aspects as an important element in their decision-making processes. Different tools have been developed based on holistic life cycle management approaches to assess the entire product life cycle, from concept development, to design and implementation, further to marketing, finally, to end-of-life issues. The Eco-efficiency Assessment often incorporates both economic and environmental aspects.

Promising products can be identified at an early stage, thus facilitating decision-making about the prime thrust of the development. Major R&D projects are to be accompanied by eco-efficiency analyses during the following development phases: mini-plant, pilot plant, and basic design of a production facility, and the projects are evaluated at each milestone.

So Eco-efficiency Assessments are powerful and supporting tools for shifting product developments, optimization of products along the whole supply chain, and the definition of new opportunities in a direction, where significant improvements of sustainability can be achieved.

This chapter introduces different ways of conducting Eco-efficiency Assessments and using the results in different situations, mainly product development, improvement, and marketing.

Keywords Eco-efficiency Assessment • Eco-efficiency analysis • ISO 14040 • ISO 14044 • ISO 14045 • Life cycle assessment • LCA • Life cycle costing • LCC • Sustainability evaluation • Sustainability management tools

---

P. Saling (✉)
Sustainability Strategy, BASF SE, ZZS/S-C104, D-67056 Ludwigshafen, Germany
e-mail: peter.saling@basf.com

M. Finkbeiner (ed.), *Special Types of Life Cycle Assessment*, LCA Compendium – The Complete World of Life Cycle Assessment,
DOI 10.1007/978-94-017-7610-3_4

Acronyms

ADP	Abiotic resource depletion
AOX	Adsorbable organic halogens
AP	Acidification potential
ASUE	Arbeitsgemeinschaft für sparsamen und umweltfreundlichen Energieverbrauch e.V. Germany: Comparison of heating costs in new developments, 2009
BOD	Biological oxygen demand
CFC	Chlorofluorocarbons
CFL	Fluorescent lamp
COD	Chemical oxygen demand
CSI	Cement sustainability initiative
ECF	Elemental chlorine free
ECM	Eco-Care-Matrix
EDP	Ecosystem damage potential
EEA	Eco-efficiency analysis
EPD	Environmental Product Declarations
GefStoffV	Hazardous substances regulation act
GHG	Greenhouse gases
GLS	General lighting standard
GWP	Global warming potential
HAL	Halogen lamp
HCs	Hydrocarbons
HMs	Heavy metals
ISIC	International Standard Industrial Classification of All Economic Activities
LCC	Life cycle costing
NPV	Net present value
NSF	National sanitation foundation
ODP	Ozone depletion potential
OSHA	Occupational safety and health act
PEMFC	Proton exchange membrane fuel cell
POCP	Photochemical oxidation creation potential
SD	Sustainable development
SEEBALANCE	Social-eco-efficiency analysis, trade name SEEBALANCE
VOC	Non-methane volatile organic compounds
WBCSD	World Business Council for Sustainable Development

1 Introduction: What Is Eco-efficiency?

1.1 Definitions

Sustainable development (SD) has been defined as the balance of economic success, ecological protection, and social responsibility. To effectively manage sustainability, companies must be able to measure or otherwise quantify sustainability in each of these pillars.

Eco-efficiency is generally measured by the ratio of a useful output divided by a useful input. The monetary value created by the business can be a useful output but is not limited to a useful input. The created monetary value, justified by environmental external costs, can be the useful output (Steen 2009). It is a common concept designed to drive the decoupling of economic growth from environmental deterioration (Carlson 2009). But it has to be pointed out that eco-efficiency has no direct link to sustainability per se. There is still a need for a consistent definition of the term. A "strong" eco-efficiency definition implies improvement in both dimensions – environment and economy (Finkbeiner 2008).

The empirical facts and developments that have been achieved in the field of the eco-efficiency are able to support decision-making on micro and macro level. Relations between economy and environment are not self-evident. Clarifying the "why and what," eco-efficiency is a first and important step toward decision-making support including relevant aspects of SD (Huppes and Mansanobu 2007).

1.2 Link to LCA Principles

1.2.1 Basic Principles and Definitions

The eco-efficiency analysis (EEA) is closely linked to ISO 14045 (ISO 14045:2012) and subsequently linked as well to ISO 14040 (ISO 14040:2006a) and 14044 (ISO 14044:2006b). Eco-efficiency Assessment is defined there as a quantitative management tool which enables the consideration of life cycle environmental impacts of a product system alongside its product system value to a stakeholder. Within Eco-efficiency Assessment, environmental impacts are evaluated using life cycle assessment (LCA) as prescribed by other international standards (ISO 14040, ISO 14044). Consequently, Eco-efficiency Assessment shares with LCA many important principles such as life cycle perspective, comprehensiveness, functional unit approach, iterative nature, transparency, and priority of scientific approach.

Eco-efficiency is defined as aspect of *sustainability relating to the environmental performance of a product system to its product system value*. Within this context, the environmental performance relates to measurable results concerning environmental aspects. Environmental aspects are defined as elements of an organization's activities, including products and services that can interact with the environment.

A product system value is worth, or desirability ascribed, to a product system and may be expressed in monetary terms or other value aspects.

A product system reflects a collection of unit processes with elementary and product flows, performing one or more defined functions, and which models the life cycle of a product.

The international standard helps to establish clear terminology and a common methodological framework for Eco-efficiency Assessment. It enables the practical use of Eco-efficiency Assessment for a wide range of product (including service) systems. It provides clear guidance on the interpretation of Eco-efficiency Assessment results and encourages the transparent, accurate, and informative reporting of Eco-efficiency Assessment results.

The standard focuses on the life cycle perspective, the iterative approach, transparency, comprehensiveness, and the scientific approach. An Eco-efficiency Assessment considers the entire life cycle from raw material extraction and acquisition, to energy and material production and manufacturing, to use and end-of-life treatment and final disposal.

Through such a systematic overview and perspective, the shifting of a potential impact between life cycle stages and individual processes can be identified and assessed with a view to an overall eco-efficiency.

1.2.2 Iterative Process and Comprehensiveness

Eco-efficiency Assessment is an iterative technique which contributes to the comprehensiveness and consistency of the Eco-efficiency Assessment and the reported results. It supports the improvement of the study by focusing on the update of the most relevant life cycle steps and improving the data quality and input information for those steps.

An Eco-efficiency Assessment considers all attributes or aspects of environment and product system value. By considering all attributes and aspects within one Eco-efficiency Assessment, potential trade-offs can be identified and assessed (Fig. 4.1).

Decisions within an Eco-efficiency Assessment are preferably based on scientific data, methodology, and other evidence. If this is not possible, decisions based on international conventions may be used. If neither a scientific basis exists nor international conventions can be referred to, then decisions may be based on value choices.

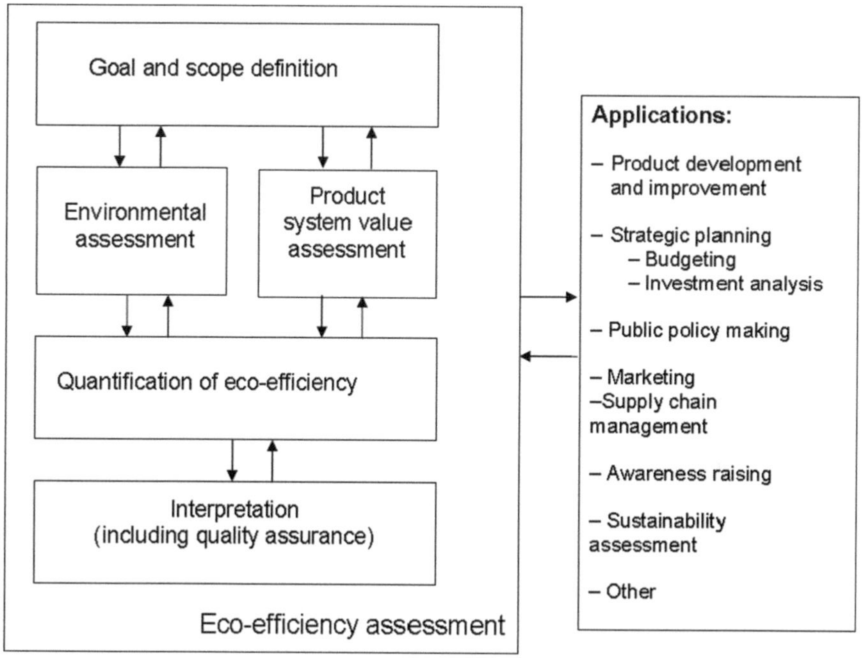

Fig. 4.1 Phases of an Eco-efficiency Assessment

1.2.3 Transparency

Due to the inherent complexity in Eco-efficiency Assessment, transparency is an important guiding principle in executing an Eco-efficiency Assessment, in order to ensure a proper interpretation of the results. It allows the reader a better understanding of the underlying principles, datasets, and information. Furthermore, the interpretation of results can be understood much better and the comparison with other studies in the same field as well.

1.2.4 The Product System Value

In ISO 14045 an approach is published how to accurately define and describe a product system by different stakeholders. They may encounter different values for the same product system. For instance, the product system value to the consumer may be different from the product system value to the producer and in turn different to the investor.

It shall be described which stakeholder's value(s) and which type of value(s) and the methods to determine the product system value(s) to be used in the assessment. The value(s) shall be quantifiable with reference to the functional unit according to the goal and scope of the Eco-efficiency Assessment. The types of product system values may be a functional value, a monetary value, or other values.

A good functional value of a product system reflects a tangible and measurable benefit to the user and other stakeholders. The functional value is a numerical quantity representing functional performance or desirability of a product system and is subject to improvement.

In the Eco-efficiency Assessment, the functional value is different from the functional unit. The functional value should be measured and related to the functional unit in a quantification of the product system performance. The functional unit provides the reference to which the input and output data are normalized (in a mathematical sense). Therefore, within an Eco-efficiency Assessment, the functional value may change, e.g., because of product improvement, whereas the functional unit remains the same.

Different types of functional values can be used in eco-efficiency studies. One type is a monetary value. This value may be expressed in terms of costs, price, willingness to pay, added value, profit, future investment, etc. Changes in costs for a single company may represent changes in the product system value over the entire life cycle. If other parts of the product system are affected, for example, if the price from suppliers is negotiated to be lower or the price to the customer is raised for the same products, then there is no net change in the product system value.

Other values may include intangible values such as aesthetic, brand and cultural and historical values. These values may be determined by means of interviews, surveys, market research, etc. This type of values must be explained accurately, so that misunderstanding and decisions into wrong directions will be avoided. In the life cycle assessment area, this is a quite unusual way of expressing sustainability, but it will be accepted in the future, if reasonable results will be derived from this approach.

The subsequent quantification of the product system value shall be carried out by using relevant product system value indicators, as defined in the goal and scope definition of the eco-efficiency study.

1.2.5 Indicators

Several types of eco-efficiency indicators can be chosen to express a quantitative statement on eco-efficiency. Quantification is a key step to evaluate different systems with eco-efficiency methods. The indicators to be used in the assessment need to be described. Additionally, the methods can be described separately, making it available to third-party evaluations with a published comment about it and subsequently used as basis for a set of evaluations. The evaluation method (s) and the presentation format of the Eco-efficiency Assessment must be defined and published to keep the whole evaluation process open and transparent. The specific BASF method, called eco-efficiency analysis (EEA), was validated by TÜV Rheinland (TÜV 2002) and the American organization National Sanitation Foundation (NSF) (Uhlman et al. 2013).

For the choice of eco-efficiency indicators, the following requirements of ISO 14045 should be considered and modeled to show higher eco-efficiencies:

- Increasing efficiency at the same product system value shall represent improved environment
- Increasing efficiency at the same environmental impact shall represent improved product system value

Life cycle impact category indicator results, as determined according to ISO 14044, may be used for Eco-efficiency Assessments. Such data will typically result in an eco-efficiency profile, where several environmental aspects are considered in parallel. All single results should be reported separately with meaningful information. They should be displayed in a way that life cycle impacts can be shown in an easily understandable way, main impacts can be detected, and improvement potentials can be identified as well.

Eco-efficiency indicators address both an environmental and a value aspect, and there are potential trade-offs between changes in environmental and product system value performance. The interpretation of results shall be done transparently and with proper justification.

1.2.6 Interpretation

To finalize an Eco-efficiency Assessment, a meaningful interpretation of the results is key for the support of decision-making as well as for the publication of results to an external audience. Therefore, the identification of significant issues based on the results of the environmental and product system value assessment phases and aspects and description of completeness, sensitivity, uncertainty, and consistency of data and results is needed. This will support the formulation of conclusions, limitations, and recommendations to the recipient or client of such a study. The comparison of Eco-efficiency Assessment results in another subsequent step of the whole evaluation process is important for the use of results and the consultancy linked to it. To derive messages and initiate improvements of product systems, the interpretation is a very important element.

The multicriteria LCA approach delivers different types of information in an eco-efficiency study. This makes it comprehensive but often not easy to interpret, even for specialists.

Therefore, for a better understanding of the outcome of an eco-efficiency study, it is very helpful to aggregate single information to an overall result. But, the aggregation of a set of indicators to one single result needs normalization, aggregation, and weighting steps. Weighting is subjective and bears potentials for the possible manipulation of results. Therefore weighting schemes must be balanced and reflect opinions of important stakeholder groups. The generation of weighting schemes can be done by assessing questionnaires sent to important stakeholder groups with a high level of representativeness. The benefit of the aggregation of single results to an overall single scoring is that decision-makers as well as an interested audience are able to easier understand the outcome of a study. The weighting should follow specific guidance which is given in ISO 14040 and ISO 14045 as well. However, weighting is not allowed in the case of comparative assertions, according to ISO 14040 and ISO 14044. In such cases, the critical review has to be performed according to the panel method.

The interpretation phase of an Eco-efficiency Assessment comprises the following elements, according to the goal and scope of the study:

- Identification of significant issues based on the results of the environmental and product system value assessment phases
- Evaluation that considers aspects of completeness, sensitivity, uncertainty, and consistency
- Formulation of conclusions, limitations, and recommendations

The requirements and recommendations in ISO 14044 should also apply for the interpretation of Eco-efficiency Assessment. In addition, the interpretation shall consider the relation between environmental results and product system value results.

1.2.7 Sensitivity and Uncertainty Analysis

Sensitivity analysis is a procedure to determine how changes in data and methodological choices affect the results of the Eco-efficiency Assessment. A sensitivity analysis may provide additional information on data choice(s). In an Eco-efficiency Assessment, several different methods for the determination of environmental and product system value indicators may be used. Therefore, a sensitivity analysis should be conducted to assess the consequences on the Eco-efficiency Assessment results of different choices of methodology and data.

An uncertainty analysis should be conducted to determine how uncertainties in data and assumptions affect the reliability of the results of the Eco-efficiency Assessment.

An analysis of results for sensitivity and uncertainty shall be conducted for Eco-efficiency Assessments to be used in comparative eco-efficiency assertions disclosed to the public.

1.2.8 Comparison of Eco-efficiency Assessment Results

When comparisons of Eco-efficiency Assessment results between product systems or within the same product system are made, they shall be based on the same eco-efficiency indicator. The comparative environmental assessment results and the product system value assessment results shall then be separately included in the Eco-efficiency Assessment report.

When improvements of Eco-efficiency Assessment results are identified or comparisons based on an Eco-efficiency Assessment results are performed, the following cases should be differentiated:

- Improvement or superiority in both aspects (environmental performance and product system value)
- Improvement or superiority in just one of both aspects
- No improvement or superiority in any of the two cases

The first and the third case do not contain trade-offs between the two dimensions. In the first case, improvement/superiority of the Eco-efficiency Assessment can be unambiguously determined.

In the third case, improvement/superiority can be unambiguously denied.

The second case is the most challenging, because of the trade-off between environmental and product system value aspects. In this case, an improvement or superiority shall only be reported, if the trade-off is clearly communicated and the underlying product system value assumptions are documented and justified.

If a claim of improvement or superiority is disclosed to third parties for the purpose of comparative eco-efficiency assertions, the Eco-efficiency Assessment results shall demonstrate an equal or better environmental performance.

1.2.9 Weighting

Following the standard, weighting shall not be used in Eco-efficiency Assessments for comparative eco-efficiency assertions to be disclosed to the public. A comparative eco-efficiency assertion is defined as a claim regarding the superiority or equivalence of one product versus a competitor's product that performs the same function.

If weighting is used in an Eco-efficiency Assessment, additional requirements similar to those in ISO 14044 apply. Those are defined as basic principles:

- Weighting principles
- Weighting factors
- How the weighting factors were determined including:
  - Methodology
  - Which stakeholder values have been used to determine the weighting factors

1.2.10 Reporting

The eco-efficiency results shall be reported as defined in the goal and scope definition phase of the study.

The results and conclusions of the Eco-efficiency Assessment shall be completely and accurately reported without bias to the intended audience. The results, data, methods, assumptions, and limitations shall be transparent and presented in sufficient detail to allow the reader to comprehend the complexities and trade-offs inherent in the Eco-efficiency Assessment. The report shall also allow the results and interpretation to be used in a manner consistent with the goals of the Eco-efficiency Assessment.

The results of the environmental assessment and product system value assessment shall be documented separately.

For Eco-efficiency Assessments used in comparative assertions to be disclosed to the public, the following issues shall additionally be addressed by the report:

For the *environmental assessment*, the following issues shall be addressed:

(a) Analysis of material and energy flows to justify their inclusion or exclusion
(b) Assessment of the precision, completeness, and representativeness of data used
(c) Description of the equivalence of the systems being compared
(d) Description of the critical review process
(e) Evaluation of the completeness of the LCIA
(f) Statement as to whether or not international acceptance exists for the selected LCIA category indicators and a justification for their use
(g) Explanation for the scientific and technical validity and environmental relevance of the LCIA category indicators used in the Eco-efficiency Assessment
(h) Results of the uncertainty and sensitivity analyses
(i) Evaluation of the significance of the differences found

For the *product system value assessment*, the following issues shall be addressed:

(a) Assumptions made in the product system value assessment phase shall be clearly reported and justified
(b) Methodologies and product system value indicators used in the product system value assessment phase shall be clearly reported and justified
(c) Assessment of precision, completeness, and representativeness of data used
(d) Description of the critical review process
(e) Evaluation of the completeness of the product system value assessment
(f) Results of the uncertainty and sensitivity analyses
(g) Evaluation of the significance of the differences of results and final conclusions found

If results from an Eco-efficiency Assessment are intended to be used in comparative assertions disclosed to the public, neither the environmental nor the Eco-efficiency Assessment results shall be reported as a single overall score or number.

Table 4.1 Light source life cycle example

Terms	Example	Value indicator (unit)
Product system	Light source life cycle	
Function	Illumination	
Functional value	Brightness	Luminous flux (lumen)
Monetary value	Market price	Price (euro/piece)
Other values	Shape	Consumer ranking (numerical value from 1 to 5)

1.3 Examples

1.3.1 Lighting Systems

To get a better idea in which way an eco-efficiency value can be defined and used in an assessment, the ISO 14045 published several examples. One example (B.1) shows the definitions of light sources with an illumination function behind it. Table 4.1 shows which different values can be defined, if it is a functional value or a monetary value.

The further evaluation of the system shows eco-efficiency results for the following application.

The functional unit was defined for illumination as the illumination of the same luminous flux during 1000 h of use with lamps. In the calculation, the environmental assessment considered each stage of product life cycle. It included material acquisition, parts production, manufacture of lamps, packaging staffs, domestic distribution, and use with a reference electricity mix. For product system value assessment, the use stage was chosen to represent the product system value. The study compared the use of two different lighting systems; whereas Product A was an incandescent light bulb, Product B was a bulb-shaped fluorescent lamp.

The life cycle assessment was carried out according to ISO14040 and 14044 by using the process analysis method based on the JEMAI-LCA1.10 database for each product. The materials and parts only used in the final products were considered. Domestic distribution of "1000 km by using 4-ton trucks" was assumed. In the manufacturing stage, primary and average data were collected and used. For use, the "rated electricity consumption" through the product lifetime was adopted, so that the power change in the same duration was ignored for calculation. The lifetime was 13,000 h for Product B and 1000 h for Product A. Product system value indicators of Products A and B = 8.10 E + 05 [l m·hour]. As a result of the assessment, it was found that 98 % or more of life cycle GHG (greenhouse gas) emissions were emitted in the use stages for both products. Other impacts showed almost the same results.

The total amount of the life cycle GHG emissions was presented in the units of [kg-CO_2e] to form the environmental impact indicator.

The total amount of the life cycle GHG emissions for Product B was quite a bit larger than that of Product A due to its long lifetime. However, as the indicator for

Product B must be calculated according to the functional unit, its numerical quantity became smaller than that of Product A in this study.

The indicators of two products were calculated as follows:

- Environmental impact indicator of Product A $= 2.32\mathrm{E} + 01[\text{kg-CO}_2\text{e}]$
- Environmental impact indicator of Product B $= 4.66\mathrm{E} + 00[\text{kg-CO}_2\text{e}]$

The eco-efficiency indicator was calculated by dividing the product system value indicator by the environmental impact indicator for each in the units of [l m·h/kg-CO_2e].

The indicators of two products were calculated as follows:

- Eco $-$ efficiency indicator of Product A $= 3.49\mathrm{E} + 04[\mathrm{lm} \cdot \mathrm{hour/kg\text{-}CO_2e}]$
- Eco $-$ efficiency indicator of Product B $= 1.74\mathrm{E} + 05[\mathrm{lm} \cdot \mathrm{hour/kg} - \mathrm{CO_2e}]$

The ratio of the eco-efficiency indicator of Product B compared to that of Product A is used to clarify the difference of the eco-efficiencies between the two products assessed.

The factor result (eco-efficiency indicator of Product B/eco-efficiency indicator of Product A) was 4.98. This means the eco-efficiency indicator of Product B (bulb-shaped fluorescent lamp) is about five times larger than that of Product A (incandescent light bulb). The decrease of power for illumination and the prolongation of lifetime are significantly contributing to the improvement of eco-efficiency, because the GHG emissions derived from electricity consumption in the use phase is critical to the environmental assessment results. Since several assumptions and simplifications were made in environmental and product system value assessments, this conclusion should be understood with a couple of limitations. For example, if other functional values and indicators focusing on the different aspects were adopted, the Eco-efficiency Assessment might reach the different results.

Figure 4.2 shows the product development paths. For the presentation of eco-efficiency results, often a two-dimensional graph, a portfolio is used. It shows different sectors to position different products, technologies, or scenarios of product applications. The portfolio allows an easier understanding of complex results and shows target areas as well as undesired areas. Different approaches define the different sectors of a portfolio differently, so a clear description of the portfolio and its logic must be transparently explained. In the portfolio of Fig. 4.2, the X-axis is linked with the environmental impact of a product in its application. The lower E is, the better the product is due to its environmental impacts. The Y-axis shows the improvement of the product's functions. The higher this number is, the more favorable the product. The goal area for an improved eco-efficiency is in this case the area A. When a present product is on the point of "benchmark product," its eco-efficiency is increasing the eco-efficiency, and the tanX expresses the eco-efficiency of the development target. If the target is "eco-efficiency (EE) >2," area A is the goal, and B is better in the environmental aspect. Area C shows that the product is steadily developing toward area A.

Progress of technologies may take paths and sometimes involves a decrease of environmental performance on the way to the goal. D seems to be a bad area due to

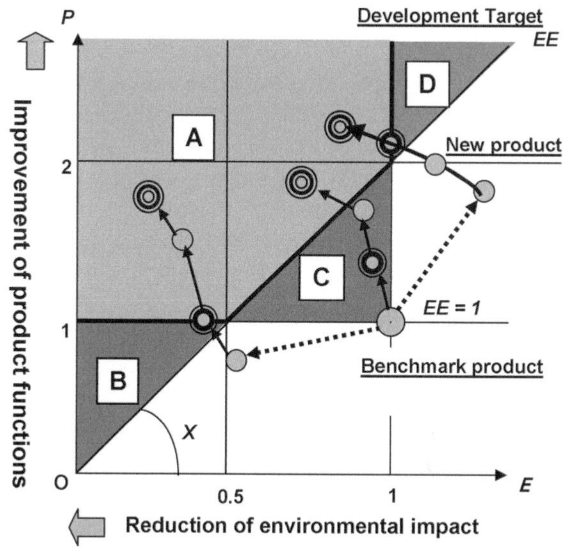

Fig. 4.2 Product development paths, evaluated with eco-efficiency indicators

the inferiority of environmental impact but may be an inevitable position to the goal by adopting the best available technology. In this context, when the product system value is increased much more than the decrease of environmental friendliness, the eco-efficiency may be reported as "improvement" in a series of product development (ISO 14045:2012; Japan Eco-Efficiency Forum 2009; Shibaike et al. 2008).

In this context, other comparisons have been made to integrate the eco-efficiency evaluation in decision-making processes.

Example: Eco-Care-Matrix

Siemens is using the "Eco-Care-Matrix" (ECM), another type of a portfolio graph, to analyze the eco-efficiency of new products compared to their predecessor or, in case of new business, to the installed base versus best available technology (BAT) (Wegener and Walachowicz 2009, 2011). Two main focuses are covered: to boost eco-efficient product development and to improve customer communication. The methodology comprises a holistic approach regarding the whole life cycle covering manufacturing, supply chains, use phase, and end of life.

Most of these case studies show that the environmental impact of the applications are mainly influenced by the energy consumption in the use phase with a share mostly higher than 90 %.

For electrical or electronic products, therefore, energy efficiency is in most cases the main driver for environmental improvements. Changes in the performance for different environmental impact categories show a linear correlation to the function of power grid mix.

Based on full-scale LCAs for representative products, system boundaries for ECM can thus be cut down to the use phase for the covered product families, if the environmental impact of the use phase is highly dominant (>90 % of impacts in all impact categories), and the differences between the compared product systems for all other life cycle stages than the use phase are evaluated to be irrelevant (<1 % for the whole life cycle for each environmental impact category).

For a transparent communication of the environmental performance in customer communication, the results of all relevant environmental impact categories are represented in a separate ECM instead of adding them up to a single score.

Global warming potential will always be addressed. Other environmental impact categories are only shown if their development is different from GHG development.

Product value in general is represented by life cycle costs. Life cycle costs include all affected cost elements during lifetime.

Objects of comparison are four different domestic lighting products including their complete life cycle. At first an incandescent lamp 40 W (General Lighting Standard, GLS) is to be compared with a halogen lamp (HAL), a compact fluorescent lamp (CFL) and an LED lamp, all featuring similar lumen outputs. To ensure comparability of the different lamp types, a lifetime of 25,000 h has been taken into account as a reference parameter. This reference flow can either be provided by one LED lamp with an average lifetime of 25,000 h or several pieces of the other lamp types with shorter lifetimes.

In the visualization of ECM results, the incandescent lamp (GLS) is chosen to be the reference. Indicator results for the other three lighting systems are displayed in relative deviation to the reference value. Results and main findings are significantly dominated by the use phase for all lighting systems. Due to the main influence of environmental impacts of the underlying power grid mix nearly, all ECM results reflect the same characteristics as the Global Warming Potential (GWP), except of elementary abiotic resource depletion (ADP elements).

Figures 4.3 and 4.4 show ECM results for the lighting systems with respect to GWP and ADP_{fossil}.

Comparing the lighting systems to the reference reveals a trade-off between the environmental impact categories of global warming potential and elementary abiotic resource depletion potential. A higher elementary resource consumption in the production of novel lighting systems enables a significant decrease in energy consumption during the use phase.

In order to find out which of the environmental impacts are considerably relevant, the indicator results have additionally to be normalized to the respective overall global impacts. In Fig. 4.5 the normalized indicator results for lighting systems over their entire life cycle are shown. The spider diagram reveals that acidification potential (AP) is the most relevant impact category followed by ADP_{fossil}.

Though the elementary abiotic resource depletion is not relevant compared to global warming and acidification, it is important to consider that the significant benefit of novel lighting systems in eco-efficiency is enabled by increased material intensity in their production.

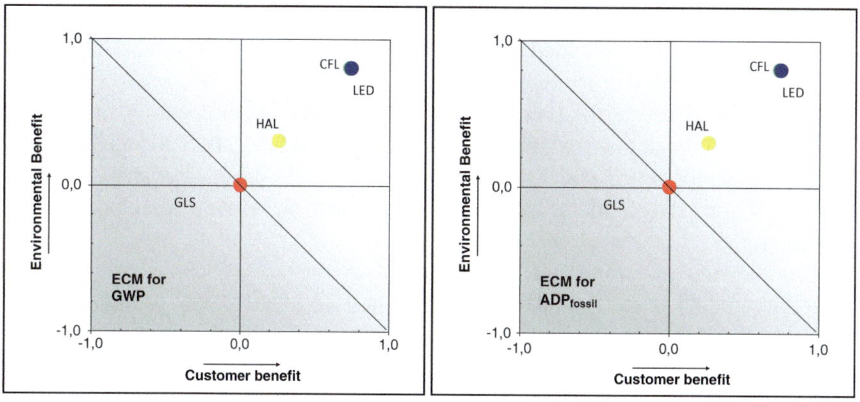

Figs. 4.3 and 4.4 ECM for GWP and ADP_{fossil}

Fig. 4.5 Normalized indicator results for alternative lighting systems over their life cycles

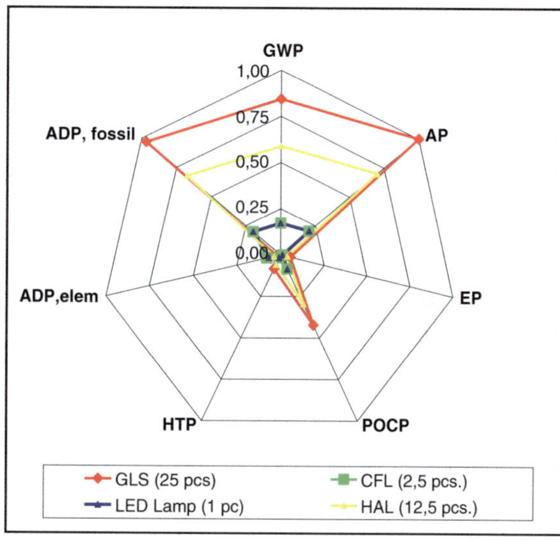

Due to the studies, key performance indicators can be extracted, summarized, and published. They can be worked out in accordance with ISO 14045. Furthermore, strategies for further improvements can be outlined.

To implement more sustainability evaluation in decision-making processes of industries will lead to more sustainable solutions (Gardner 2015).

1.3.2 Chelating Agents

With the purpose of assessing different chelating agents from environmental and financial perspectives, an Eco-efficiency Assessment was carried out for European conditions. The example of the four chelating agents was taken from Annex B.4 in ISO 14045:2012. The intended audience comprises primarily product developers but also purchasers.

The intended use is for product development and communication of product performance to business customers. The example calculates chelating agents made by four different processes, Products A, B, C, and D on an industrial scale in Europe. Main stakeholders involved are product developer, purchasers, and customers.

The function and functional unit referred to the use of chelating agents in detergents and cleaners. The goal of the study was the identification of improvement potentials along the value chain with the same detergency power of the products. The chelating agents were compared on an equal weight basis in order to make the study independent of the exact amounts used in the many detergent recipes. The functional unit is the production of 1 ton of chelating agent, linked with the same function of binding different types of metal ions, as Ca^{2+}, Fe^{3+}, Ni^{2+}, Hg^{2+}, etc.

The product system excludes the function of different detergent recipes because it is assumed to be the same for alternatives A, B, C, and D.

In eco-efficiency studies, different sets of indicators can be used. Table 4.2 shows an overview of the used indicators and elementary flows that have been assessed.

The impact categories considered in the Eco-efficiency Assessment (Saling et al. 2002) and applied for different chelating agents are primary energy consumption, resource depletion, land use, emissions, human toxicity, and risk (referring to occupational health and accidents). The impact category "emissions" is further subdivided into other impact categories (Table 4.3). Impact assessment methods used are detailed in (Saling et al. 2002).

In a further weighting process, the impact category results are aggregated into a single statement of the total strain on the environment. In the presented Eco-efficiency Assessment method, a weight that expresses the environmental importance of that impact category relative to the other impact categories for a specific region is assigned to each impact category. These weighting factors are a combination of impact category-specific "relevance factors" and "societal factors." For the European relevance and societal factors, see Table 4.3. To derive the relevance factor, the result of the alternative with the highest impact in that category is normalized against the total load of the same category in a specific region. This step yields the relative significance of the different impact category results. The societal factors express the importance of each category relative to the other impact categories as perceived by a group of people (see Table 4.3). The societal factors are based on the opinion polls in the same region chosen for the relevance factors.

The societal factors were derived through a public opinion poll (Kicherer 2005). More detailed information regarding the weighting methodology and the subsequent integration of ecological and economic data were published by Saling et al. (2002) and Kicherer et al. (2007).

The product system value in this study was assessed by using a life cycle costing (LCC) method (Bengtsson and Sjöborg 2004). Costs associated with environmental

Table 4.2 Elementary flows and indicators that have been used

Energy (MJ/FU)	**Emissions to water (mg/FU)**
Coal	COD
Oil	BOD
Gas	N-total
Waterpower	NH_4-N
Nuclear	P-tot
Lignite	Heavy metals
Recovered/other	AOX
Biomass	Hydrocarbons
	Sulfates
	Chlorides
Resources (kg/FU)	**Waste (kg/FU)**
Hard coal	Municipal waste
Oil	Chemicals waste
Natural gas	Construction waste
Lignite	Mining waste
Sodium chloride	
Sulfur	
Phosphorous	
Iron	
Lime	
Bauxite	
Sand	
Emissions to air (mg/FU)	**Land use (m^2/FU)**
CO_2	Forest
SO_X	Pasture, fallow, bio-agriculture
NO_X	Conventional agriculture
CH_4	Sealed
Non-methane volatile organic compounds (VOC)	Roads, tracks, canals
CFC	
NH_3	
N_2O	
HCl	

impacts are not covered by the LCC since, by definition, external costs are created by society and reflect the environmental impacts of the system under study (Rüdenauer et al. 2005). These impacts are covered by the LCA in the environmental assessment. In this study, the product system value for the customer, based on an equal weight basis, was the cost savings of the chelating agent for the detergent manufacturer.

In the Eco-efficiency Assessment method applied, the total costs of the studied alternatives are normalized with respect to the gross domestic product of the same region that is used in the environmental assessment.

The result of the weighting is illustrated in the bar chart and table in Fig. 4.6. They show the weighted values for each impact category and chelating agent; the

Table 4.3 Impact categories and weighting factors used in this study

Impact category	Social weighting factor S (%)	Relevance factor R (%)	Total weighting factor W (%)[a]
Resource use	20	4	11
Primary energy use	20	5	13
Area use	10	0,3	2
Toxicity potential	20	20	20
Risk potential	10	10	10
Emissions	20	61	44
Water emissions[b]	35	95	78
Solid waste	15	0	0
Air emissions	50	5	22
Greenhouse gases (GHG)	50	69	68
Photochemical oxidation creation potential (POCP)	20	8	15
Ozone depletion potential (ODP)	20	0	0
Acidification potential (AP)	10	23	17

[a]Geometric mean of S and R

[b]This impact category includes the eutrophication potential of substances emitted to the water recipient

top of the bars denotes the total and final environmental results that were integrated with economic data in the complete Eco-efficiency Assessment.

The eco-efficiency method includes a weighting of environmental impacts and costs, resulting in a two-dimensional diagram (Fig. 4.7). This overview graph enables the reader an easier understanding and a good overview on the results of a comprehensive study. It helps in the step of interpretation to derive overall conclusions and makes it easier for nonexperts to use the results. It bears the risk of unbalanced weighting, but if this process is quite openly and transparently described, it reduces those risks very much.

The eco-efficiency method takes into account the contribution of the studied alternatives' environmental impact to the total environmental impact within a specific region. In the same way the costs of the studied alternatives are compared to the gross domestic product of the same region. Hence, this is a normalization step, which yields two ratios communicating the significance of the environmental and financial impact. If the environmental impact is greater, for example, more weight will be put on the environmental performance of the studied alternatives. The axes in the diagram are inverted so that the alternative that has the lowest environmental impact and the best financial performance is found in the upper right corner. This alternative is termed the most eco-efficient alternative and is hence favored from an eco-efficiency perspective.

The result of this study indicates that the product system for chelating agent A has the lowest total environmental impact. A performs well in all important aspects compared with the other alternatives, mainly because it is based on renewable raw

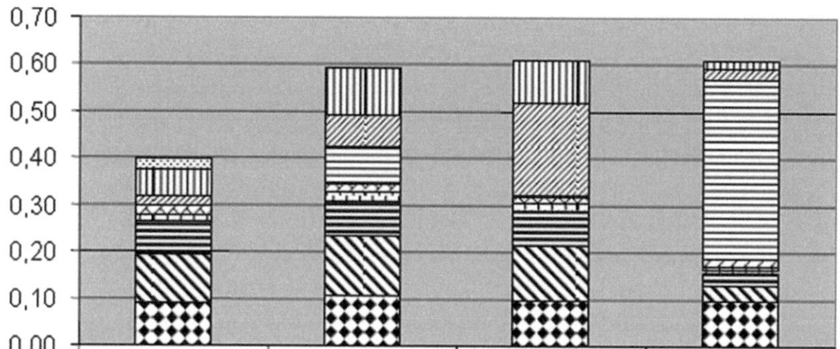

	A	B	C	D
Area use	0,02	0,0002	0,0002	0,0001
Risk potential	0,06	0,10	0,09	0,02
Toxicity potential	0,02	0,07	0,20	0,02
Water emissions	0,0009	0,08	0,003	0,38
AP	0,02	0,02	0,01	0,02
POCP	0,01	0,02	0,02	0,006
GWP	0,07	0,08	0,07	0,04
Energy use	0,11	0,13	0,12	0,03
Resources use	0,09	0,11	0,10	0,10

Fig. 4.6 Weighted values for the different impact categories and chelating agents

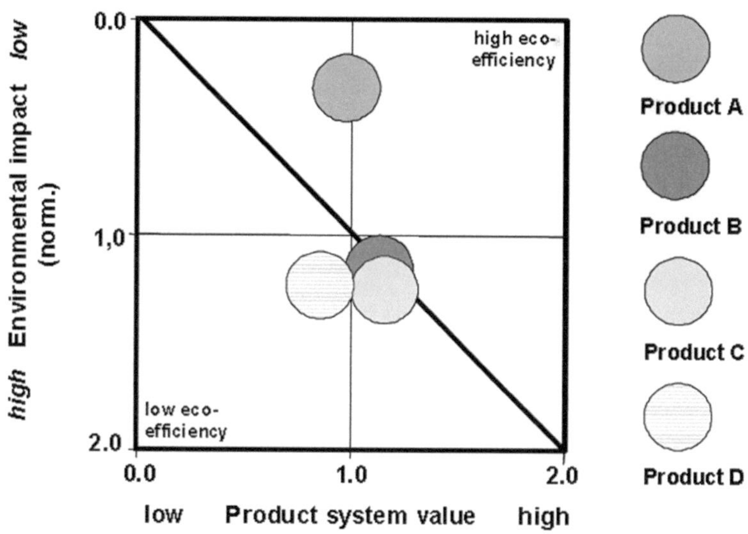

Fig. 4.7 Weighting of environmental impacts and costs, resulting in a two-dimensional diagram

materials and is readily biodegradable. Another advantage of A and B is that (unlike D and B) they do not give rise to any phosphorus emissions to water, and hence the eutrophication potential, which is included in the water emissions of A, is minor as shown in Fig. 4.6. The most significant impact of chelating agents is water emissions, according to the applied weighting methodology. This is due to the fact that a lot of the eutrophication is caused by the use of phosphorous in detergents. More than 60 % of the environmental impact of chelating D is due to eutrophication, which is the single impact category that makes this chelating agent have a higher environmental impact than agent A.

With respect to the toxicity potential, A scores much better than especially C, for which there is limited evidence of carcinogenic effects from exposure. For these reasons it can be concluded that on an equal mass basis, A is the most environmentally preferred product system. A sensitivity analysis also showed that this result is robust with regard to the region (continent) that is chosen for the weighting (Borén et al. 2009).

1.4 Pulp Industry Example

Eco-efficiency Assessment is used within the pulp industry and may be used to support decision-making. It might increase the development of products or processes with higher eco-efficiency, if applied early in the development or design phase.

Examples of research material within the topic may be found in Gäbel (2001) and Löfgren (2009). The need for flexible LCA models has grown, yet the knowledge and use of LCA during the development phase of products and processes are limited within companies. The companies have started to realize that the use of LCA during the development phase of products and processes may be beneficial. However, it is important to realize that the environmental impact of a product can be stimulated more easily and continuously during the development phase of a product or process in order to make LCA useful.

Optimization of existing products and processes is also an area where LCA may be of great interest for companies and producers. Since most of product and process development is performed using computer-based programs, the output parameters have to be connected to the LCA tool in order to make it useful. Material and energy balances are example of parameters that are outputs from design or optimization tools that may be used and combined with LCA. It is of great importance that it is possible to make changes in the design or optimization program and quick and easily implement those changes to an LCA. Thus, it is possible to investigate if and how changes during the design process influence the environmental performance (Tegstedt 2011).

Furthermore, the environmental impact of a product or a process may be minimized through optimization, using a simulation tool. It may thus be possible to include environmental optimization as a design parameter when making strategic decisions of new or existing products and processes. Environmental performance may be one of many parameters when performing multi-objective optimization of a product or a process.

Combining process simulation and LCA is under development. One possibility would be to investigate if it is possible to combine process simulation and Eco-efficiency Assessment in order to identify efficient working procedures. The idea is to study literature within the subject and try to conclude, or at least investigate, if it is possible to apply on Eco-efficiency Assessments.

The goal of the case study is to assess and compare the eco-efficiency of alternative concepts for producing bleaching chemicals for the elemental chlorine-free, ECF, pulp industry. The following four production concepts are compared from an environmental and economic perspective:

1. Chlorine dioxide generator-type SVP-SCW
2. Chlorine dioxide generator-type SVP-LITE
3. Chemical Island, CI
4. Integrated Plant, IP

The assessment and comparison are first carried out for the supply of bleaching chemicals to a specific pulp mill in Russia. Second, possible key parameters and assumptions are varied in order to identify flexible parameter settings significantly affecting the eco-efficiency. Third, the geographic location of the pulp mill is varied in order to find out how this affects the eco-efficiency.

The intended audience of the Eco-efficiency Assessment is Eka Chemicals. It is Eka Chemicals who has commenced the project in the first place. Furthermore, the study may be of interest for the pulp and paper industry, environmental authorities, and NGOs interacting in regions where the products and services of Eka Chemicals may be implemented.

The Eco-efficiency Assessment is performed in order to provide Eka Chemicals with information about the environmental and financial performance of the alternative concepts to provide a pulp mill with ECF bleaching chemicals. The knowledge will be used for educational purposes by the organization working with process development. It will give employees at Eka Chemicals a better understanding of environmental and economic issues related to their processes and products. Eka Chemicals also aims for using Eco-efficiency Assessments:

- In strategic business decisions such as investments
- For providing information to a customer about a specific production site and to identify the most optimal production concept for the specific site

Figure 4.8 shows how a simulation can be used starting from a base case and integrating new impact figures or a different scaling. The portfolio graph shows clearly what the improvement potential is, if indicators or input tables are changed. In this scenario, Indonesian electricity was used. It shows that the two systems with stand-alone chlorine dioxide generators are most eco-efficient, followed by the systems with IP and CI. When it comes to environmental impact, the differences appear to be quite large. In the technologies that are assessed in this example,

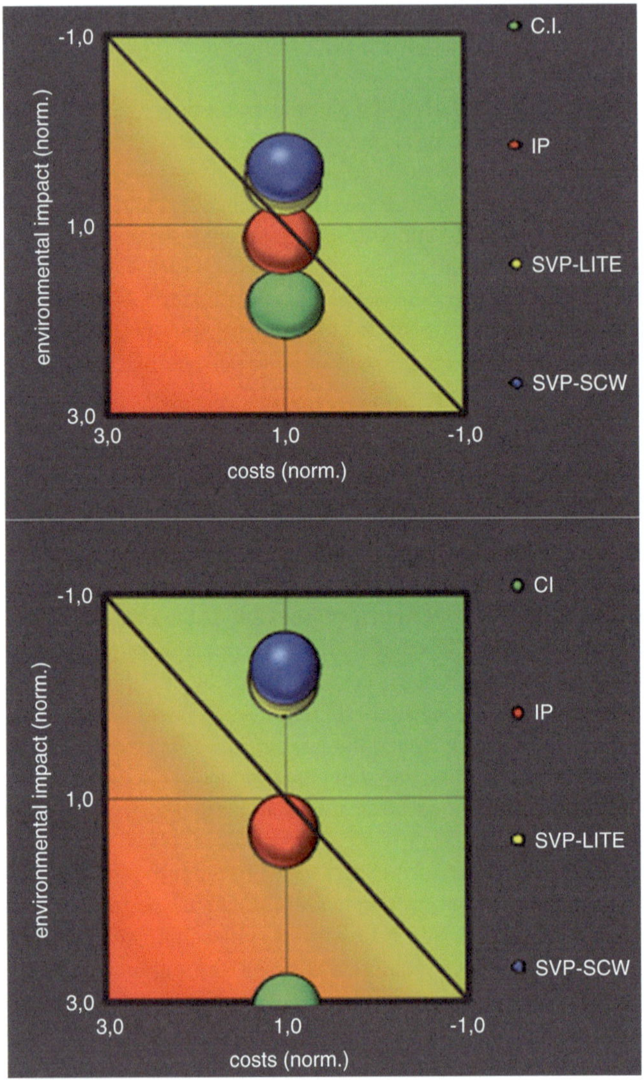

Fig. 4.8 The eco-efficiency portfolio comparison (*upper*, base case; *lower*, scenario with Indonesian electricity mix)

electricity plays an important role and influences the results significantly. The CI system has the highest total environmental impact, more than twice the environmental impact caused by the systems with stand-alone chlorine dioxide generators. The air emissions generated from the system with SVP-SCW has the smallest impact from air emissions of the four production systems compared. Its impact is 43 % of the corresponding impact of the CI system and 64 % of the corresponding impact of the IP system. However, the POCP is highest for the systems with

SVP-LITE and SVP-SCW. This result supports strongly the decision-making process of which technology should be used in which regions. They support the improvement of technologies and the identification of important levers of the production system.

1.5 Cement Industry Example

The production of concrete, notably its most important ingredient, cement poses several sustainability issues that need to be managed: Cement production emits CO_2 and other air emissions, and the quarrying of raw materials produces local impacts such as noise and dust. Also, water use needs to be carefully looked at in locations where water is scarce. The industry is well aware of these impacts and addresses them both collectively, via the CSI (Cement Sustainability Initiative) or regional and national trade organizations, and individually as producers within their sphere of influence (WBCSD 2015). In this context, eco-efficiency will be able to support process improvements from a holistic point of view.

Based on the needs and requirements of the cement industry, a life cycle process model has been designed and built. The purpose of the model is to provide a tool for supporting decisions made on product and process development options. The requirements on a flexible model that generates information on product performance, economic cost, and environmental performance, from a life cycle perspective, were seen as important (Gäbel 2001).

The chosen approach, which entailed dividing the cement manufacturing process into a foreground and a background system and modeling the two subsystems with different techniques and levels of detail, proved to be successful. The modeling approach used to build the foreground system model, a calculation non-causal model, physical modeling, and an object-oriented modeling language can enhance the flexibility, modularity, and comprehensiveness of the model. Together with an appropriate simulation tool (for this application, ASCEND IV was used), the approach provided a flexible and general purpose model. The chosen software also supports nonlinear and dynamic elements for future use. By making use of physical and object-oriented modeling, all physical entities in the manufacturing process are kept together and described independently of each other. This satisfies the requirement for flexibility, in terms of future modifications to represent and simulate process development options and other manufacturing plants.

The foreground system model has been validated and shows satisfactory agreement with the real system's properties, with the exception of metals.

The result is a flexible and valid life cycle model of the current manufacturing process at the site. The model can be used to solve many different problems. So different development options can be simulated and information on potential environmental, product, and economic performance generated. Thus, the same model can be used for a number of different purposes. The model has been used to explore the potential for reducing the negative environmental impact of cement

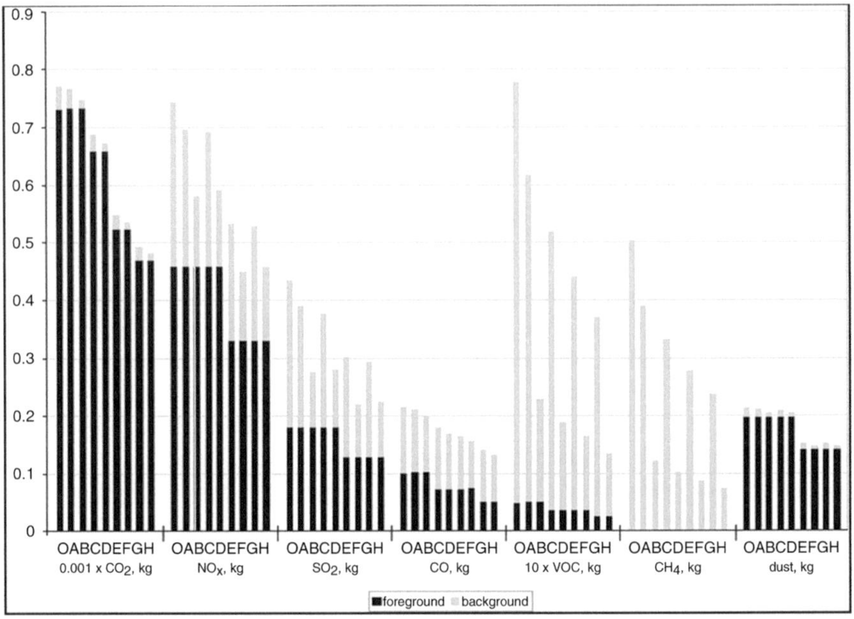

Fig. 4.9 Emission to air, per 1000 kg cement, from background and foreground systems, scenarios O to H. The bar for each scenario is divided into the emission from the foreground system and the emission from the background system

manufacturing through an increase in the use of recovered material and alternative fuel. It has been shown that the model can simulate the desired development options differently. The desired information is generated and assessed in relation to current requirements on product performance. The generated information can be used to give indications for development options for further investigation and study. The nine simulations show that the use of recovered material and alternative fuel can be increased with no negative effect on product performance. The use of resources and the studied emission to air can be substantially reduced (Fig. 4.9).

The simulations show that emissions of CO_2, NO_x, SO_2, CO, VOC CH_4, and dust can be reduced with an increase in the use of recovered material and alternative fuel (see Fig. 4.9). Even if the emission of CO_2 from alternative fuel is valued the same as the CO_2 emission from fossil fuel, the total CO_2 emission can be reduced from about 780 to about 480 kg per ton cement.

The emission of small amounts of toxic substances, such as metals, dioxins, and furans, has not been modeled. These emissions depend, to a large degree, on minor variations in the raw material and fuel chemical composition.

2 The Sustainability Assessment Toolbox Developed by BASF

2.1 Overview

BASF has worked out methods for the assessment of the sustainability of chemical products and production processes through the development and use of its EEA as well as SEEBALANCE® or AgBalance™ analysis. Next to the environmental impact, which is assessed based on ISO14040 and ISO14044 standards, economic factors are taken into account and implemented in the Eco-efficiency Assessment, following ISO 14045. The SEEBALANCE (social-eco-efficiency analysis, trade name SEEBALANCE) and AgBalance (specifically for agriculture processes based on SEEBALANCE) also consider social impacts of products and processes.

The tools are used by BASF and its customers to assist strategic decision-making, facilitate the identification of product and process improvements, enhance product differentiation, as well as support the dialogue with opinion makers, NGOs, and politicians.

Both detailed in-depth results of individual impact indicators and aggregated results and a sustainability evaluation score are output of the sustainability evaluation methods. Different types of footprinting in combination with other information tools can support decision-making efficiently. Specific developed Quick Scan tools can give a basic overview of alternatives in this context. Sensitivity analyses can be employed to study the robustness of the model results and to investigate trade-offs or the response to external influences. Scenario analyses can model different situations by simulating new sets of inputs followed by an assessment of improvement potentials of the analyzed system.

At the United Nations Conference on Sustainable Development (UN 2015), the United Nations system was requested to mainstream the economic, social, and environmental dimensions of sustainable development throughout its work.

Improvements of products and processes should be supported that consider the three dimensions of sustainability, i.e., economy, ecology, and society, which can be covered with the updated toolbox of BASF.

An overview about the toolbox with its various opportunities is given in Fig. 4.10. The more complex the studies are, the more comprehensive information can be generated. Often it is not needed to get the full picture of a sustainability assessment, so in the early phases of a product development. More simple tools as the Sustainability Check or a Quick Scan can be applied to give general directions for further development to be updated during the development process. For reporting life cycle inventories together with specific technical information, the Environmental Product Declarations (EPD) can be worked out and can support the LCA of complete systems. Today, it is mainly used in the building and construction industry. In this article, the Eco-efficiency Assessment, SEEBALANCE, and AgBalance will be highlighted as well as the web-based assessment which needs a sustainability study as basis. This effort is not included in the position of the web-based method in Fig. 4.10.

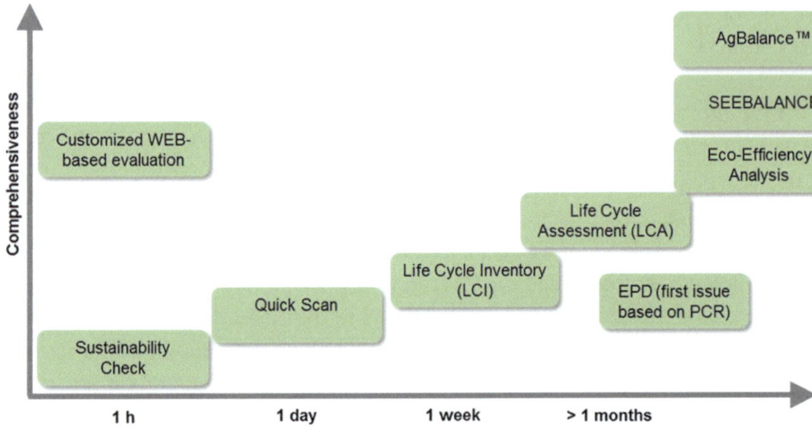

Fig. 4.10 Overview of sustainability evaluation and LCA-based tools showing efforts and comprehensiveness

2.2 *The Eco-efficiency Analysis of BASF: Methodological Aspects*

The method was initially developed in the late 1990s by BASF in cooperation with Roland Berger Consulting, Munich. The method is seen as a life cycle management tool and can be involved in assessments of the entire product life cycle, from concept development, to design and implementation, further to marketing, finally, to end-of-life issues. This “marketing life cycle,” in different steps, can be linked with the “physical life cycle” of LCA. The analysis method may incorporate both economic and environmental aspects and lead to a comprehensive evaluation of products and processes over their entire life cycle.

In the method of eco-efficiency analysis, results are presented as aggregated information on costs and environmental impacts and show the strengths and weaknesses of a particular product or process. A comparative assertion with competitors’ products as it is described above is done only in some selected cases because this is not intended in the current standards. In those cases, all disaggregated figures are shown as well, and the aggregation is transparently described in the report.

The ecological calculations of the single results in each category are following the ISO-standards 14040 and 14044, and the assessment method in total follows ISO 14045.

It will be shown, in several examples, how this methodology can be used to support and to evaluate chemical products and processes over the whole life cycle.

Every eco-efficiency analysis passes several key stages (Uhlman et al. 2013). This ensures consistent quality and the comparability of different studies. Environmental impacts are determined by LCA, and economic data are calculated using the usual business or, in some instances, national economic models.

The basic preconditions in eco-efficiency analysis are (Saling 2009):

- Products or processes studied have to meet the same defined customer benefit.
- The entire life cycle is considered.
- Both an environmental and an economic assessment are carried out.

The methodology has been approved by the German "Technischer Überwachungsverein" (TÜV, English: "Technical Inspection Association") (TÜV 2002) and by the American National Sanitation Foundation (NSF) (Uhlman et al. 2013).

For the calculation and comparison of the environmental position of each alternative, data from the different production methods are assimilated and analyzed to provide a value for energy consumption, raw material consumption, emissions, and use of area, risk potential, and toxicity potential. All raw materials, required in the process and how these are derived, are factored in the study, as these are the steps required to bring the product to the end user.

In the same manner, economic data from the life cycle chain of a product application or process evaluation may also be calculated and summarized. The rationale behind this assessment tool has been described by Saling et al. (2002) and by Landsiedel and Saling (2002).

Practical examples can show how the metrics for sustainability can support decision-making processes answering different questions.

The process for performing an EEA has been previously published (Saling et al. 2002), and it involves measuring the life cycle environmental impacts and life cycle costs for product alternatives for a defined level of output. In other words, a BASF EEA evaluates both the economic and environmental impacts that products and processes have over the course of their life cycle. The eco-efficiency methodology is a comparative analysis and thus does not determine the sustainability of a product by itself, rather how it compares with other alternatives considered. Thus a product which was deemed most eco-efficient in one analysis may be a less eco-efficient alternative in another study when a different application is considered.

The basic preconditions of an eco-efficiency analysis are defined as follows: (1) products or processes studied have to meet the same defined functional unit or customer benefit, (2) the considered alternatives should cover at least 90 % of the relevant market, (3) the entire life cycle is considered, and (4) both an environmental and an economic assessment have to be carried out. The eco-efficiency method is based upon the ISO 14040 and 14044 (ISO 2006a, b) standards for life cycle assessments and fulfills both the required and optional phases. However, in addition to the requirements found in this standard, it includes additional enhancements that allow for the expedient review and decision-making at all business levels. The newly developed ISO standard 14045 (ISO 14045:2012) for eco-efficiency is, furthermore, a relevant standard that gives guidance for performing studies for internal and external uses. The general process for conducting an EEA follows a specific and defined way of calculation:

1. Calculation of total cost from the customer viewpoint.

2. Preparation of a specific life cycle analysis for all investigated products or processes according to the rules of ISO 14040, 14040, and 14045. The water footprint standard 14046 is as well implemented. For carbon footprinting within the method, the Greenhouse Gas Protocol is applied.
3. Determination of impacts on the health, safety, and risks to people.
4. Assessing the use of land over the whole life cycle.
5. Calculation of relevance factors for specific weighting.
6. Weighting of environmental factors with societal factors.
7. Determination of relative importance of the environment versus the economy.
8. Creation of an eco-efficiency portfolio.
9. Analyses of appropriateness, data quality, and sensitivities.
10. Scenario analysis for further interpretation of results.
11. Interpretation of results.
12. Identification of improvement potentials.
13. Summary and conclusion.
14. Optional: Transfer to a web-based evaluation tool for further assessment of scenarios.

2.2.1 Definition of Customer Benefit, Alternatives, and System Boundaries

The first step of the analysis is the definition of the goal and scope of the study. During this step, the customer benefit or functional unit of comparison as well as the alternatives are identified. The functional unit provides the reference point from which the economic and environmental inputs and outputs for each alternative are compared. The functional unit or customer benefit should be consistent with the goals of the study and clearly state criteria with both performance attributes and spatial and temporal limits. The definition of a functional unit can cover a wide range of options and allow the comparison of different solutions for a defined functional unit. For example, a functional unit could be the production of 100 m^2 of a flooring system in an office building for the use of 5 years. Alternative solutions for this functional unit can be wooden materials, carpets, PVC tiles, ceramic tiles, etc. The main criterion is to fulfill the functional unit with a comparable service and use by customers. As the eco-efficiency method is a comparative analysis, as many possible alternatives represented in the marketplace or in development which can perform the same function need to be considered. Finally, the scope of the EEA is defined by the specific elements within the production, use, and disposal phases of the product's life cycle that will be considered. The whole calculation considers relevant and significant system boundaries. In some cases, a so-called cradle-to-grave or, in other cases, a so-called cradle-to-cradle might be the selected boundary conditions. Depending on the goal and scope, both approaches can be sufficient to deliver good and acceptable results for decision-making. Specific cutoff rules helping to define the system boundaries to be accurate are the basis for the generation of meaningful and reliable results. The same life cycle stages must be considered for each alternative, and as the study is comparative in nature, any life cycle stage that is identical for each alternative can be excluded from the analysis as its impact across each alternative will be the same. However, in these cases, a complete calculation will be made to see how relevant the remaining calculation is compared to the complete system.

2.2.2 Determination of Economic Impacts

The EEA assesses the full economic impact of a product or process over its life cycle in order to determine an overall total cost of ownership for the defined customer benefit. The specific approaches for conducting a life cycle cost (LCC) analysis vary from study to study and are strongly dependent upon the customer benefit selected and the system boundaries and alternatives considered. Cost accounting needs to take into account both initial costs and all future cost impacts or benefits of the products as well as any costs having an environmental aspect (e.g., disposal of hazardous waste). Either constant (real) or nominal monetary values can be used for cost accounting but they cannot be mixed in the analysis. In addition, the final cost analysis can be calculated as a point in time or can account for the time value of money. If the analysis is performed to consider the time value of money, then a net present value (NPV) or similar metric shall be calculated. A representative but not all inclusive list of economic metrics normally considered for each alternative includes the costs of raw materials, labor, energy, capital investment, maintenance activities, transportation, illnesses and accidents, and waste disposal. When all the costs are identified and accounted for, they are summed and combined in appropriate units (e.g., $ or €) without additional weighting of individual financial amounts. This rigorous accounting of life cycle cost impacts can help identify not only the economic benefits of a product in its use and application that can help manufactures better understand their economic value proposition but also uncover hidden costs and cost intensive areas of the life cycle that can become the focus of optimization efforts.

2.2.3 Determination of Environmental Impacts

The eco-efficiency method applies a consistent and comprehensive approach to assessing the potential environmental burden of a product as opposed to focusing on just a few specific metrics or considering only part of a product's life cycle. As described previously, the method goes beyond the basic requirements of the ISO standards, and measures, at a minimum, 11 environmental impacts in 6 main categories: energy, resource consumption, emissions (to air, water, or land), land use, toxicity potential, and risk potential. In a testing phase, a seventh environmental impact was introduced, namely, the consumptive water use.

The process for data acquisition and calculation is done according to the requirements of ISO 14040, 14044, and 14045 and is collected for each impact category as defined by the study's scope, boundary conditions, and customer benefit. This process is known as the inventory analysis stage. It is followed by the impact assessment, and this starts by applying a systematic approach to classifying and characterizing all the information identified in the inventory analysis. The environmental impacts are then aggregated using normalization and weighting schemes for each impact category. The specific steps of the impact assessment are discussed and illustrated below.

2.2.4 Environmental Categories, Classification, and Characterization

Energy Consumption and Raw Materials

The cumulative amount of primary energy, normally expressed in terms of megajoule(MJ)/customer benefit, consumed during the life cycle of each alternative, is measured and converted back to its primary energy source such as oil, gas, coal, lignite, nuclear energy, biomass, and hydroelectric. Fossil fuels are included before production and renewable energy before its harvest or use. The consumption of the individual primary energy sources is included as well in the raw material or resource consumption category.

Key materials consumed during the life cycle of each alternative are considered in the resource consumption category. The usage amounts for the different raw materials considered are aggregated together into a common unit of consumption, say, kg, by applying a weighting factor to each material which takes into consideration its exploitable reserves as identified, for example, by the US Geological Survey and its current level of consumption. This allows a higher weighting to be applied to materials that are either scarce or have a very high consumption rate in society. Renewable raw materials if produced through sustainable management practices are viewed as having a theoretically infinite reserve and thus would have an applied weighting factor of zero. In principle, other evaluation systems as CML[1] or ReCiPe[2] can be used as well and might be adopted, if the standard development will make further progress in the future.

---

[1] Within the European debate on the development of a standard LCA methodology, the Center of Environmental Science, Leiden University, the Netherlands (Centrum voor Milieukunde Leiden: CML), soon conquered a position of hegemony regarding the "agenda setting" for further research on LCA. Most European experts today agree that the methodology published in 1993 by CML marked a breakthrough in the scientific foundation of LCA methodology.

[2] Goedkoop M, Heijungs R, Huijbregts MAJ, De Schryver A, Struijs J, Van Zelm R (2009) ReCiPe 2008 A life cycle impact assessment method which comprises harmonised category indicators at the midpoint and the endpoint level. Report I: characterisation, 1st edn. 6 Jan 2009. http://www.lcia-recipe.net

Water Footprint

The importance of one resource in particular, fresh water, and understanding how we impact the quality and availability of this valuable and in some areas limited resource is a topic of key interest and political debate in our society. The eco-efficiency method is currently piloting some advanced methods to incorporate a more rigorous approach to assessing the use and impacts of water consumption. The method proposed by Pfister et al. (2009) which assesses damages to three areas of protection, namely, human health, ecosystem quality, and resources, is currently being applied and evaluated in ongoing studies. The finalization of the ISO Standard 14046 in 2014 for assessing water footprint supported the integration of this impact category to the method.

Air Emissions

The emissions category is further subdivided into emissions to air, water, and land (soil). In emissions to air, the inventory analysis is classified into four subcategories: global warming potential (GWP), acidification potential (AP), ozone depletion potential (ODP), and photochemical ozone creation potential (POCP), also known as summer smog. Some air emissions, such as methane, can be included in more than one classification. Through characterization, weightings are applied to each emission to enable aggregation within each classification. For example, global warming potential is measured as the total amount of CO_2 emitted into the atmosphere. Thus, the global warming potential of a substance will then be measured in relation to CO_2. The 100-year global warming potential of methane is 2511, meaning that for every 1 kg of methane emitted to the air, it is equivalent to 25 kg of CO_2 emissions. Correspondingly, the POCP of methane is only 0.007 meaning that it is over 140 times less potent than a similar quantity of ethene, the comparative unit of measure for POCP.

Water Emissions

For emissions to water, the inventory analysis will include, for example, COD (chemical oxygen demand), BOD (biological oxygen demand), total nitrogen, hydrocarbons, heavy metals, chloride, sulfates, ammonium, and phosphates. The method of critical volumes or critical limits for discharge is used for characterization. As described in the German AbwasserVerordnung from 1997 (AbwV 1997), each pollutant emitted into the water contaminates sufficient water, until the critical load or statutory limit, as defined by waste water regulations, for the substance is reached. The greater the water hazard posed by a substance, the lower its limit. The amount of uncontaminated water needed to dilute the water emission to the statutory limit is the determined regulation. For example, if the legal limit for COD is 75 mg/L, the factor is 0.013 or the inverse of 75. For a more potent contaminate such as heavy metals, if the legal limit is 1 mg/L, the factor will be 1. Through this approach, one can multiply all the identified water emissions by their corresponding dilution factor, and then through aggregation of these values over the life cycle, a single number or critical water volume can be determined for each alternative. The statutory or maximum threshold limits utilized in the model are based on the German waste water ordinance. They consider how severe effects of different pollutants to the environment are and put them in a defined order. These thresholds have been worked out by experts considering the different importance and potency of the emissions they have in the environment linked with a realistic management of those emissions to acceptable levels. So in general the thresholds can be applied to other regions as well. There are no provisions in the current model to regionalize or localize these values as they are common waste water constituents with well-established toxicity and limits should be quite similar across various geographies.

Wastes

The results of the inventory analysis partitions solid wastes into the various categories of materials that will end up in a landfill (cradle-to-grave). Thus materials that are recycled (cradle-to-cradle) are not counted in this category. Wastes are categorized as either municipal, hazardous, construction, or mining with a weighting factor to account for their varying impact potentials applied to each type based on their representative costs of disposal. All weightings are normalized toward the municipal waste category which is assigned a value of 1. Examples of classification of various wastes streams include:

- Municipal: household trash
- Hazardous: follows RCRA[3] definition of hazardous waste
- Construction: nonhazardous waste materials generated during building or demolition activities
- Mining: nonhazardous earth or overburden generated during raw material extraction activities

The impacts are then summed up to obtain an overall impact amount which is expressed as kg/customer benefit.

Land Use

The importance of how the products we produce, use, and dispose impact the land around us is becoming more prominent in life cycle assessments. Though there is

[3] The Resource Conservation and Recovery Act by EPA governs the management of hazardous wastes.

much debate on how to incorporate land use as impact category in LCA, and while no singular approach or methodology has been universally adopted, the EEA has adopted a comprehensive land use evaluation which quantifies the damages a series of land transformations, land occupations, and land restorations having on specific biodiversity indicators. As adapted from the approach proposed by Köllner and Scholz (2007, 2008), this is a more robust approach than the previous methodology utilized which was based on the hemeroby or biodiversity of an area and only considered the specific use of the land. The characterization factor proposed by Köllner and Scholz (2007, 2008) for land use is labeled the ecosystem damage potential (EDP). This approach is in line with the UNEP/SETAC framework on land use assessment, and inventory data is available in many LCA databases.

Toxicity Potential

The remaining two environmental categories are often not assessed in LCA, but the EEA has included them in order to provide a more comprehensive approach to assessing the environmental impact of products. The toxicity potential category focuses on human toxicity potential. This is often seen very critical by different stakeholders due to the fact that the chemicals industry often uses products which are labeled as harmful or toxic. To integrate that in decision-making processes, this is important, in particular for industries with harmful substances. The life cycle-based approach focuses not only on the final products but also on all the reactants and chemical precursors required during manufacture and ultimate disposal. The general framework for performing this analysis is described by Landsiedel and Saling (2002). To score the toxic properties of a substance, a classification of its possible adverse human health effects is needed. The system adapted into this model is based on the European classification of different toxic effects and the assignment of H-phrases of the Globally Harmonized System (GHS),[4] or risk phrase, as defined in Annex III of the European Union Directive 67/548/EEC. In the evaluation of human toxicity in the applied model, only H-phrases of the 300 series were used (University Münster[5]). They express health hazards, where the 200 series contains physical hazards and the 400 series contains environmental hazards. After extensive research and surveying knowledgeable experts in the area (e.g., toxicologists), the H-phrases were grouped into six categories which reflected the relative severity of each toxic effect relative to one another. Less hazardous effects such as "may cause drowsiness or dizziness" (H336) or "causes skin irritation" (H315) are appropriately scored lower with an evaluation factor of 100 or 300 points than much more hazardous effects such as "toxic if inhaled" (H331) and "may cause cancer" (H350) or "may cause cancer by inhalation"

---

[4] Globally Harmonized System of Classification and Labelling of Chemicals (GHS).

[5] Source: https://www.uni-muenster.de/imperia/md/content/physikalische_chemie/praktikum/h_p_phrases.pdf

(H350i). They are linked with 750 or 1000 points. The range of these figures starts with 100 points and ends with a maximum of 1000 points. Comparable ranges have been published in the official "Technische Regel für Gefahrstoffe," Germany (TÜV 2002) and TRGS 600 (TRGS 2014). There was the same system applied but with a higher range until 50,000 points.

If there is only one H-phrase for the substance, it will be assigned to the appropriate group; however, if there are multiple H-phrases, the substance will be upgraded by one group level. Weak effects and local effects (group 1 and group 2, respectively), and if the same effect is caused by an additional exposure route (e.g., oral and dermal), will not lead to an upgrade. In general, there is only one upgrade for a substance, irrespective of how many additional H-phrases are present. In the case where H-phrases are not specifically identified for a chemical but a material safety data sheet exists, health effect information obtained from the data sheet can be used to estimate the appropriate H-phrases. For data sheets which follow the OSHA[6] recommendations for hazardous communications, relevant information can be found in the following sections: hazard identification, toxicological information, or regulatory information. However, in some cases, limited or no toxicological information may exist for a substance. In these cases, valuable toxicological data can be obtained from related substances, structure-activity relationships, and even data from the results of preliminary tests. The estimation of possible toxic effects of substances utilizing any of these data sources requires expert judgment and consultation with toxicologists.

However, different systems can be applied here based on a differentiation of substances linked to their human toxicity potential. In an ideal case, real exposure factors will be applied and linked with the evaluation system. This is often very difficult to realize and to generate data, so a risk-based system is applied instead of a hazard-based system. Other systems like USEtox[7] try to integrate a risk-based approach but have the same problem to get real data. They replace it by calculation models, which enables practitioners in an easier way to apply such systems but which have a high level of uncertainty. Further developments are ongoing which try to implement information, the REACH program generated, and where more realistic risk evaluations are behind it, into the LCA or eco-efficiency methodology (Kalberlah et al. 2015).

After the inventory analysis phase accounting for all the chemicals utilized, an overall toxicity score is developed across each life cycle stage by multiplying the quantities of each material by their respective toxicity score. During this process,

---

[6] Occupational Safety and Health Act (OSHA), United States Department of Labor, is the main federal agency charged with the enforcement of safety and health legislation.

[7] USEtox is a model which can be used to calculate characterization factors for human and ecotoxicity impact categories used in life cycle assessment (Rosenbaum RK, Bachmann TM, Gold LS, Huijbregts MA, Jolliet O, Juraske R, Koehler A, Larsen HF, MacLeod M, Margni M (2008) USEtox—the UNEP-SETAC toxicity model: recommended characterisation factors for human toxicity and freshwater ecotoxicity in life cycle impact assessment. Int J Life cycle Assess 13: 532–546).

weighting factors are applied to each chemical's toxicity score based on several factors related to safety standards present during manufacturing, whether the materials are handled in an open or closed system.

Risk Potential

The final impact category is risk potential and focuses on both quantitative and semiquantitative data. The quantitative data focuses mainly on the impact categories of working and fatal working accidents and occupational illnesses and diseases. By correlating the type of materials identified in the inventory analysis with statistical data from the type of industries they were produced in (e.g., farming, textile, electronics, chemicals), a quantitative relationship can be established. Products are linked to the statistical industry data through the ISIC codes. ISIC[8] codes classify all industries into different branches (e.g., mining and quarrying of products or extraction of crude petroleum and natural gas). By linking the industrial production volumes with the statistical data (e.g., fatal working accidents), a defined correlation is developed (e.g., number of fatal working accidents per kg of material produced).

2.3 *Normalization, Weighting, and Aggregation*

After having classified and characterized all the environmental impacts in each of the six or, respectively, seven categories for each alternative over the defined life cycle, it is then necessary to manipulate the data in a way that will facilitate clear understanding and comparison. The methods applied in the EEA are described below and include normalization, weighting, and aggregation.

The first normalization is quite simple and applies to all the main environmental categories except emissions, as the emissions' category requires some additional weighting steps since it is composed of several subsections of emissions – e.g., air[9] (AP, GWP, POCP, ODP), water, and solid waste – which need to be combined together in a logical approach. This weighting will be discussed in more detail later. For the areas other than emissions, the summed life cycle consumption data (e.g., total MJ/customer benefit for energy usage) or scores (e.g., risk and toxicity potential) are normalized relative to the alternative which had the highest impact in this area. The least favorable alternative or the one with the highest impact is assigned a value of 1, and the other alternatives would have values less than one

[8] ISIC: International Standard Industrial Classification of All Economic Activities, Revision 4, United Nations, New York 2008.

[9] AP, acidification potential; GWP, global warming potential; POCP, photochemical ozone creation potential; ODP, ozone depletion potential.

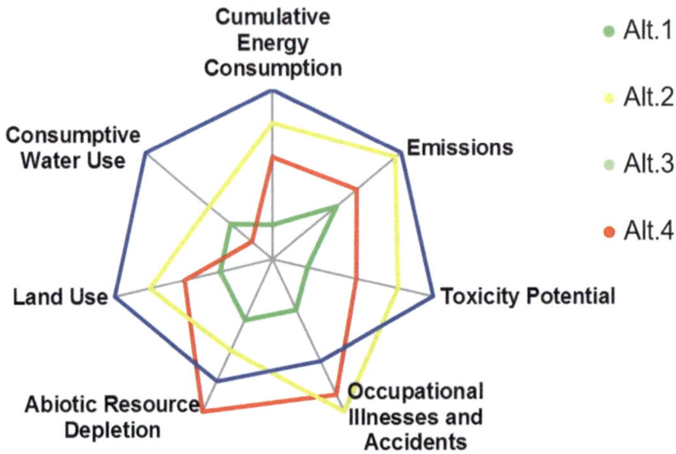

Fig. 4.11 The environmental fingerprint gives an overview about the impacts to different indicators in a comparison system of four alternatives

(closer to the center of Fig. 4.11). After normalization it is possible to plot, in a graphical form, the relative environmental results for each alternative. This diagram is called environmental fingerprint and is depicted in Fig. 4.11 with each colored line reflecting a different alternative.

The fingerprint makes it possible to easily visualize the trade-offs between alternatives by clearly showing where certain alternatives performed well and where they had less desirable results. However, to clearly understand which alternative had the overall lowest or highest environmental impact and thus realize which impact categories were important in driving the results of the study, an additional weighting procedure is required in order to combine the normalized results reflected in the environmental fingerprint into one single score. This weighting process, and analogous to the procedure used for the emissions category, involves incorporating both scientific relevance factors with societal weighting factors. The relevance factors help put into context how important or significant an environmental impact is for each individual eco-efficiency analysis. The relevance factors calculated are unique for every study and are different depending on the specific results of the study and on which region of the world the study applies. Advantages of this approach are that high environmental burdens are more heavily weighted than relatively low ones, and changes in society's view, relative to each individual impact category, can be included.

More specifically, the normalization factors reflect the level to which the emission or energy consumption determined in a specific study contributes to the total emissions or energy consumption in a given geographic region. To determine the specific factor, the impact of each category as determined during the inventory, classification, and characterization phase is divided by the total environmental impact in the region. This impact data is normally found in various, publically available statistical databases.

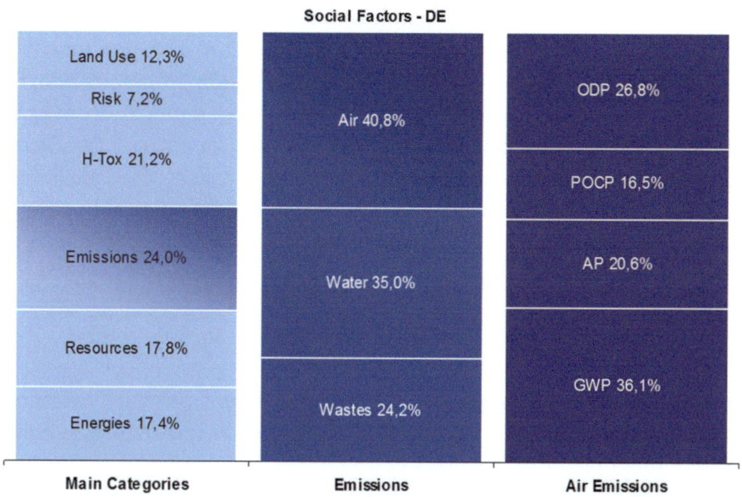

Fig. 4.12 Overview about societal weighting factors for Germany

The societal weighting factors used in conjunction with the environmental relevance factors account for society's opinion on the importance of each of the environmental impacts. These values are derived from the results of third party market research and polling and are constant for each study but should be updated on a periodic basis, as society's view can change over time. For example, currently GWP is grabbing all the attention as the key air emission. However, not too long ago, air emissions related to ozone depletion or acid rain were gaining more notoriety. Figure 4.12 shows an example of typical social weighting factors used in the EEA with each column reflecting how society viewed the importance of each impact category relative to the others.

Relevance or Normalization Factor

- What does the emission (energy use, etc.) contribute to the overall emissions (energy use, etc.) of the region/country?
- Based on statistics.
- Focus is on the weaknesses of the product analyzed.
- Changes from analysis to analysis depending on product.

Societal Factor

- What value does society attach to the reduction of the different environmental impact categories?
- Based on representative polls
- Dedicated factors for different regions
- Independent of customer benefit

An overall weighting factor is determined for each impact category by taking the geometric mean of the environmental relevance factor and the social/societal weighting factor. The results of the normalization step (environmental fingerprint) are multiplied by these overall calculation factors and summed up over the six categories to represent the final environmental impact for each of the alternatives studied. Though the environmental impact assessment and cost calculations are separate steps of the eco-efficiency analysis, the goal of the study is to present both findings in a balanced method that supports clear understanding and facilitates strategic decision-making (Eq. 4.1).

$$\text{Calculation Factor} = \sqrt{\text{Relevance Factor*Societal Factor}} \tag{4.1}$$

The environmental position with the aggregated figure for the environment is accomplished through the eco-efficiency portfolio. After one final weighting step, described in more detail by Kicherer et al. (2007), which takes into consideration whether the environmental or cost impacts were more influential in driving the results of the study, the environmental impact score is combined with the normalized economic impact defined earlier and plotted on a biaxial graph for each alternative (Fig. 4.13).

The costs are shown on the horizontal axis and the environmental impact is shown on the vertical axis. The graph reveals the eco-efficiency of a product or process compared to other products or processes. As both environmental impact and costs are equally important, the most eco-efficient alternative is the one with the largest perpendicular distance above the diagonal line in the direction of the upper right-hand quadrant. Less eco-efficient alternatives are located in the lower left-hand section and reflect an area of higher environmental burden and higher life cycle costs. Scenario and sensitivity analyses can be effectively conducted on the study parameters through the dynamic nature of the eco-efficiency model with results easily depicted in the associated graphs and diagrams and decision-making supported through the revised portfolios.

2.3.1 Case Study for Heating Systems

The reason for carrying out the study is to compare alternative systems for providing space heating and hot water for domestic buildings (detached houses, new developments), examining both renewable and nonrenewable fuels. The EEA was based on the ASUE study[10]; the EEA study examines a subset of the systems in the ASUE study and adds some additional heating systems.

The comparison focused on one hand on established commercially available technologies based on fossil fuels (gas or fuel oil condensing boiler both with

[10] ASUE: Arbeitsgemeinschaft für sparsamen und umweltfreundlichen Energieverbrauch e. V. Germany: Comparison of heating costs in new developments, 2009.

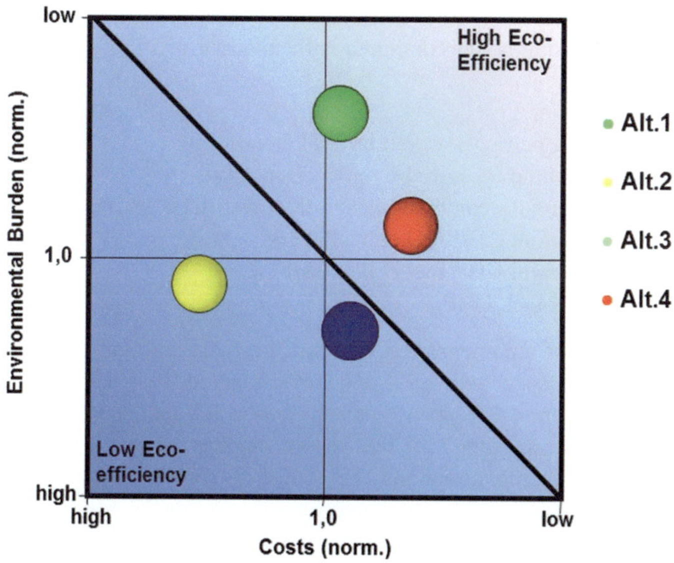

Fig. 4.13 Summary of an eco-efficiency study plotted in a portfolio graph

additional solar water heating, district heating block heat, and power plant) and renewable sources (wood pellet boiler, split logs boiler) and, on the other hand, on new commercially available technologies (brine-water and air-water heat pump, micro combined heat and power Stirling engine) as well as new, not yet commercially available technologies (fuel cells).

The goal of the EEA was to compare the different alternatives for their environmental and economic advantageousness as of now and in the near future. The results were created to support decision-makers in industry but as well as in the society who plan investments in heating systems. Often the comprehensive and life cycle-based information are not available because decisions are made based on single information and costs. To initiate improvements for more sustainable heating systems, a holistic analysis is needed. The results have been transparently published and support the investment in new technologies. They help additionally to identify improvement potentials of those technologies and the research activities in this field. Researchers can improve their systems more efficiently and their improvements can be assessed, based on the eco-efficiency model, very easily and fast. That reduces development times and costs and considers as well sustainability aspects at the very beginning of innovation processes.

The functional unit (customer benefit) of this EEA study is the provision of space heating and hot water for a single-occupancy detached house (floor area 150 m^2) during 1 year, corresponding to annual space heating requirements of 7500 kWh/a and annual hot water requirements of 1875 kWh/a. The reference flows of space heating and hot water are provided by the following alternatives:

1. Natural gas condensing boiler combined with solar thermal hot water
2. Fuel oil condensing boiler combined with solar thermal hot water
3. Split log boiler
4. Wood pellet boiler
5. Brine-water heat pump (powered by grid electricity)
6. Air-water heat pump (powered by grid electricity)
7. Natural gas-fed micro combined heat and power (CHP) Stirling engine
8. Natural gas-fed mini CHP fuel cell (SOFC)
9. Natural gas-fed mini CHP fuel cell (PEMFC)
10. District heating block CHP

The calculation of the energy requirements per heat and hot water output was performed in accordance with the applicable standard DIN 4701-10 (as in the ASUE study). Different life spans were assumed for the systems, discounting emissions and costs accordingly. It was shown that the results are not particularly sensitive to the life span assumptions.

All in all, these data are deemed suitable for compiling the life cycle inventory (LCI) of this product system in accordance with ISO 14040, 14044, and 14045.

Portfolio Result

Systems based on natural gas used the more eco-efficient solution than the natural gas condensing boiler combined with solar thermal hot water and the district heating block CHP.

The CHP Stirling engine and heat pumps were found to be less eco-efficient due to higher costs and similar (or only slightly lower) environmental impacts. Fuel cells were considered to offer potentials in term of future cost reductions. The split logs boiler was found to perform somewhat less eco-efficient than the above. The study concluded that the worst alternatives from an eco-efficiency point of view are the fuel oil condensing boiler (with solar thermal hot water) due to environmental impacts and the wood pellet boiler due to capital costs. The result is shown in an eco-efficiency portfolio in Fig. 4.14. The whole study was checked with a peer review worked out by the German DEKRA.

Life cycle Assessment Results

LCA delivers different single results without any weighting factors behind. The reader can follow every single impact that was calculated in a cradle-to-grave or, if needed and helpful, also in a cradle-to-cradle manner. All environmental criteria considered are shown in Fig. 4.15. The evaluation covers important criteria from common LCA but also some additional criteria like risk potential, showing impacts of accidents and occupational diseases, water emissions from heavy metals, COD or other specific emissions, as well as toxicity potential for humans.

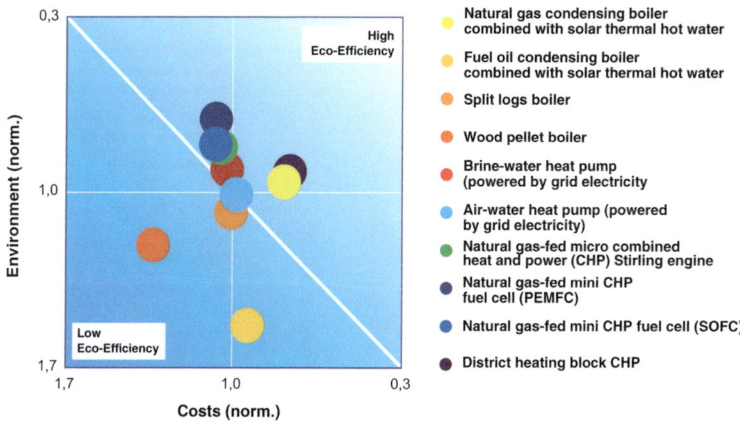

Fig. 4.14 The portfolio graph showing more eco-efficient solutions in Germany

All the single graphs can be normalized and show the best and the worst alternative in each category. With this overview it will be possible to get a good overview of the environmental impacts of the alternatives. This would be the basis for further developments and improvements.

The best energy consumption is linked to the proton exchange membrane fuel cell (PEMFC), alternative to the high utilization factor and the bundled production of heat and electricity.

The best resource consumption is linked with the split logs boiler due to the use of wood as an energy containing resource with a relatively good resource factor of renewable materials.

The risk potential is positively evaluated mostly for the natural gas alternatives due to the low statistical numbers for working accidents of gas delivery compared to alternatives. Due to the avoided production of electricity from the grid because of the intrinsic electricity production of these alternatives, the mining of coal with a certain amount in the grid with relatively high-working accident numbers can be avoided. That credit leads to reduced working accidents in the whole life cycle of these alternatives.

The toxicity potential is mainly determined in the use phase by the NO_x and CO emissions. In this category, the alternatives using natural gas have the most positive results. The alternative that uses oil as raw material has the worst position due to the higher emissions, dust emissions, and the toxicity of the oil itself.

All the results were determined by quantitative methods generating numbers that can be compared directly. After that, a normalization step is done to put the alternatives on the axis that are shown in Fig. 4.15.

The next step is an aggregation and weighting step using statistical numbers and "relevance factors." To clearly understand which alternative has the overall lowest or highest environmental impact and thus which impact categories is important in driving the results of the study, an additional weighting procedure is required in order to combine the normalized results reflected in the environmental fingerprint

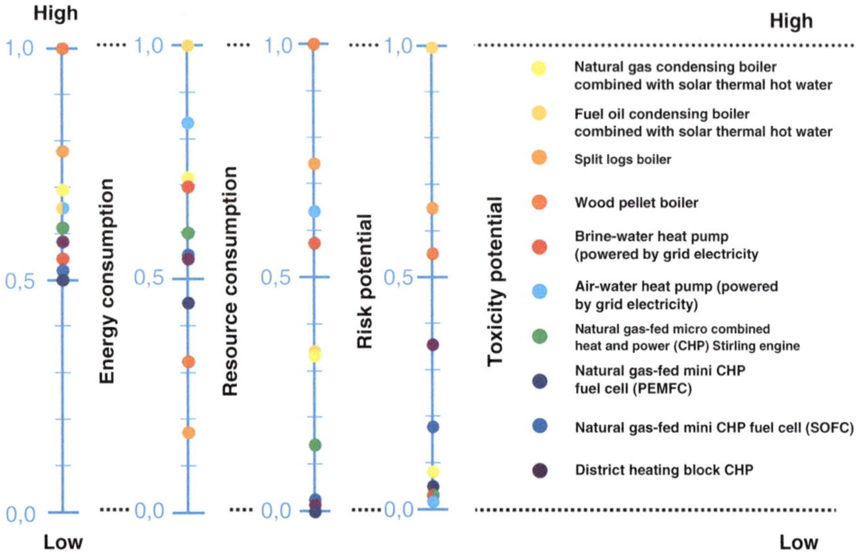

Fig. 4.15 Overview of single LCA results, normalized

into one single score. This weighting process, and analogous to the procedure used for the emissions category, involves incorporating both scientific relevance factors with societal weighting factors. The aggregated figures supporting decision-makers in industry but much more the end users who want to find out which heating systems should be installed for the realization of a more sustainable solution.

Costs

Additionally, costs information along the whole life cycle were worked out and integrated with the environmental factors to the final results in the portfolio.

The specific approaches for conducting a life cycle cost (LCC) assessment varies from study to study and is strongly dependent upon the customer benefit selected and the system boundaries and alternatives considered. Cost accounting needs to take into account both initial costs and all future cost impacts or benefits of the products as well as any costs having an environmental aspect (e.g., disposal of hazardous waste). Either constant (real) or nominal monetary values can be used for cost accounting, but they cannot be mixed in the analysis. In addition, the final cost analysis can be calculated as a point in time or can account for the time value of money. If the analysis is performed to consider the time value of money, then a net present value (NPV) or similar metric shall be calculated. A representative but not all inclusive list of economic metrics normally considered for each alternative includes the costs of raw materials, labor, energy, capital investment, maintenance activities, transportation, illnesses and accidents, and waste disposal. When all the costs are identified and accounted for, they are summed and combined in

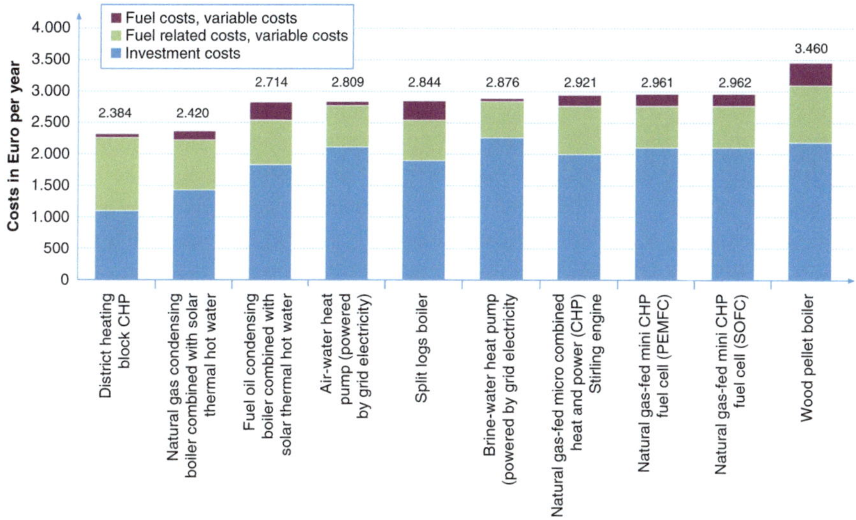

Fig. 4.16 Overview of life cycle costs of the alternatives

appropriate units (e.g., $ or €) without additional weighting of individual financial amounts (Uhlman and Saling 2010).

In the case studies, it was shown that the best alternative due to costs was the district heating block CHP based on natural gas use. The relation of investments costs and maintenance costs per year is here around 50–50 %, whereas in other alternatives, the investment costs are significantly higher leading to higher total costs. In the total, yearly costs of the cheapest alternative are 2400 Euro and of the most expensive alternative were 3500 Euro per year (Fig. 4.16).

The benefit of this study was the creation of a good overview about important impact factors and improvement potentials of a complicated product system. Without knowing all background details, it is possible to learn very quickly and easily which heating systems are the most sustainable solutions. A final result is needed because decision-making processes need clear and easily understandable information. In a typical sustainability evaluation, there are often huge ranges of single data which cannot be understood in a defined objective way without the help of a methodology that aggregates all the information to a final result by using defined algorithms.

The quantitative weighting steps to get the ecological fingerprint and the portfolio are additional features.

2.3.2 Life cycle Assessment for the Production of Various Carotenoid Additives for Use in Poultry Feeds

Within animal feed, among the inputs being most closely scrutinized are the carotenoids, a group of pigments applied in animal rations to impart color to poultry, fish, and crustacean products.

Canthaxanthin, a chemically synthesized red carotenoid used in poultry and salmon feeds, is also found in nature in shrimps, bacteria, algae, birds, insects, fungi, and fish (Shahidi et al. 1998).

As companies strive to comply with ethics guidelines, naturalness is promoted over anything man-made. This is especially true in the food industry and the animal feed sector. Given the reported importance of eco-friendliness to the customer perception of food and feed products, we conducted a study to evaluate the ecological impact of red and yellow carotenoids used in the pigmentation of egg yolks. For this, the EEA has been used, assessing a majority of the life cycle inputs and outputs of a carotenoid production system, from cradle to grave.

The aim of the present study was to assess the ecological impact of the use of various pigmenting carotenoids (or xanthophylls) included in rations for laying hens. In this way, chemically synthesized carotenoids, canthaxanthin, and ß-apo-8′-carotenoic acid ethyl ester (C30-ester) were compared to their plant-derived counterparts, paprika oleoresin and marigold oleoresin, respectively.

Canthaxanthin, or paprika oleoresin (containing capsanthin and capsorubin), may be used to provide red colors to the egg yolk. C30-ester and marigold oleoresin (supplying lutein and zeaxanthin) both yield a yellow egg yolk color. When targeting a brighter colored yolk, mixtures of red and yellow pigments are used. The example makes up a part of a larger study examining the overall eco-efficiency of pigment use in this application field (Baker 2002).

Data Used in the Comparative Study

In the first step of assessment, the system boundaries were defined. These include all relevant steps in the whole life cycle that must be considered in the final calculation and are shown in Fig. 4.17, with some key assumptions for calculation in Tables 4.1, 4.2, and 4.3. For the chemically synthesized pigments, data from BASF's commercially available products were used (Lucantin® Red and Lucantin® Yellow) (Saling et al. 2006).

All results were expressed relative to units of customer benefit. In this case we opted for the impact of a given production method per 100,000 pigmented eggs produced to display a specified yolk color.

Realistically, practical product feed doses were taken (3 mg chemically synthesized carotenoid per kg feed or equivalents from plant-extracted products), as well as published efficacy rates for the pigmentation of the carotenoid sources. In layer applications, paprika oleoresin is published to be only 30 % as efficient as Canthaxanthin (Blanch and Hernandez 2000; Steinberg et al. 2000a), and similarly, marigold oleoresin is approximately 30 % as efficacious as C30-ester (Steinberg et al. 2000b).

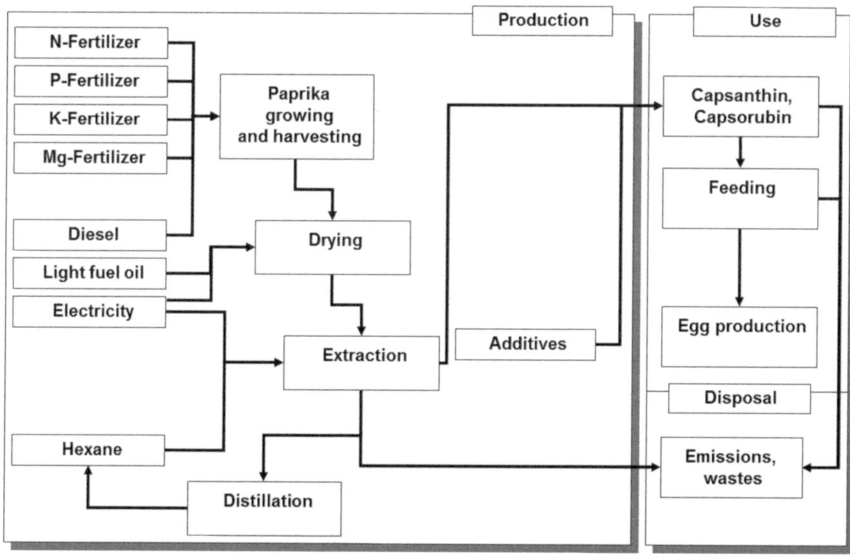

Fig. 4.17 System boundaries for the production of pigments from paprika

Environmental Burdens

Primary energy consumption was determined over the entire life cycle. Fossil energy media used in preproduction steps and renewable energy media during harvest or production phase were included in this appraisal. This captures conversion losses from electricity and steam generation. In the case of the chemical-synthesis processes, BASF's own, company-specific data were used. In the case of processes from the comparison products, energy data of defined geographical regions were used. Marigold processes and conditions were calculated for China, while the paprika processes and conditions were assumed as for Mexico (Luck 2004, personal communication), reflecting the locality of main production. Feedstock energies of the biomass were also included in the overall calculations. Additionally, energies for growing and harvesting of the plants, for drying and extracting, and for work-up of solvents and materials were included in the calculation of the extraction processes of plant materials.

In the chemical processes, the energies for the preparation of basic materials (precursors), and of using energy in the processes and work-up, were considered over the whole life cycle, cradle to grave.

The mass of raw materials necessary for each alternative was determined. The individual materials were then weighted according to a factor incorporating the life span and the quantified consumption of that material (Hargreaves et al. 1994; Römpp Chemie Lexikon 2004; U.S. Geological Survey 2004; World Resources Institute 1996).

In the case of renewable raw materials, sustainable farming methods were assumed, in that the resource that was removed was supposed to have been replenished in the period under consideration. This means an endless life span and thus a weighting factor of zero. Of course, in the case of renewable raw materials from non-sustainable farming (e.g., rainforest clearance), an appropriate (nonzero) weighting factor was used for the calculation. High energy consumption can be correlated with low material consumption if renewable raw materials such as wood or hydroelectric power are used. What therefore appears to be a double-accounting of raw material and energy consumption does not occur with these two categories.

Emission values were initially calculated separately as air, water, and soil emissions (waste). The calculation included not only values from electricity and steam production and transport but also values from direct emission from the process.

For the air emission category, GWP, POCP, ODP, and AP were calculated.

Unlike the "air emissions" category, for emissions to water, there is currently no available comparable standardized, scientifically documented method for calculating the impact potentials. For the inventory of emissions to water (COD; BOD; N-tot, total nitrogen; NH_4^+, ammonium; PO_4^{3-}, phosphate; AOX, adsorbable organic halogens; HMs, heavy metals; HCs, hydrocarbons; SO_4^2, sulfate; Cl^-, chloride), the method of critical volumes or critical limits for discharges into surface waters was therefore used (BUWAL 1991). Each pollutant emitted into water contaminates a sufficient volume until the statutory limit for this substance is reached (critical load). The limits used for the respective emission to water were the limits listed in the schedule of the German wastewater regulations (AbwV 1997). The greater the water hazard posed by a substance, the lower its discharge concentration limit.

The results of the inventory on solid wastes were combined to form four waste categories: special wastes, wastes resembling domestic refuse, building rubble material, and construction waste. In the absence of other criteria, impact potentials for solid wastes were estimated on the basis of the average costs for their disposal. Wastes designated as containing dangerous materials were assessed with higher factors than nonhazardous waste. The toxicity potential was determined using the assessment method described in the method part (Landsiedel and Saling 2002) based in that time on the preparation of the study on the R-phrases of the Hazardous Substances Regulation Act (GefStoffV).

The life cycle consists of construction, operation, and demolition and is put in relation to the overall capacity of the system. In the case of nonrenewable resources, the recultivation time is taken into account.

The risk potential in the EEA was established using BASF-assigned risk matrices that were developed by the BASF's safety department and validated by independent experts employing ABC analysis to assess where the risks are higher or lower compared to other alternatives (Krems 2004). These matrices consider variables like temperatures and pressures of chemical reactions and processes.

In the risk potential category, the focus is always on the severity of potential damage that operations can cause, multiplied by its probability. In the risk potential category, mainly accident statistics for various industries or occupations may be included, such as safety data on various types of reactions in the chemical industry. Risk potentials are calculated values. In order to be able to estimate a risk actually occurring to humans, additional calculations and estimates were required. Furthermore, accident statistics from different industry sectors in different countries were used. Transportation risks were estimated where longer transportation distances resulted in higher risk evaluations.

Environmental Fingerprint

Results of the calculation of environmental data reveal differences between the studied alternatives.

On calculation of every impact category, results were normalized, the worst alternative receiving a nominal value of 1, and others ordered relative to this worst-case alternative.

An ecological fingerprint shows the weaknesses and strengths of each alternative according to the defined environmental category.

Figure 4.18 reveals the advantages, in all environmental categories, for the chemical production of the studied carotenoids. The single graphs for every subcategory are shown in Figs. 4.19 and 4.20.

Emissions were translated, according to their ascribed impact category factors, to the overall GWP. The factors reflect the varying impact of the different emissions on GWP. The factor for carbon dioxide is 1, because this gas is expressed as the equivalent gas, or standard. Figure 4.19 shows the impact category factors and the

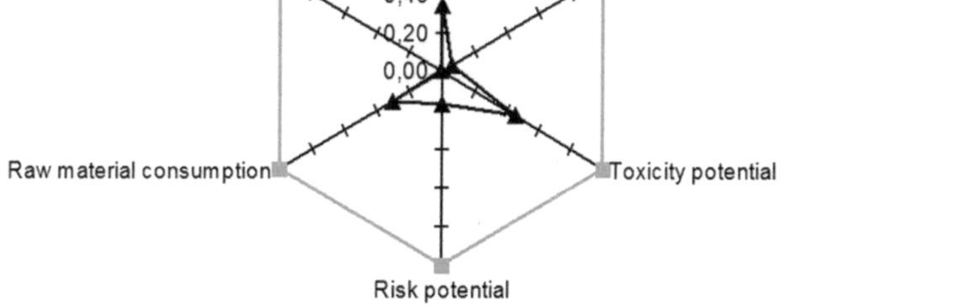

Fig. 4.18 Environmental fingerprint of the use of chemically produced or plant-derived pigments to color egg yolks. Data are represented as normalized values with 1 demonstrating the highest impact for each category

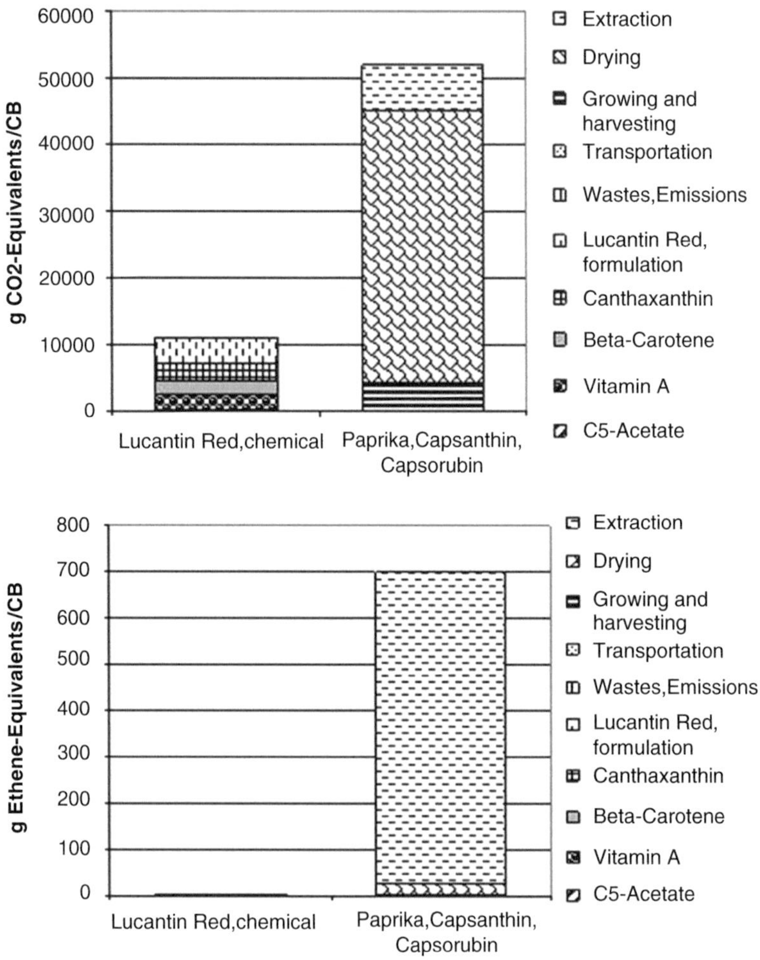

Fig. 4.19 Ecological impact of the alternatives for yellow and red pigments per unit of customer benefit: (*top*) GWP and (*bottom*) summer smog (POCP)

resulting graph for the other emissions contributing to the GWP, assuming relative impact factors of 21, 4500 and 310 for methane, CFC (chlorofluorocarbons) and nitrous oxide, respectively (Renner and Klöpffer 1995).

In most of the environmental categories, the chemical alternatives outperformed the plant-extract products. The normalized ecological results are summarized in an overall view of the environmental burden of the tested products, and the environmental fingerprint is shown in Fig. 4.20. It is obvious that products from chemical synthesis had a lesser impact on the environment, mainly due to the drying and extraction steps required for products based on plant biomass. This was valid for all of the subcategories in the calculation.

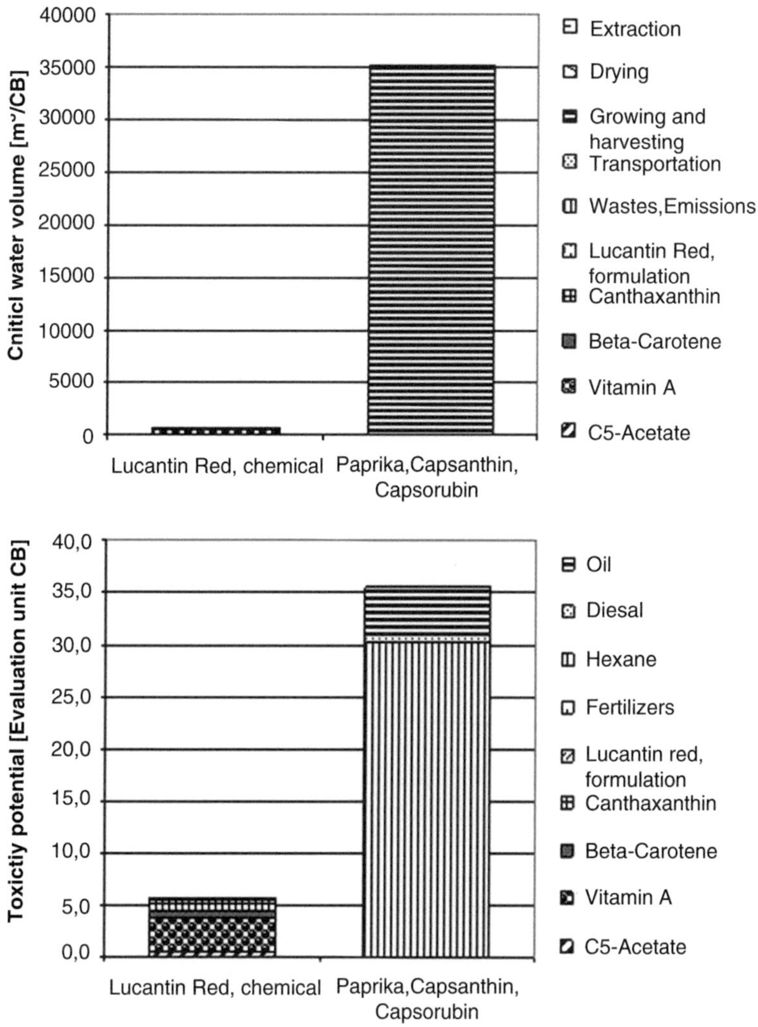

Fig. 4.20 Ecological impact of the alternatives for yellow and red pigments per unit of customer benefit: (*top*) critical water volume and (*bottom*) toxicity potential

From the above data, it is clear that plant-derived pigment products performed relatively poor because of the ecological burdens imposed during the agriculture processes, the low content of pigments, the efforts of drying and extracting the plant material, and the high demand for transportation. In areas of low rainfall, the need for irrigation also contributes to the ecological impact.

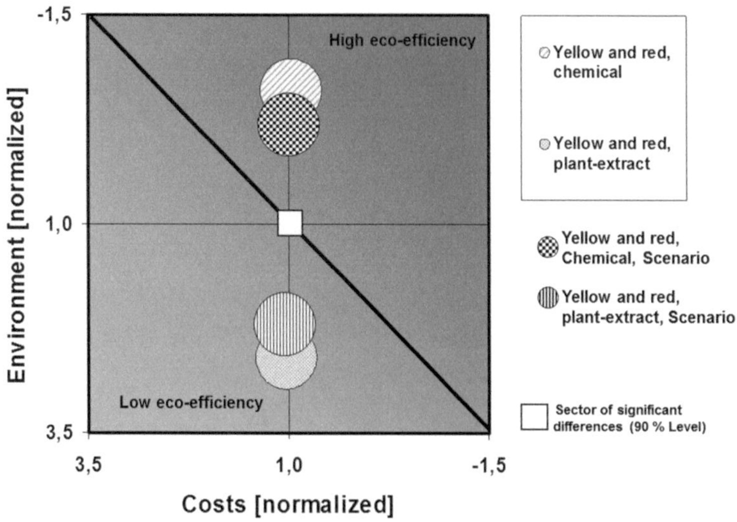

Fig. 4.21 The eco-efficiency portfolio, demonstrating a scenario where there is an assumed reduction of drying efforts for natural products by 50 % compared to the base case situation

The Portfolio Calculation

Despite the fact that economic data has been omitted from this study, it has still been possible to demonstrate the benefits of this eco-efficiency approach with regard to evaluating the relative impacts of improvements in production technology on both sides.

In this study, the alternatives were significantly different ($P < 0.05$). That is valid also for the single calculation for red and yellow pigments with a large difference in the environmental burden shown on the Y-axis in Fig. 4.21.

Evidently, the portfolio analysis may also be used to test assumptions related to production efficiency or performance changes. Scenario analysis may be performed by comparing the base-case portfolio to one where new assumptions are inserted as input data. This may answer many questions, allowing a comprehensive analysis and permitting an optimization of strategic decisions for R&D, for marketing, or for political discussions.

As an example, Fig. 4.21 shows as well a scenario analysis where the drying efforts for the plant-extracted materials are reduced by 50 %. In that case, the chemical processes are still more eco-efficient than the biological alternatives. With this method it is possible to define break-even points for further developments and optimization of the whole production and value chain.

Natural Materials Versus Chemically Produced Goods

Are **n**atural **m**aterials *always* **b**etter than **c**hemically **p**roduced **g**oods? In the present case study, it was shown that the chemical synthesis of pigments for poultry feeds is much more sustainable than extraction from natural materials. *The result contradicts the widely held belief that natural materials are always better and more sustainable than chemically produced goods.* In this specific case, the chemical products are much more sustainable. Clearly, this is because there is often a high efficiency and high yield associated with a chemical process. In addition, these products have a high-purity and a high-quality standard.

The relatively lower eco-efficiency of the solvent-extracted plant materials is logical given the use of electricity and fossil fuels in agricultural processes and downstream processing, as well as the inputs of fertilizers and crop-protection products used in the cultivation of plant crops. In many cases, there is also a need for solvents and some basic chemicals for the extraction of the target molecules. Emissions and risks from transportation of relatively low-potency plant products also contribute to the overall poorer ecological performance of these oleoresins of plant origin.

It was demonstrated, in the systems under test, that the ecological impact of chemical production is less than in the traditional plant-extraction procedures.

To assess sustainability of additives (or indeed any product), an Eco-efficiency Assessment may be performed and resulting information used in decision-making processes. Whether used by feed additive producers, animal feed manufacturers, animal growers, retailers, or legislators, this analysis can lead to better decisions with regard to product design, material utilization, product purchase, or legislation. Determination of the sustainability of products and processes will become more important in the future to distinguish between more or less sustainable solutions. It may be necessary to prove to the stakeholders that a preferred solution is more sustainable than another. This should be done employing an accepted scientific method such as our presented eco-efficiency analysis, carried out without any preconceptions regarding certain technologies.

3 Simulation Tools and Techniques

Nowadays no matter which engineering area one wants to explore, there is an abundance of knowledge available. Modeling and simulation is a very wide engineering area of which the importance has increased over the last few years. The reason is the rapid development of powerful computers to perform more advanced types of simulations. Now it is possible to digitally simulate systems that were almost impossible to deal with before. Besides, as the development goes on, even more complex systems can be calculated. As a result, more research effort has been put on finding ways to perform the modeling to render a more flexible model concerning both modularity and types of applicable simulations.

Modeling and simulation in LCI has not, however, made use of these recent findings yet, in spite of there being a great potential in doing so. That involves not only increasing the number of available evaluation techniques but also to learn and adapt nomenclature from related areas. This will expand the usability, understandability, and general acceptance of LCA as well as LCA in a wider perspective. Another aspect is that already available software systems may be used within the specific area of interest, which may result in saving of effort (Forsberg 2000).

The generation of original and reliable data will be time consuming also in the future. High quality and documentation demands and technical changes in quite a fast way are challenges of the future, even if databases and source are easier available than in the past. How "big data" can improve data availabilities will be shown in the future.

The decision on most eco-efficient production scheme for producing customized products is highly time consuming and error prone as it depends on several factors most notably the geographical locations, legal requirements, material characteristics, and process-related parameters. Furthermore, the planners must be experienced with the simulation tools as well as related terminologies, environmental laws, and standards to conduct precise assessments. The limited functionalities of decision-making tools as well as nonautomated exchange of information and data, absence of environment-related knowledge, and non-standardized interfaces between the simulation tools and planning systems make planners incapable of reaching quick decisions on eco-efficient choices. A web-based inference system seamlessly connected with the LCA simulation tools and legacy systems has been described by Minhas and Berger (2014). The web-based software automatically analyzes the input on manufacturing alternatives to produce customized products and generates automatically the corresponding LCA simulation models. These models can be used inside LCA tools.

3.1 Web-Based Simulation Tool

The web-based Eco-Efficiency Analysis Manager Online is a tool which enables the user to assess the total costs and environmental impacts created by one's company's products and processes over their entire life cycle. The Eco-Efficiency Analysis Manager Online is based on the EEA methodology (www.eeaman.com).

Benefits of the Eco-efficiency Manager The Eco-Efficiency Analysis Manager Online provides practitioners as well as in LCA methodologies unexperienced users the opportunity to enhance the sustainability of products and processes along the value chain. The user can:

- Select the most cost-efficient and environmentally sound alternative
- Calculate eco-efficiency portfolios for his/her product or process
- Identify strengths and weaknesses
- Analyze the impact of single parameters

- Simulate scenarios
- Create diagrams for presentations

With the web-based system, based on an eco-efficiency study, users can create, in very short time spans, a high number of scenarios. Via an interface which is connected to the calculation basis, comprehensive studies can be used very easily. It is not necessary that the user understands all the specific details of the technologies and calculation routines behind the interface. There is supporting material available and clearly described which underlying principles and calculations were used. They are responsible for their data input and the results they create with such a system. The web-based simulation tool provides easily accessible input fields for fast changes of input figures. There is no limitation of input fields. The interfaces can be more sophisticated for technical experts as well as very simple for other users. In parallel, several users can use the same system and can calculate scenarios. The calculation is very fast within seconds or, in complicated systems, minutes. User rights can be clearly defined and make the administration comfortable. The calculation basis can be maintained and changed by experts from a central point; the surface can be designed tailor-made linked with specific functions to support an easy handling. Different functions can be linked to the surface so that unrealistic inputs or incomplete input datasets can be detected and displayed to the user. Furthermore, ranges of data can be defined and support the data input via the internet platform.

In the system there is an automatic reporting and data export function involved for an easy documentation of a number of scenarios. Figure 4.22 shows an input table of such a web-based interface. It was applied and designed for the evaluation of different fish diets for salmon production in different regions. The different diets and their compounds can be optimized under conditions of sustainability improvement. Another nutrient optimization tool can be linked with the web-based manager tool. So different optimizations can be combined and can lead to a holistic improvement of the sustainability of fish feeding alternatives.

The generated information can be used for improving contributions of the whole supply chain. To understand where most of the contributions come from will help to work together with the responsible stakeholders and also to improve their parts of the value chain. This common understanding and improvement of products can serve as well for outside communication and for the decision-making inside a company.

In Fig. 4.23 an example is shown, as to how results from a scenario calculation can be displayed and used by the practitioners. Those graphs are part of the reporting template linked with additional information from the calculation.

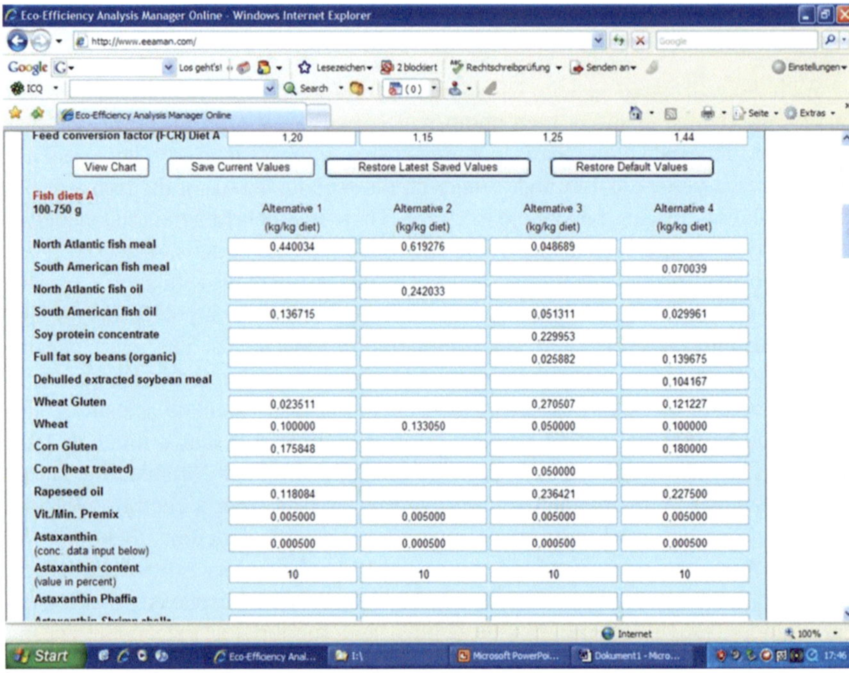

Fig. 4.22 Input table of a web-based manager tool for fish feeding diets

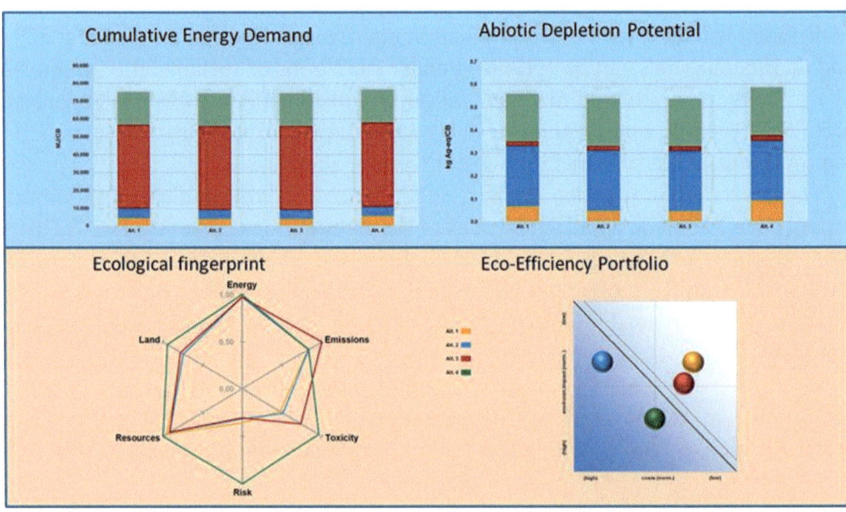

Fig. 4.23 Example of output graphs of a web-based manager tool for fish feeding diets

4 Additional Tools Based on Eco-efficiency Including Social Indicators

4.1 SEEBALANCE and AgBalance

4.1.1 Introduction

In addition to the eco-efficiency analysis, some years ago, the SEEBALANCE was developed (Kölsch et al. 2008; Schmidt et al. 2004) and used by BASF and its customers to assist strategic decision-making, facilitate the identification of product and process improvements, enhance product differentiation, as well as support the dialogue with opinion makers, NGOs, and politicians. Both EEA and SEEBALANCE analyses are comparative methods; the advantages and disadvantages of several alternatives are assessed according to a predefined customer benefit with a holistic approach. Next to the environmental impacts, which are assessed based on ISO14040 and ISO14044, all economic factors are taken into account following the ISO 14045 standard as described above. The SEEBALANCE also considers social impacts of products and processes (Uhlman and Saling 2010).

Both eco-efficiency analysis and SEEBALANCE have been employed in the food and feed value chains in order to assess the key drivers of sustainability in various production systems.

Ongoing developments on indicators, methods, and evaluation systems for social metrics show that, in the close future, more and more social indicators will be assessed and integrated in holistic sustainability evaluation systems. There are guidelines available that reflect the discussion about how to handle those indicators and methodologies (Benoît et al. 2010; UNEP/SETAC Life cycle Initiative 2009; Round Table for Product Social Metrics 2014).

4.1.2 Agriculture

It has been shown in various case studies that agriculture can have a large share of the entire sustainability profile of food and feed value chains. At the same time, logistics, transport, processing, and, not least, consumption can play a substantial role as well. In 2012, for example, BASF analyzed the CO_2 balance for veal and beef products with the client Westfleisch, supporting them in improving the sustainability of their meat production along the whole value chain (Westfleisch 2012).

Feed ingredients for a more sustainable aquaculture of salmon have been identified through EEA in collaboration with Biomar A/S (Saling et al. 2007). Three ways to produce astaxanthin as an ingredient of salmon diets were compared: chemical synthesis, fermentation, and production via fermentation of algae. In this case study, the astaxanthin derived from chemical synthesis was the most eco-efficient product (Saling et al. 2007). Other examples comprise the production of beef with Cattlemen's Beef Board and National Cattlemen's Beef Association

(NCBA 2014). The EEA portfolio shows that the present-day US beef value chain is more sustainable than in 2005.

This shows that in general EEA and the SEEBALANCE can be applied in different sectors quite successfully and are able to support decision-making processes, R&D activities, as well as marketing support with specific information on sustainability aspects (Saling et al. 2005).

However, the limitation of eco-efficiency and SEEBALANCE to cover the agricultural production level is the lack of specific indicators, among others capturing the impacts on biodiversity, soil health, and the agri-sociological context of production. For this reason, BASF has developed the AgBalance™, a holistic method for assessing sustainability in agriculture and food production for and identifying key drivers for improvement. The method was developed in an international team of different experts. It was described in detail and validated by NSF, DNV, and TÜV Süd (Schoeneboom et al. 2012).

4.1.3 AgBalance: Life cycle-Based Approach in Agriculture

AgBalance comprises a multicriteria life cycle-based approach in combination with a defined aggregation and summary of single results into a single sustainability score (Frank et al. 2012). AgBalance delivers results that enable farmers, the food industry, politicians, and society to objectively evaluate processes in terms of their sustainability profile. In doing so, a vast amount of information on individual factors can be ascertained in addition to overall statements on the sustainability of agricultural practices (e.g., plowing). AgBalance can be used to map an individual farm or the whole agricultural sector in one region, for example. The focus can either be on the agricultural production system alone or on the processes that have established themselves downstream in the value chain, such as logistics or processing.

The SLC Case Study

A case study with the holding company SLC Agricola in Brazil involved an internal benchmarking of two large farms, each with over 10,000 ha, to identify the central sustainability drivers for their crop rotation consisting of soya, maize, and cotton and to derive follow-up opportunities for their continuous improvement. An average cultivated hectare for each of the two farms, Panorama (Bahia state) and Planalto (Mato Grosso do Sul state), was compared on the basis of the operation data from the 2009/2010 season. The indicators from all three sustainability dimensions – environment, economy, and society – were investigated using a holistic approach over a section of the life cycle that starts with the raw materials used in the production (the "cradle" of the process, e.g., phosphorus extraction or oil production) and ends with the delivery of the harvested goods at the nearest port. The analysis revealed that the Planalto farm is substantially more sustainable than the

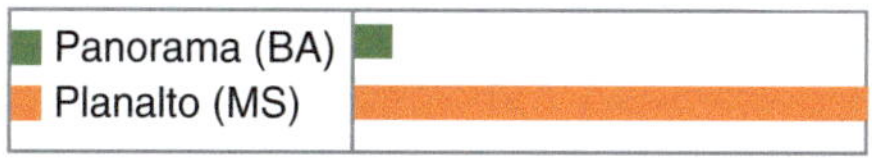

Fig. 4.24 Relative sustainability index of the two farms Panorama and Planalto. Planalto achieved a 40 % better result

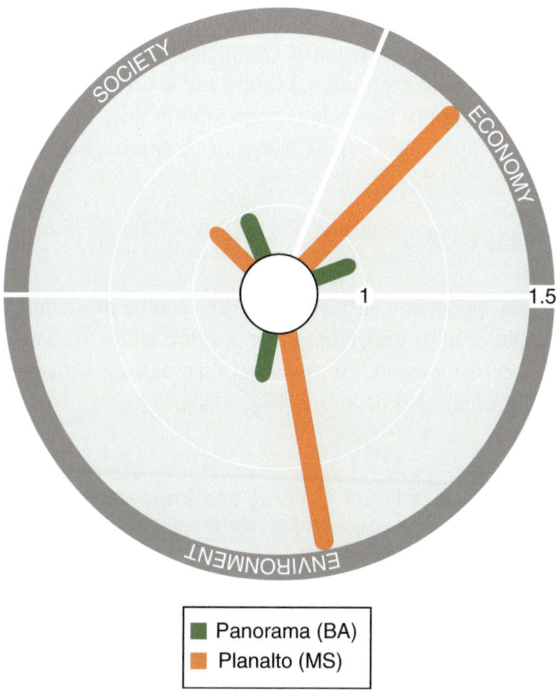

Fig. 4.25 Representation of the sustainability index in terms of the three dimensions of sustainability. The length of the bar indicates higher sustainability. Each time the worse alternative is normalized to the value 1

Panorama farm (Fig. 4.24), which is largely due to better results in the economy and environment dimensions.

In a more detailed view, all three dimensions can be shown and results can be derived from a specific spider diagram (Fig. 4.25).

Whereas the result in the social indicators did not exhibit significant differences (see Fig. 4.24), the most important drivers in terms of economy turned out to be an improved cost situation and an increased profit of the Planalto farm. In terms of the environment, the key drivers turned out to be a predicted imbalance for nitrogen and, above all, phosphorous in the soil as well as the pesticide regime. According to initial calculations, the optimization of the fertilization regime in Panorama could lead to savings of almost 15 million kWh of energy (this corresponds to the energy use of roughly 2000 households in Brazil) in addition to substantial cost savings. The CO_2 equivalents saved using AgBalance™ amount to almost 8000 tons per year. These results, together with the additional findings on pesticides, can serve as the starting point for a continuous improvement program at SLC Agricola. With its knowledge base, BASF supports a suitable product portfolio throughout the whole life cycle and works toward creating common solutions toward greater sustainability.

Measuring sustainability can be a central key to steady improvements toward sustainable agriculture. It is therefore an essential requirement that it succeeds in translating results from complicated life cycle analyses into farmers' everyday reality and to derive specific recommendations for action. However, agricultural production globally is strongly dependent upon smallholder farming, which is not easily accessible by complex and expert-based LCA approaches.

Corn Production in the USA

Another example that shows how differently results can be presented, and how they can be displayed in an easy-to-understand manner, is a study for corn production in Iowa, USA. All three dimensions of sustainability can be shown there and can finally be aggregated to an overall figure (Fig. 4.26).

Additional IT Solutions Linked with Training and Improvement Programs

Novel IT solutions are required in order to make use of life cycle-based knowledge for a more sustainable crop management on farm. This is the basic idea of the concept "AgBalance Farm." It uses key learnings of socioeconomic LCA studies for the development of web-based crop management support applications for farmers. This strategy is based upon BASF's experience with the Eco-Efficiency Analysis Manager Online (web-based application; see above) as outlined in Saling (2013).

The Samruddhi Case Study and Training Initiative

India is the fifth largest producer of soybean in the world, but soybean yields currently reach only half the global average of 2.4 mt/ha. The lack of knowledge about good farming practices comprises the key reasons for the low productivity. Through the training program "Samruddhi" (Sanskrit for "prosperity"), farmers are educated not only on the timely usage of crop protection inputs but also about correct fertilization, seed rate, and spacing to enable higher yields (GIZ 2013).

While the contribution of Samruddhi to the profitability of the Indian soybean farmers had been shown (Ramachandran 2014), its contribution to the sustainability of the production was largely unknown (Voeste 2012). Against this background, a

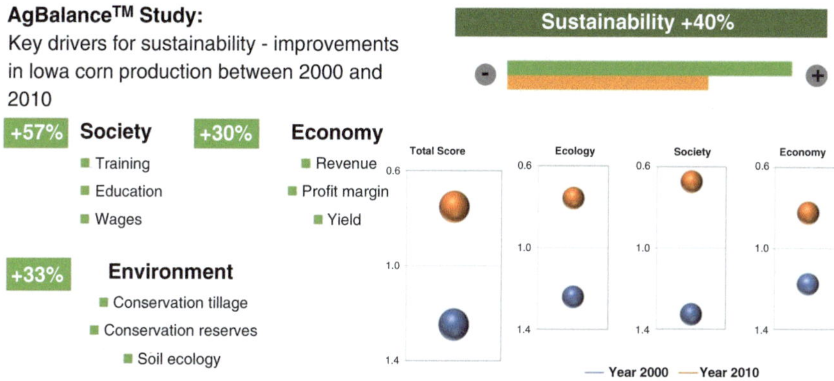

Fig. 4.26 Results of all three dimensions of sustainability in an AgBalance study

holistic socioeconomic LCA using AgBalance methodology was conducted. As a test case, soybean production under "Samruddhi" and "non-Samruddhi" in the state of Madhya Pradesh was compared. The AgBalance revealed that the "Samruddhi" production practice outperformed "non-Samruddhi" in all three dimensions of sustainability. In the economic dimension, a better cost position (fixed and variable) and higher profits per ton of soybean resulted in a better score. In the social dimension, a stronger emphasis on professional training favored the "Samruddhi" practice. In the environmental dimension, the better performance of the "Samruddhi" practice in some LCA impact categories was accompanied by the inferior scores in categories such as soil health, biodiversity potential, and emissions. As the key driver for this, the fertilizer regime of "Samruddhi" was identified.

A newly developed web-based application was generated that can be used, e.g., by "Margdarshaks" to help soybean farmers in their villages optimizing their production protocol toward higher yield, profitability, and sustainability. It basically conducts scenario analysis interactively, as demonstrated for the concept of the Eco-Efficiency Analysis Manager Online. With this "AgBalance Farm" strategy, the effective and easy use of the potential of socioeconomic LCA will support crop management decisions of individual farmers. The final goal is to achieve more sustainability in the food, feed, and nutrition sector.

4.2 *Conclusion*

Sustainability is becoming increasingly important as a key factor for growth and value creation. Customers along supply chains want more sustainable products and system solutions. There is a need to integrate sustainability much more closely into businesses and decision-making processes. To manage this in an effective way and to support decision-making processes, *a sustainability evaluation toolbox is needed* which can be applied to assess products and processes in a holistic manner. Both detailed in-depth results of individual impact indicators, as well as aggregated results, and a single sustainability evaluation score are output of the sustainability evaluation methods. Holistic, LCA-based, and scientifically sound methods to measure sustainability are key success factors for the realization of more sustainable production systems.

Different types of footprinting methods in combination with other information can support decision-making efficiently. The coverage of the dimensions economy, ecology, and social indicators is a key for a realistic, holistic, and complete overview of positive and negative impacts along the whole supply chain.

Establishing Eco-efficiency Assessment systems that were developed by different groups is an important approach to improve the sustainability of products and processes. A single indicator expressing stakeholders' value(s) or product system value(s) is one option to integrate eco-efficiency in decision-making processes. The value(s) shall be quantifiable with reference to the functional unit according to the goal and scope of the Eco-efficiency Assessment. The types of product system values may be a functional value, a monetary value, or other values. The most relevant standard to describe the processes and requirements to work out meaningful study results is the ISO 14045. Based on this standard, which allows a relative wide range of applications, other examples were established. Often the eco-efficiency is expressed in matrices or portfolio diagrams to support the easier understanding and interpretation of results which is crucial to implement it in a higher number of decision-making processes.

Often the different assessment approaches are designed to fulfill specific needs of industries, e.g., the electronic industry, chemicals industry, or agriculture industry. Different indicators can be used in an Eco-efficiency Assessment in combination with economic factors but also with other value indicators. In several approaches, sets of indicators are aggregated to a single score for a better understanding of the outcomes of a comprehensive study. In this case, the single results are available but a weighting step is applied as well. This must be done carefully and transparently to avoid misinterpretation. The communication of those results is much easier and will be understood by nonexperts as well. This is important to foster the consideration of sustainability aspects in decision-making processes. Based on those information, the willingness and the opportunities to improve the sustainability of products and applications in the market and the acceptance of product selections will be much higher.

To use Eco-efficiency Assessments more intensively, good case studies in different fields of industries and services should be available. It should be shown how it is possible to generate meaningful results in a reasonable time frame with reasonable amounts of data needed for reasonable costs. An additional opportunity to use this type of assessments is to generate specific tools which can generate a high number of scenarios in a short time frame based on automatic systems which support practitioners very efficiently. Different types of such tools were developed and successfully applied.

The engagement of all relevant stakeholders in supply chains but linked to actors along the whole life cycle of products based on meaningful information for the decision-making processes is needed, to improve the sustainability significantly. Eco-efficiency Assessments will support this engagement in a very powerful manner.

Glossary

ASUE Arbeitsgemeinschaft für sparsamen und umweltfreundlichen Energieverbrauch e.V. Germany: Comparison of heating costs in new developments, 2009

Eco-efficiency Eco-efficiency is defined as aspect of sustainability relating to the environmental performance of a product system to its product system value

Eco-efficiency Analysis (EEA) Eco-efficiency Analysis is closely linked to ISO 14045 (ISO 14045:2012) and subsequently linked as well to ISO 14040 (ISO 14040:2006) and 14044 (ISO 14044:2006). BASF has developed the eco-efficiency analysis tool to address not only strategic issues but also issues posed by the marketplace, politics and research. The goal was to develop a tool for supporting decision-making processes, which is useful for many of applications in the chemical and other industries. A part of the eco-efficiency analysis involves the evaluation of the toxicity and the ecotoxicity potential

Eco-efficiency Assessment Eco-efficiency Assessment is defined as a quantitative management tool which enables the consideration of life cycle environmental impacts of a product system alongside its product system value to a stakeholder

OSHA Occupational Safety and Health Act, United States Department of Labor, is the main federal agency charged with the enforcement of safety and health legislation

Resource Conservation and Recovery Act The Resource Conservation and Recovery Act by EPA governs the management of hazardous wastes

Sustainable development Sustainable development is defined as the balance of economic success, ecological protection, and social responsibility

USEtox USEtox is a model which can be used to calculate characterization factors for human and ecotoxicity impact categories used in life cycle assessment. Rosenbaum RK, Bachmann TM, Gold LS, Huijbregts MA, Jolliet O, Juraske R, Koehler A, Larsen HF, MacLeod M, Margni M (2008) USEtox—the UNEP-SETAC toxicity model: recommended characterisation factors for human toxicity and freshwater ecotoxicity in life cycle impact assessment. Int J Life Cycle Assess 13: 532–546

References

AbwV (Abwasserverordnung) (1997) Verordnung über Anforderungen an das Einleiten von Abwasser in Gewässer. Regulation on requirements for the discharge of wastewater into surface waters, March 27, 1997. http://www.gesetze-im-internet.de/bundesrecht/abwv/gesamt.pdf

Baker RTM (2002) Canthaxanthin in aquafeed applications: is there any risk? Trends Food Sci Technol 12(7):240–243

Bengtsson S, Sjöborg L (2004) Environmental costs and environmental impacts in a chemical industry – eLCC and LCA on two colorants, Akzo Nobel. http://www.dantes.info/Publications/Publication-doc/Environmental%20costs.pdf

Benoît C, Norris GA, Valdivia S, Ciroth A, Moberg A, Bos U, Prakash S, Ugaya C, Beck T (2010) The guidelines for social life cycle assessment of products: just in time! Int J Life Cycle Assess 15:156–163. doi:10.1007/s11367-009-0147-8
Blanch A, Hernandez JM (2000) Use of red carotenoids for yolk pigmentation. Int Poult Prod 8 (2):25–29
Borén T, Ludvig K, Halldén KA, Seetz J (2009) Eco-efficiency analysis – applied to different chelating agents. Int J Appl Sci Personal Care Deterg Spec 135:2–10
BUWAL (1991) Bundesamt für Umwelt, Wald und Landschaft, Schriftenreihe Umwelt Nr. 132, Bern, 1991
Carlson R (2009) Eco-efficiency – the conceptual model, the concept model and an operational data structure. Chalmers. http://life cyclecenter.se/wordpress/wp-content/uploads/2011/01/2009_2.pdf. Accessed 9 July 2012
Finkbeiner M (2008) On the sustainability of eco-efficiency. LCA VIII – Seattle 2008. TU Berlin, Sustainable Engineering. http://lcacenter.org/LCA8/presentations/ss-ecoefficiency-Finkbeiner.pdf. Accessed 9 July 2012
Forsberg P (2000) Modeling and simulation in LCA. http://www.cpm.chalmers.se/document/reports/00/Modelling_and_Simulation_in_LCA_CPM_Report_2000_1.pdf
Frank M, Schöneboom J, Gipmans M, Saling P (2012) Holistic sustainability assessment of winter oilseed rape production using the AgBalance™ method – an example of 'sustainable intensification'? In: Corson, MS, van der Werf HMG (eds) Proceedings of the 8th international conference on life cycle assessment in the agri-Food Sector (LCA Food 2012), 1–4 Oct 2012, Saint Malo, France. INRA, Rennes, pp 58–64
Gäbel K (2001) A life cycle process model – simulation of environmental, product and economic performance in cement production. Chalmers University of Technology, Gothenburg
Gardner S (2015) The state of sustainability 2015. In: Liam Dowd L (ed) Ethical corporation. FC Business Intelligence Ltd, London. http://www.ethicalcorp.com
GIZ (2013) Partnering progress. BASF's Samruddhi Project. In: Responsible Corporate Investment in Rural India, pp 11–17. http://de.scribd.com/doc/183777721/Responsible-Corporate-Engagement-in-Rural-India-A-Compendium-of-Good-Practices. Accessed 10 Feb 2014
Hargreaves D, Eden-Green M, Devaney J (1994) World index of resources and population. Dartmouth Publishing Company, Aldershot, 417 pp. ISBN 1855215039
Huppes G, Mansanobu I (2007) Eco-efficiency in industry and science, quantified eco-efficiency: an introduction with applications, vol 22. Springer, London
ISO 14040 (2006a) Environmental management—life cycle assessment—principles and framework. Geneva
ISO 14044 (2006b) Environmental management—life cycle assessment—requirements and guidelines. Geneva
ISO 14045 (2012) Environmental management—eco-efficiency assessment of product systems—principles, requirements and guidelines. Geneva
Japan Eco-Efficiency Forum (2009) Guidelines for standardization of electronics product eco-efficiency indicators ver. 2.1. http://lca-forum.org/english/eco/pdf/01.pdf
Kalberlah F, Saling P, de Hults Q, Grahl B, Schmincke E (2015) Approach for a human and eco toxicity indicator for construction products and works (ToxScale). SETAC Europe, 25th annual meeting, Barcelona (2015), Proceedings
Kicherer A (2005) BASF eco-efficiency analysis methodology seminar. BASF AG, Ludwigshafen
Kicherer A, Schaltegger S, Tschochohei H, Pozo BF (2007) Eco-efficiency – combining life cycle assessment and life cycle costs via normalization. Int J Life CycleAssess 12(7):537–543
Köllner T, Scholz R (2007) Assessment of land use impacts on the natural environment, part 1: an analytical framework for pure land occupation and land use change. Int J Life Cycle Assess 12:16–23
Köllner T, Scholz R (2008) Assessment of land use impacts on the natural environment, part 2: generic characterization factors for local species diversity in Central Europe. Int J Life Cycle Assess 13:32–48

Kölsch D, Saling P, Kicherer A, Grosse-Sommer A, Schmidt I (2008) How to measure social impacts? A socio-eco-efficiency analysis by the SEEBALANCE® method. Int J Sustain Dev 11 (1):1–23

Krems B (2004) In: Online-encyclopedia of administration. http://www.olev.de

Landsiedel R, Saling P (2002) Assessment of toxicological risks for life cycle assessment and eco-efficiency analysis. Int J Life Cycle Assess 7(5):261–268

Lexikon Römpp Chemie (2004) Online version 2.5. Thieme Verlag, Stuttgart. http://www.roempp.com/prod/index1.html

Löfgren B (2009) Capturing the life cycle environmental performance of a company's manufacturing system. Master Thesis, Gothenburg: Chalmers University of Technology, Göteborg

Luck T (2004) BASF Plant Science Holding GmbH, Germany. Personal communication

Minhas SUH, Berger U (2014) A web-based smart inference system to support automated generation of LCA simulation models. 47th Hawaii international conference on system science

NCBA (2014) Cattlemen's Beef Board and National Cattlemen's Beef Association; sustainability, 2014, Maday J, Benchmarking beef sustainability. http://www.cattlenetwork.com/drovers/Benchmarking-beef-sustainability-195679321.html

Pfister S, Koehler A, Hellweg S (2009) Assessing the environmental impacts of freshwater consumption in LCA. Environ Sci Technol 43(11):4098–4104

Ramachandran R (2014) From emerging to multinational: leaders shaping new models to meet the BoP, Business Call to Action Annual Forum. http://www.theguardian.com/sustainable-business/2014/oct/31/from-emerging-to-multinational-leaders-shaping-new-models-to-meet-the-bop

Renner I, Klöpffer W (1995) Untersuchung der Anpassung von Ökobilanzen an spezifische Erfordernisse biotechnischer Prozesse und Produkte. UBA-Texte 23/95. UBA-FB 000713

Round Table for Product Social Metrics (2014) Handbook for product social impact assessment. In: Fontes J et al (eds) PRé sustainability. The Netherlands

Rüdenauer I, Gensch CO, Grießhammer R, Bunke D (2005) Integrated environmental and economic assessment of products and processes. J Indian Ecol 9:105–116

Saling P (2009) "Metrics for sustainability" as part of RSC Green Chemistry No 4. Sustainable solutions for modern economies. In: Höfer R, Clark J (eds) The Royal Society of Chemistry, Green Chemistry Series 25–37

Saling P (2013) User handbook for web-based eco-efficiency analysis. https://croplife.org/wp-content/uploads/2014/04/User_Manual_-_Web-based_Eco-efficiency_Analysis.pdf

Saling P, Kicherer A, Dittrich-Krämer B, Wittlinger R, Zombik W, Schmidt I, Schrott W, Schmidt S (2002) Eco-efficiency analysis by BASF: the method. Int J Life Cycle Assess 7(4):203–218

Saling P, Maisch R, Silvani M, König N (2005) Assessing the environmental-hazard potential for life cycle assessment, eco-efficiency and SEEbalance. Int J Life Cycle Assess 10(5):364–371

Saling P, Baker R, Gensch CO, Quack D (2006) Differences in life cycle assessment for the production of various carotenoid additives for use in poultry feeds. J Sustain Agric 29(1):15–41

Saling P, Grosse-Sommer A, Alba-Perez A, Kölsch D (2007) Using eco-efficiency analysis and SEEBalance in the sustainability assessment of products and processes. In: Sustainable neighbourhood – from Lisbon to Leipzig through research (L2L). https://www.bmbf.de/pub/l2l_conference.pdf

Schmidt I, Meurer M, Saling P, Kicherer A, Reuter W, Gensch CO (2004) SEEbalance – managing sustainability of products and processes with the socio-eco-efficiency analysis by BASF. Greener Manage Int, Greenleaf publishing Sheffield, S. Seuring S (guest ed), Issue 45, Spring 2004, pp 79–94

Schoeneboom J, Saling P, Gipmans M (2012) AgBalance™, technical background paper. http://www.agro.basf.com/agr/AP-Internet/en/function/conversions:/publish/upload/sustainability/AgBalance/307736_BASF_Tech-E_Paper-AgBalance.pdf

Shahidi F, Metusalach A, Brown JA (1998) Carotenoid pigments in seafood and aquaculture. Crit Rev Food Sci 38:1–67

Shibaike N, Shinomura Y, Ichinohe M, Tanaka M, Takata N, Miyake K, Takeyama N (2008) Activity of Japanese electronics industry on environmental performance indicator toward future standardization. Electronics goes green 2008+. Proceedings, pp 473–477

Steen B (2009) Eco-efficiency of a wind farm. A case study. Environmental systems analysis CPM – center for environmental assessment of product and material systems, Chalmers. http://www.cpm.chalmers.se/document/reports/09/2009_1%20EE%20of%20Borkum.pdf. Accessed 9 July 2012

Steinberg W, Devaud A, Schierle J, Klünter A-M (2000a) Comparative pigmentation efficacy of canthaxanthin and capsanthin/capsorubin in egg yolks. Proc Soc Nutr Physiol 9:86

Steinberg W, Grashorn MA, Klünter A-M, Schierle J (2000b) Comparative pigmentation efficacy of two products containing either apo-ester or tagetes extracts in egg yolks and liquid eggs. Arch Geflügelkunde 64(4):180–187

Tegstedt F (2011) Eco-efficiency assessment. Production of bleaching chemicals for the Elemental Chlorine Free, ECF, pulp industry. Master Thesis, Chalmers University of Technology, Göteborg

TRGS (2014) Technische Regeln für Gefahrstoffe. http://www.baua.de/de/Themen-von-A-Z/Gefahrstoffe/TRGS/pdf/TRGS-600.pdf?__blob=publicationFile

TÜV ("Technischer Überwachungsverein", engl.: "Technical Inspection Association") (2002) ID-Nr. 5711150561: BASF Aktiengesellschaft, Ökoeffizienz-Analyse, Geprüfte Ökoeffizienz-Analysenmethode. http://www.certipedia.com/quality_marks/5711150561/public_documents?locale=de

U.S. Geological Survey (2004) Mineral commodity summaries. Government Printing Office, Superintendent of Documents, P.O. Box 371954, Pittsburgh. http://minerals.usgs.gov/minerals/pubs/mcs/index.html

Uhlman BW, Saling P (2010) Measuring and communicating sustainability through eco-efficiency analysis. Chem Eng Prog (CEPE) 106(12):17–26

Uhlman B, Laginess T, Green D, Steinmetz D, Saling P (2013) NSF protocol 352, validation and verification of eco-efficiency analysis, part A, BASF's eco-efficiency analysis methodology, USA. http://www.nsf.org/services/by-industry/sustainability-environment/claims-validation/eco-efficiency/

UN (2015) Mainstreaming of the three dimensions of sustainable development throughout the United Nations system, report of the secretary-general, general assembly, seventieth session, Item 20 of the preliminary list, A/70/50

UNEP/SETAC Lifecycle Initiative (2009) Guidelines for social life cycle assessment of products. http://www.oeko.de/oekodoc/908/2009-023-en.pdf. ISBN: 978-92-807-3021-0

Voeste D (2012) Experience on gathering meaningful data for life cycle analyses: BASF'S ecoefficiency tool in Indian agriculture. In: A sustainability challenge: food security for all. National Academies Press, Washington, DC, pp 46–48

WBCSD (2015) http://www.wbcsdcement.org/index.php/en/key-issues/sustainability-with-concrete. Dec 2015

Wegener D, Walachowicz F (2009) Comprehensive approach to energy and environment in the Eco Care Program. Risoe international energy conference 2009

Wegener D, Walachowicz F (2011) Improving energy efficiency in industrial solutions – walk the talk. Risoe international energy conference 2011

Westfleisch (2012) Annual sustainability report, p 22. http://www.westfleisch.de/fileadmin/Bilder/02_Unternehmen/02.07_Archiv/02.07.01_Geschaeftsberichte/2012/Westfleisch_eG_GB_2012_EN.pdf

World Resources Institute (1996) A guide to global environment, 1996–1997. World Resources Institute, United Nations Environment Program, World Bank. Oxford University Press, 400 pp. ISBN: 0-19-521161-8

Chapter 5
LCA Perspectives for Resource Efficiency Assessment

Laura Schneider, Vanessa Bach, and Matthias Finkbeiner

Abstract Efficient use and consumption of natural resources is an important strategy in sustainable development. This chapter discusses available methods and indicators to assess "resource efficiency" beyond the assessment of the quantities of materials used and toward available indicators in life cycle assessment (LCA). According to the classical definition in LCA, natural resources encompass input-oriented environmental interventions (e.g., extraction of abiotic resources, such as oil, ore deposits, fossil, and fresh surface water, as well as biotic resources such as fish and trees). LCA and existing life cycle impact assessment (LCIA) methods are seen as a good basis for measuring resource efficiency. Despite several models to assess resource use and depletion within LCA, important challenges remain. Available models do not fully evaluate resource use and availability in the context of their functional relevance for human purposes. For the efficient use of resources, all dimensions of sustainability need to be addressed. Environmental, economic, and social implications of material use and availability have to also be considered. Assessment of resource utilization and efficiency associated with product systems needs to shift toward life cycle sustainability assessment (LCSA).

Keywords AoP • Area of protection • LCA • LCIA • LCSA • Life cycle assessment • Life cycle impact assessment • Life cycle sustainability assessment • Natural resources • Resource availability • Resource consumption • Resource depletion • Resource efficiency • Resource use • Sustainable development

Acronyms

AADP	Anthropogenic stock extended abiotic depletion potential
ADP	Abiotic depletion potential
AoP	Area of protection

L. Schneider • V. Bach • M. Finkbeiner (✉)
Chair of Sustainable Engineering – Institute of Environmental Technology, Technische Universität Berlin, Office Z1, Strasse des 17. Juni 135, 10623 Berlin, Germany
e-mail: matthias.finkbeiner@tu-berlin.de

M. Finkbeiner (ed.), *Special Types of Life Cycle Assessment*, LCA Compendium – The Complete World of Life Cycle Assessment,
DOI 10.1007/978-94-017-7610-3_5

CF	Characterization factors
CML-IA	CML impact assessment
DMC	Domestic material consumption
DMI	Domestic material input
EMC	Environmentally weighted material consumption
ESP	Economic resource scarcity potential
GDP	Gross domestic product
LSFs	Life-support functions
MSY	Maximum sustainable yield
NPP	Net primary production
PEF	Product environmental footprint
PGMs	Platinum group metals
REE	Rare earth elements
TMI	Total material requirement

1 Introduction

Mankind depends on natural resources, such as metals, minerals, oil, water, land, wood, fertile soil, or clean air for survival, and these resources constitute vital inputs for a stable economy and society (see, e.g., European Commission 2012; Meadows et al. 2004). Their exploitation is strongly linked to the supply of products and services within society (UNEP 2010a, 2011). "Human well-being and its improvement, now and for a still growing world population in the future, is based upon the availability of these resources" (UNEP 2011). Natural resources are important for their structural properties and as energy carriers to humans and machines. In recent decades, the demand for many resources has increased significantly due to the growing importance of these resources for industrial and technological development (Behrens et al. 2007; European Commission 2013b; Gordon et al. 2006; Kleijn 2012; Petrie 2007; UNEP 2011; Weterings et al. 2013). In Fig. 5.1, two examples are shown to emphasize the whole extent of resource use and its increase in the past years. In the left part of the figure, an overview of the cumulated world production of different metals is provided. Around 80 % of the cumulative mine production of platinum group metals (PGMs) or rare earth elements (REE) has occurred over the last 30 years (Hagelüken and Meskers 2010; USGS 2014). The right part of the figure shows the amount of captured fish in the last decades in relation to the stock status. Not only is the overall quantity of captured fish increasing, but also the stocks within biological unsustainable levels are rapidly increasing (FAO 2014b).

The availability of natural resources as input for production processes is not infinite nor is the capacity of natural resources to absorb pollution (UNEP 2010a). The accelerating pace at which resources are exploited and increasing pollution burdens are the basis of growing concern. Current patterns of resource consumption could lead to irreversible damage to the planet's natural environment and

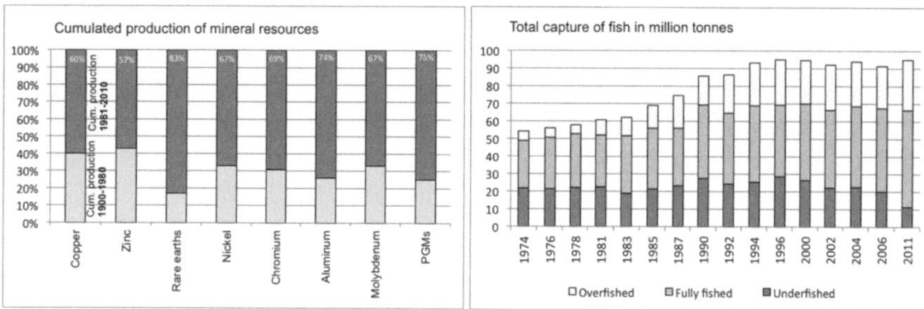

Fig. 5.1 Trends in resource production – cumulated production of mineral resources (*left*) and total capture of fish (*right*) (Based on data published by FAO 2006, 2012, 2014b; USGS 2014)

jeopardize its very ability to provide resources and ecosystem services that humankind is dependent on (BIO Intelligence Service 2012).

In the context of growing economic activity and population, resource use in line with the requirements of sustainable development is one of the most important challenges faced by our society (European Commission 2013b; Giljum and Polzin 2009). Natural resource use is prominently featured on the environmental policy agenda in Europe and in other world regions, and increasing resource efficiency has been a topic of intensive public discussion and legislation over the past years (Giljum et al. 2008; Klinglmair et al. 2013). Resource efficiency refers to the sustainable use of the Earth's limited resources and is one of the commonly mentioned strategies for sustainable development. Key goals of resource efficiency include ensuring security of supply of resources and to decrease environmental impacts of resource use (BIO Intelligence Service 2012; European Commission 2013a; Giljum and Polzin 2009). For assessing the efficiency of resource use, companies need reliant and applicable indicators and methods (European Commission 2005, 2011c, 2012). Resource efficiency is commonly defined as the relation of economic output (added value) and required resource input (Eq. 5.1) (Demurtas et al. 2014; ISO 14045 2012; UNEP 2011). The concept characterizes how efficiently resources are used to add economic value.

$$\text{Resource efficiency} = \frac{\text{added value}}{\text{resource input}} \tag{5.1}$$

The assessment of resource efficiency can be a good starting point for evaluating the pressure put on the natural environment. The added value refers to the socio-economic benefit of resource use and is often expressed in terms of economic market value, for example, gross domestic product (GDP) or revenue. The determination of the resource input depends on the classification of resources as such and needs to be measured by means of appropriate indicators that can adequately capture the goals of sustainable development. For achieving sustainable development, resources need to be managed on a corporate level under consideration of their availability for products and production processes and the impacts of their extraction on environment and society. Currently, there is the tendency to adopt indicators simply based on mass aggregation (Klinglmair et al. 2013). Several material-flow-based indicators are available for assessing resource needs from an economy-wide perspective. Indicators like the domestic material input (DMI)[1] or the total material requirement (TMR)[2] can be used to assess material efforts associated with economic activities (Bringezu and Schütz 2010; Ritthoff et al. 2002; van der Voet 2005). At the EU level in the *Roadmap to a resource efficient Europe*, such a material-flow-based indicator, the domestic material consumption (DMC),[3] has been adopted and is recommended for measuring resource efficiency on national and regional level (European Commission 2011c). However, these concepts are based on the idea to use less resource input per unit of economic output. Neither reflects the availability or environmental relevance of the resources, as such, nor the pressure arising from resource consumption. A simple aggregation of resources on the basis of mass (or energy) has little informative value. Such indicators suggest that resources "are exchangeable and equally important with respect to their mass or energy content" (Steen 2006). Those measures are not sufficient for assessing impact as no relation to potential scarcity is provided. Besides, the problem with measuring resources solely on a mass (or energy) basis is that the environmental and economic relevance are disregarded and that no link to the availability or potential scarcity of resources is provided (European Commission 2011c). These indicators do not take into account material specific aspects (e.g., the potential for specific environmental damages), as all flows are accounted for in unweighted units (Behrens et al. 2007). For example, in mass-based evaluations, the use of sand, gravel, or copper is of the same value. Approaches for linking the mass to an environmental impact score have been proposed. Van der Voet (2005) developed a method to compare and evaluate the environmental impacts associated with the DMC of an economy (see environmentally weighted material consumption (EMC) (van der Voet 2005)). However, this method has been developed and is applied independently from evaluating resource use on a product level. Similar to the economy-wide analysis of resources, resource utilization on a product level is often reduced to an evaluation of mere quantities of resource input. For example, the Wuppertal Institute developed an indicator focusing on the quantity of resources used for a specific product or service (material input per service unit, MIPS) (Ritthoff et al. 2002). However, based on these indicators, no statement can be made with regard to the effects of resource use on the environment, human health, or availability of these resources in nature. Today, the topic of resource security is more pressing than in previous times as it is a high-priority issue

[1] The DMI accounts for the environmental resource used in domestic production and consumption, including imports (Wuppertal Institute for Climate 2013).

[2] The TMR comprises the use of all domestic and foreign primary materials for production and consumption (Wuppertal Institute for Climate 2013).

[3] The DMC refers to the amount of resources used for domestic consumption (excluding exports and unused extraction) (Wuppertal Institute for Climate 2013).

for economic development and implementation of technologies (Kesler 1994; Valero et al. 2013; Wäger et al. 2011). In the context of ever-increasing demand, the evaluation of natural resources becomes more and more important. Indicators should not only relate to the quantities of resources used but consider resulting impacts on the environment and their availability (see, e.g., BIO Intelligence Service 2012). Even though sustainability is a macroeconomic problem, concerned with all resources required to sustain production and the well-being of current and future generations (Mikesell 1994), sustainable development is mainly promoted at the microeconomic level, for example, by reducing material inputs. Thus, appropriate indicators and measures are needed for evaluating the resource input and resource efficiency on a product level.

Life cycle approaches are necessary to avoid shifting impacts and to capture all potential effects associated with resource use. Life cycle assessment (LCA) is considered suitable to provide support in integrating sustainability of resources into design, innovation, and evaluation of products (Klinglmair et al. 2013; Sala et al. 2013), and the evaluation of resource use has been a common practice in life cycle impact assessment (LCIA) for several years (see, e.g., European Commission 2011c; ISO 14040 2006; ISO 14044 2006; Lindeijer et al. 2002). For a meaningful assessment of resource efficiency, all impacts associated with resource use need to be accounted for. Thus, LCA and existing LCIA methods are seen as a good basis for the evaluation of resource efficiency and for shifting the definition of the "resource input" away from mere quantities toward a more comprehensive assessment in the context of sustainable development.

This chapter focuses on the determination of appropriate indicators for evaluating "resource input" in order to assess resource efficiency. An overview of available methods and approaches for the analysis of resource use in LCA is provided and existing shortcomings are discussed. Furthermore, potential enhancements for addressing resource use more comprehensively are outlined, and future research needs are summarized, including moving toward life cycle sustainability assessment (LCSA) for a comprehensive assessment of resource use. In the following section, as a first step, the classification of natural resources is presented and relevant characteristics of abiotic and biotic resources are described.

2 Classification of "Resource" and "Problem Definition" in the Context of Life Cycle Assessment

The evaluation of resource use and efficiency depends on the understanding of the "resource" as such. In this section, the different types of natural resources and the underlying problem definition in the context of LCA are outlined.

The European Commission defines resources as all natural resources including physical resources and environmental media such as air, land, and water (European Commission 2005). This implies that "resource input" for the calculation of resource efficiency should comprise any input from the environment, beyond materials. In the context of LCA, different areas of protection (AoPs) are distinguished. These AoPs are used to express what is of value to human society and what needs to be sustained for achieving human welfare (see, e.g., Dreyer et al. 2006).[4] In LCA, a distinction between "natural resources" and "natural environment" is made. The AoP "natural environment" refers to the protection of functional and nonfunctional values of nonhuman life, the regulation of climate, generation or regeneration of soil, or the protection of biodiversity and natural landscapes (Lindeijer et al. 2002). Consequences of activities on the environment need to be assessed and avoided in order to protect these functions. Contrary, the AoP "natural resources" encompasses mainly input-oriented environmental interventions (e.g., extraction of abiotic resources, such as oil, ore deposits, fossil, and fresh surface water, as well as biotic resources such as fish and trees). In this context, natural resources are defined as "extractable entities with implications for their present, but mainly future availability" (Lindeijer et al. 2002) and refer to assets that can be used for economic production or consumption (United Nations 2005).

For a comprehensive assessment of resource use, the removal of a resource from nature and the consequences of the extraction on the availability of resources as well as impacts caused by the extraction process as such need to be assessed (see Fig. 5.2). The extraction of resources reduces the quantity or the quality of resource stocks and can result in impacts on the natural environment (Lindeijer et al. 2002). In line with the key goals of resources efficiency, two different dimensions need to be addressed: the security of supply of resources and the reduction of the impacts of resource extraction and use.

While the assessment of consequences or impacts of extraction activities (extraction processes) on the "natural environment" is already established and several models and indicators to describe environmental impacts are available and generally agreed on,[5] the assessment of the consequences of natural resource extraction (removal from nature) is not yet well defined (Udo de Haes et al. 2002). For the assessment of natural resources in LCA, only the environmental intervention *extraction* as such is considered, representing the removal of a certain quantity of a resource from nature (Lindeijer et al. 2002). Basically, three different environmental interventions associated with natural resource use are distinguished (following Lindeijer et al. 2002):

- Extraction of abiotic resources
- Extraction of biotic resources
- Allocation of land areas to man-controlled processes

[4] The societal values can be grouped into different areas of protection (AoPs). Within LCA, mainly three AoPs are differentiated: "human health," "natural environment," and "natural resources."

[5] Several impact categories are available in LCA that model the effects of the resource extraction process on the natural environment. As an example, for climate change, which is caused by several greenhouse gases, methods for quantifying the global warming impacts of activities are available and well established (see, e.g., Guinée et al. 2002).

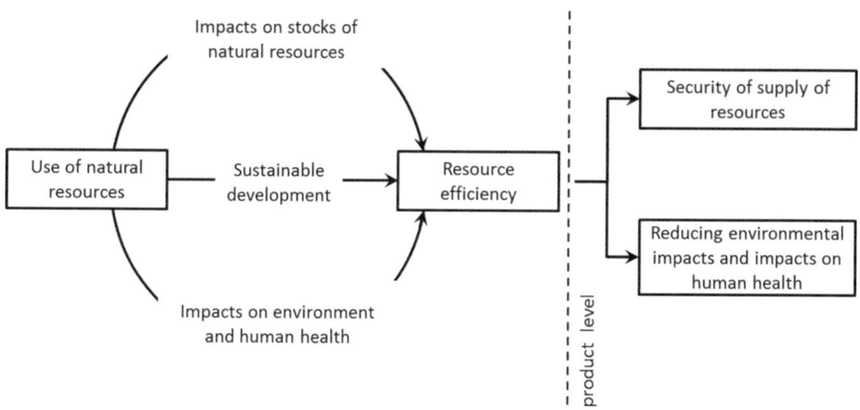

Fig. 5.2 Implementation of resource efficiency on product level

Extraction of resources leads to decreasing availability of resources for current or future generations. The process of exhausting the abundance or availability of resources can be defined as *depletion* (Guinée and Heijungs 1995; Guinée et al. 2002; Lindeijer et al. 2002; Radetzki 2002). The effect of decreasing availability is somewhat different for defined types of resources. In the assessment of natural resources, commonly a distinction between land use, water use, and the use of biotic or abiotic resources is made. Water and land can be included in the category of abiotic resources, but are often seen as resource classes of their own, and (impact) assessment separate from other abiotic resources is common practice (e.g., Berger et al. 2012; Curran et al. 2011; Gontier et al. 2006; Henzen 2008; Koellner et al. 2013; Milà i Canals 2003, 2007; Milà i Canals et al. 2007; Millennium Ecosystem Assessment 2005; Schenck 2001; Souza et al. 2013; Steinbach and Wellmer 2010; UNEP 2010c; Watson et al. 2005). The analyses of water use[6] as well as water footprinting[7] are addressed in separate chapters in the "LCA Compendium." Availability of flow resources such as sunlight and wind is so far not addressed within LCA as the process of "extraction" is somewhat different and does not lead to depletion or limited availability.[8]

In the following sections, characteristics of abiotic and biotic resources are outlined in more detail focusing on the relevance and problem definition in the context of LCA.

---

[6] See Chap. 12 on "Water Use" by Stephan Pfister, LCA Compendium, the volume on "Life Cycle Impact Assessment" (Pfister 2015).

[7] See this volume, Chap. 3 on "Water Footprinting" by Berger et al. (2016).

[8] Commonly, different types of resources are identified: stocks, funds, and flows. Stocks are irreversibly depletable, while funds are temporarily or locally degradable. Contrary, flows are nondegradable, but with a limited availability at a certain time (see Lindeijer et al. 2002).

2.1 Abiotic Resources

Abiotic resources are chemical elements and minerals from the Earth's crust. There is no other source from which they can be obtained (Kesler 2007). Abiotic resources such as metallic or energy minerals are a real challenge to modern civilization because they are formed by geologic processes that are much slower than the rate at which they are exploited and are thus classified as nonrenewable (Kesler 1994; UNEP 2010a). Natural stocks of mineral resources are irreversibly depleted by extraction processes as they cannot regenerate within human lifetimes.[9]

Minerals have no intrinsic values in themselves while locked up in ore bodies buried in the earth and are considered to be "outside the biosphere" as a reduction of natural stocks has no direct influence on ecosystems (Petrie 2007; Udo de Haes et al. 2002). It is generally agreed that the interest of mankind is not the abiotic resource as such or its value in the natural environment but predominantly the function it fulfills in the economic system to achieve human welfare (see also Jolliet et al. 2004; Stewart and Weidema 2005; Udo de Haes and Lindeijer 2002; van Oers et al. 2002; Weidema et al. 2005; Wellmer and Becker-Platen 2002; Yellishetty et al. 2009). Thus, emphasis is placed on the evaluation of abiotic resources in the context of their potential to fulfill functions in products and to create value by meeting human needs (Lindeijer et al. 2002; Yellishetty et al. 2011).

The removal and use of abiotic resources diminishes the availability of natural stocks in the environment, that is, every consumption equates to a reduction of the natural stocks and decreasing availability for future use (Brentrup et al. 2002; Petrie 2007). The direct impact related to the use of abiotic resources is denoted as the *depletion* of resources (UNEP 2010a). In current literature, depletion of abiotic resources is assessed by means of:

- Purely physical aspects, referring to the decreasing availability of resource stocks (deposits) in the context of current (and future) extraction of these stocks
- Increasing expenses of resource extraction associated with decreasing resource stocks, assuming that costs of producing minerals will rise to a point where they are no longer affordable (see, e.g., Tilton 2003)

The first notion is based on the growing consumption in the context of finite resources, implying the approach of a physical limit and exhausting the resources in an absolute sense. This could have negative impacts on a global scale and is certainly opposing the principles of sustainable development. However, the definition of available physical stocks leaves a large room for interpretation and is often

[9] In some cases, natural resources are a mix of biotic and abiotic components (mineral nutrients) (Lindeijer et al. 2002). In this context, it is often referred to as an abiotic factor (a nonliving component of a habitat), rather than an abiotic resource. Nutrients from soil minerals are *fund* resources that are temporarily or locally degradable but that can be regenerated within human lifetime (Lindeijer et al. 2002). Soil is considered within the impact category "land use" as soil quality parameters are adequate indicators to express consequences of land use changes.

related to economic considerations and assumptions. This leads to the second notion, which is based on the increasing expenses of resource extraction. Increasing costs associated with the extraction of resources can be related to decreasing ore grades and deposit size (Skinner 1976; Vieira et al. 2012).[10] High demand of energy and increasing costs can lead to constrained availability of mineral resources, even before the stock has been exhausted. Such parameters are often assumed to determine the limits to growth and constrain supply in the end (see, e.g., Bardi 2011; Meadows et al. 1972; Turner 2008).

There is consensus in regarding abiotic resources as something that is subject to depletion or decreasing availability and scarcity (Steen 2006). The steadily increasing demand for abiotic resources means that natural resources are depleted at an ever-increasing rate. Meeting present needs of mankind without abiotic resources is inconceivable. Abiotic resources are essential for the quality of life that modern society is accustomed to, and play a key role in underpinning the prosperity of future civilizations (Auty 1993; Giurco and Cooper 2012; Reuter et al. 2005; Science and Technology Committee 2011; Verhoef 2004). "Easy access to [abiotic] resources is often seen as a precondition for economic development" (UNEP 2010a) as minerals and fossil fuels are crucial inputs for most production processes and are the starting material for the production of almost all manufactured products (Azapagic 2004; UNEP 2010b). Metals, for example, provide unique features and are a relevant input into most products and production processes due to electrical and thermal conductivity and good processability. Energy supply, high-tech products, and emerging clean technologies are highly dependent on several mineral resources and will thus significantly raise and change the demand for these in the future, putting additional pressure on supply (Achzet et al. 2011; Angerer et al. 2009a; Bardi 2013; Kleijn 2012; Wäger et al. 2011).

2.2 *Biotic Resources*

Biotic resources are living objects such as fish or wood removed from the natural environment by human activities and play a vital role for sustaining the livelihood of many people. Over two billion people use fish as an important part of their daily animal protein source, and 12 % of the global population rely on fishery as their principal source of income (FAO 2014a). The rainforests as one of the most important and remaining forest resources play a key role in sustaining biodiversity and storage of carbon (Olson et al. 2011; Steffan-Dewenter et al. 2007). According to FAO (2014a) billions of people depend on forests and the services they supply (e.g., food and shelter). Considering recent developments, like rising fish yields and decreasing stocks as well as the transformation of forests into agricultural areas, the

[10] Low-grade deposits are likely to be more difficult to mine than high-grade deposits. As a result, energy required could be one or two orders of magnitude greater when metals would need to be extracted from low-grade deposits, causing a significant increase in costs (Skinner 1976).

pressure on these resources will increase. Thus, to preserve natural biotic resources for current and future generations, it is widely accepted to evaluate resource use against depletion.

Biotic resources are classified as renewable, due to their general ability to regenerate within human lifetime. Depletion of biotic resources occurs if their use exceeds their replenishment rates resulting in decreasing availability (Lindeijer et al. 2002). Furthermore, distinctions need to be made between temporal and permanent depletion. Whereas impacts can be reversed when temporal depletion occurs (e.g., extraction of fish below their recovery rate), effects are irreversible by permanent depletion (e.g., extinction of species) (Klinglmair et al. 2013; Lindeijer et al. 2002; Stewart and Weidema 2005). Contrary to the depletion of abiotic resources, which is determined by available stocks and extraction rates or increasing costs of extraction, depletion of biotic resources is conditioned by the relation between the extraction and the replenishment rate (Lindeijer et al. 2002) and can also be affected by additional impacts on the ecosystem like extreme weather events.

All common definitions within an LCA context exclude biotic materials that are products of a man-made culture since they are not considered a depletable resource. Agri-, silvi- and aquacultural products, such as crops, farm animals (including fish farms), or grown wood, are not considered as biotic resources but as products derived from resources such as land, water, solar energy, and nutrients (Guinée et al. 1993; Heijungs et al. 1997; Klinglmair et al. 2013; Lindeijer et al. 2002).

As part of the ecosystem, biotic resources have an intrinsic value represented by biodiversity as well as a functional value by maintaining life-support functions (LSFs).[11] Thus, contrary to abiotic resources, their removal not only reduces the natural stock of the resource and affects availability for human purposes but also exerts a change on ecosystems resulting in, for example, loss of biodiversity (Guinée et al. 1993; Lindeijer et al. 2002). Furthermore, external factors (beyond extraction as such) can influence and decrease the availability of biotic resources. For example, as plants are highly dependent on soil nutrients and soil availability, acidifying substances can cause a decline in soil quality and thus decrease the availability of resources.[12]

In Fig. 5.3, the differentiation in the cause-effect chains of abiotic and biotic extractions is highlighted. While availability is the main concern associated with abiotic resource extraction, biotic resource extraction not only decreases[13] the

[11] The functional values of the natural environment are represented by life-support functions. Examples of LSFs include nutrient dispersal, climate regulation, or purification of water.

[12] Within the LCA framework, soil quality is addressed in the impact category "land use" (Lindeijer et al. 2002; Milà i Canals et al. 2006; Núñez et al. 2012) and is thus not further considered here. See chapter 11 on "Land Use" by Milà i Canals and de Baan, in the volume LCIA of the "LCA Compendium" (Milà i Canals and de Baan 2015).

[13] However, compared to abiotic resources, biotic resources can also increase during human lifetime when no material is extracted. Increasing stocks are so far not considered within LCIA methods.

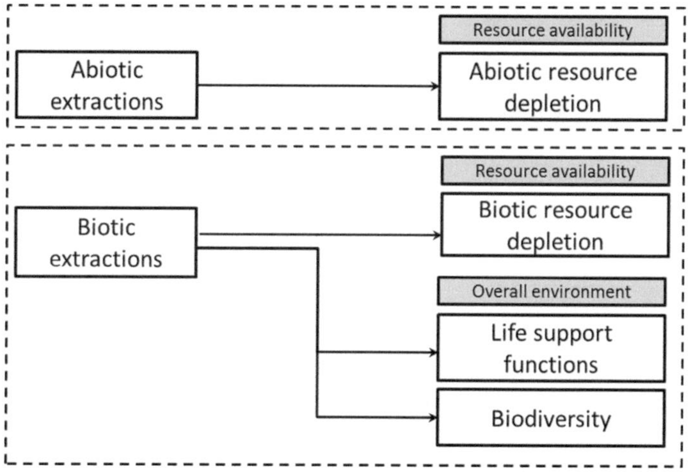

Fig. 5.3 Cause-effect network of abiotic and biotic resource extraction (Based on Lindeijer et al. 2002)

availability of biotic resources but also amplifies the risk of harming biodiversity and life-support functions (Lindeijer et al. 2002). Thus, two different types of approaches need to be used for the evaluation of abiotic and biotic resources (see Fig. 5.3):

– Focusing solely on resource availability
– Focusing next to resource availability on the overall environment, including effects on biodiversity and LSFs

The necessity of sustainable use and preservation of natural resources for current and future generations is widely accepted, and resource efficiency requires extracting and using natural resources within the planet's long-term boundaries (European Commission 2011a).

3 Assessment of Resource Use in LCA

This section captures the state of the art of evaluating natural resource use in LCA and provides an overview of existing approaches that can be used as a basis for the assessment of resource efficiency.

As outlined in the previous section, extraction of natural resources, such as metal ores, fish, or wood, needs to be assessed in the context of degrading quality, decreasing availability, and, for biotic resources, associated damage to nature. Evaluation of resource use and availability is common practice within LCA (European Commission 2010c; ISO 14040:2006; ISO 14044:2006). As of today, a wide variety of LCIA methods have been developed (especially for abiotic resources) to assess resource use by quantifying the contribution of resource extraction to the potential depletion of resources. Several authors propose ways to integrate the effects of natural resource use and the assessment of resource availability into the LCA framework. Available methods address different aspects of decreasing resource availability based on varying perceptions of the underlying problem. In general, two levels to model impacts of resource extraction can be distinguished: *midpoint* approaches and *endpoint* approaches. While midpoint approaches start at the environmental intervention and describe the mechanism of depletion, the endpoint approaches address the damage level and attempt to capture the consequences of resource extraction (see, i.a., Klinglmair et al. 2013; Udo de Haes and Lindeijer 2002). The distinction between mid- and endpoint approaches is not very clear for resource depletion, as the impact category (resource depletion) and area of protection (natural resources) are congruent.

In the next sections, the state of the art for the assessment of abiotic and biotic resources in LCA is outlined.

3.1 Abiotic Resources

Modeling the impacts of abiotic resource use is a major topic of debate in the LCA community. The ILCD Handbook, a reference handbook on best practice in life cycle impact assessment, provides an overview of existing models for the assessment of abiotic resource use and availability in LCA (European Commission 2010b, c). The identified models relate to energy and mass of a resource used, exergy or entropy impacts, future consequences of resource extraction (e.g., surplus energy, surplus cost), or diminishing geologic stocks (see, i.a., Bösch et al. 2007; BUWAL 1998; Dewulf et al. 2007; European Commission 2010c; Finnveden et al. 2009; Finnveden and Östlund 1997; Goedkoop and Spriensma 2000; Guinée et al. 2002; Hauschild and Wenzel 1998; Klinglmair et al. 2013; Lindeijer et al. 2002; PE International 2013; Steen 2006; Stewart and Weidema 2005; van Oers et al. 2002). In Table 5.1, an overview of the different LCIA methods is provided with regard to available mid- and endpoint metrics and the underlying concepts (based on the European Commission 2010b; Klinglmair et al. 2013; Pennington et al. 2004).[14] All indicators have in common that they aim at expressing decreasing availability of resources either based on the physical finiteness of resources in the geosphere or with regard to future consequences on the extraction of a resource (see also Klinglmair et al. 2013). The shortcomings of the individual methods in the context of depicting sustainable resource use are discussed in more detail in Sect. 4.1.

[14] See "LCA Compendium," the volume on "Life Cycle Impact Assessment," Chapter 13 by Swart et al. for more information on LCIA methods (Swart et al. 2015).

Table 5.1 LCIA, methods for abiotic resource depletion

Method	Characterization factors				References
	Midpoint		Endpoint		
	Unit	Concept	Unit	Concept	
ADP (CML 2002)[a]	kg Sb-eq.; MJ	Reserves and annual extraction rates	–	–	Guinée et al. (2002) and van Oers et al. (2002)
Eco-indicator 99	–	–	$MJ_{surplus\ energy}$	Surplus energy	Goedkoop and Spriensma (2001)
Ecoscarcity 2013[b]	UBP/g kg Sb-eq.; UBP/MJ oil eq.	Distance-to-target (weighting method)	–	–	Frischknecht et al. (2009)
EDIP 1997[c]	Person reserve	Reserves and annual extraction rates	–	–	Hauschild and Wenzel (1998)
EPS 2000[d]	–	–	$\$_{WTP}$	Willingness-to-pay	Steen (1999)
Exergy (CEENE, CExC)[e]	MJ_{exergy}	Loss of exergy	–	–	Finnveden and Östlund (1997), Dewulf et al. (2007), and Bösch et al. (2007)
IMPACT 2002+[f]	kg Fe-eq.; MJ	Mineral extraction	$MJ_{surplus\ energy}$	Surplus energy	Jolliet et al. (2003) and Humbert et al. (2012)
LC-IMPACT[g]	%/kg	Decrease in ore grade due to increasing extraction	$/kg	Surplus cost increase as response to lower ore grade	Vieira et al. (2011)
LIME	Sb-eq. kg; MJ	Consumption energy, reciprocal of recoverable reserves	¥	User costs (displaying costs of overuse)	Itsubo and Inaba (2012)
ReCiPe[h]	kg Fe-eq.; MJ	Decreased concentration	$\$_{surplus\ cost}$ ($/kg)	Surplus cost/ damage to resource cost	Goedkoop et al. (2008)

[a]*ADP* abiotic depletion potential. In current LCA, resource depletion is commonly assessed by means of the abiotic depletion potential (ADP), which is the differentiation between fossil depletion and element (metals/minerals) depletion

[b]Distance-to-target approaches set environmental interventions against predefined targets. The characterization factors in this method are based on Guinée et al. (2002) but are transferred to EcoPoints (EP) via weighting of the characterization factors

[c]Method involves normalization and weighting. Amount of the resource extracted is normalized to the average annual consumption of one world citizen and weighted according to the static lifetime of available economic reserves (European Commission 2011b; Hauschild and Wenzel 1998)

[d]The method of Environmental Priority Strategies (EPS) accounts for the monetary cost of avoiding damages to natural resources

(continued)

Table 5.1 (continued)

[e]Exergy of a resource or a system is defined as the maximum work potential of a resource (Dewulf et al. 2008). *CEENE* cumulative exergy extraction from the natural environment, *CExC* cumulative exergy consumption

[f]Damage CFs are taken directly from Eco-indicator 99. The midpoint characterization factors (CF) are obtained by dividing the damage CF of the considered substances by the CF of the reference substance (iron). However, midpoint indicator are not recommended for use (Humbert et al. 2012)

[g]The midpoint method is not applied, as no linear relation to the endpoint exists. The endpoint is calculated as the ratio of predefined critical flow to the actual flow of a resource

[h]The method uses increased costs as endpoint indicator and "the slope (relation grade yield) divided by availability" as midpoint indicator (Goedkoop et al. 2008)

Most methods described in Table 5.1 focus only on abiotic resources. Impacts from biotic resource depletion are generally excluded, except for EPS2000 (Steen 1999). Steen (1999) considers in his method, next to mineral and fossil depletion, the production capacity of ecosystems like wood and fish (see also Hauschild et al. 2013).

3.2 Biotic Resources

Unlike the assessment of abiotic resource availability, so far no generally applicable methods or indicators are available for the assessment of biotic resource extraction. Current analyses are predominantly focusing on fishery aspects, which cannot be transferred to other biotic resources like terrestrial animals without adaptation. In the following, existing approaches are reviewed with regard to their ability to assess impacts of biotic resource extraction. The extraction of biotic resources affects not only the availability of the resources as such (AoP "natural resources") but also the ecosystem (AoP "natural environment"). Thus, impact assessment methods for the evaluation of biotic resource use have to consider both areas (Lindeijer et al. 2002).

For assessing the depletion of biotic resources, Lindeijer et al. (2002) suggest a marginal approach based on Heijungs et al. (1997) and Klöpffer and Renner (1994) to evaluate the extraction of wild fish and wood. The category indicator is calculated considering the scarcity of the biotic resource and the quotient of extracted resources and their annual replenishment to their current stocks. For determining the scarcity of the resource, first ideas are proposed (e.g., recovery rate) which might vary for different species. Similarly, Guinée et al. (2002) propose the consideration of stocks of species in relation to their production and renewability rates. Derived from the ADP model, results are linked to a reference unit, in this case African elephants. Both approaches have not been used in LCA case studies so far due to lack of applicability because species-specific data and geographically explicit inventory data regarding the supply chain of products are missing (Avadí and Fréon 2013; Emanuelsson et al. 2014; Pelletier et al. 2006; Vázquez-Rowe et al. 2012a).

For fishery, several methods have been proposed and applied in case studies. Most commonly, indicators based on NPP are used to quantify biotic resources (Aubin et al. 2006; Efole Ewoukem et al. 2012; Hornborg et al. 2012; Jerbi et al. 2012; Libralato et al. 2008; Papatryphon et al. 2004a, b; Pelletier et al. 2006; Vázquez-Rowe et al. 2012a). The NPP expresses the biological production capacity based on total carbon fixed and aims at determining the extracted biomass due to direct uptake – here extracted fish – as well as used biomass for upstream and downstream processes (e.g., for fish fodder). NPP-based indicators are used to measure the overall biomass needed within the system to produce a certain amount of fish. Therefore, comparison of production systems, as well as definition of hot spots with regard to the biotic resource use, is possible. In some approaches, even the removal of lower trophic levels (Libralato et al. 2008) or effects of discard and by-catch (Hornborg et al. 2012; Vázquez-Rowe et al. 2012a) are addressed. Instead of NPP-based indicators, Ziegler and colleagues (Ziegler et al. 2011, 2003; Ziegler and Valentinsson 2008) propose to determine catch rates, discard rate ratios, and amounts of catch for noncommercial use to account for discard and by-catch in fishery.

Other methods addressing biotic resource extraction quantify aspects of overfishing in relation to renewability using indicators based on maximum sustainable yield (MSY) (Emanuelsson et al. 2014; Langlois et al. 2012, 2014). The MSY represents the highest catch that can be taken from a species stock in the long term without depletion. While Langlois and colleagues (Langlois et al. 2012, 2014) evaluate the potential regeneration time of the resource, Emanuelsson et al. (2014) quantify the overexploitation of existing stocks by comparing the ratio of MSY and current yields.

Steen (1999) addresses biotic resource use by taking the production capacity of wood and fish into account using the weight of the extracted resources as an indicator. However, he does not clarify if the resources are extracted from ecosphere or technosphere. The concept of production capacity is based on the assumption that the extraction of resources is ideal, when high amounts are achieved. Even though this is the case for agricultural products, where harvesting is the main purpose, for wild biotic resources, only small amounts should be extracted to avoid overall depletion of the species as well as related impacts on the environment.

Regarding the evaluation of biotic resource availability, the field of fishery is the most advanced. So far no impact assessment methods evaluating biotic resource depletion of wild forests exist. Managed forests and the extraction of wood, for example, for biofuel production or furniture, have been solely assessed regarding emitted greenhouse gases (Hansen et al. 2013; Müller-Wenk and Brandão 2010; van Zelm et al. 2014; Vogtländer et al. 2013) or in relation to land use (Doka et al. 2006; Goedkoop et al. 2008; Helin et al. 2014), but not regarding impacts on availability. Furthermore, so far no applicable methods for evaluating depletion of other biotic resources (e.g., terrestrial animals) are available.

Next to the availability of biotic resources, effects of biotic resource extraction on the natural environment need to be assessed. Biotic resources often have an important function in the ecosystems, and their extraction can affect LSFs and biodiversity (Lindeijer et al. 2002). The assessment of those aspects in LCA is still challenging due to missing inventory data and lack of applicable impact assessment methods (Finkbeiner et al. 2014).

Lindeijer et al. (2002) propose a characterization factor for assessing the loss of biodiversity caused by biotic extraction based on marginal damage. Here, the potentially affected fraction of different species can be aggregated to have one species-unrelated result for biodiversity loss caused by biotic resource extraction. However, different threat levels for species[15] are not included and have to be considered by means of additional factors. This is important, as the same amount of extracted species can lead to no impacts but also to extinction depending on the threat level of the extracted species. So far this method has not been applied in LCA case studies.

Only a few approaches for evaluating biodiversity loss due to extraction of fish have been proposed, assessing the effects of fishing on the fertility of the ecosystem (e.g., due to disruption of food webs) and the resulting decrease of biomass (Langlois et al. 2012, 2014). Next to the evaluation of biotic resource extraction, NPP has also been used for measuring loss of biodiversity within different frameworks due to the high correlation between NPP and biodiversity (Catovsky et al. 2002; Costanza et al. 2007; Itsubo and Inaba 2012). NPP has been used to determine how many biotic resources are extracted and to quantify the decrease of the population, which is assumed to correlate with biodiversity loss. Biodiversity loss due to biotic resource extraction of wild forest is so far partly evaluated in the impact assessment category land use but no separate methods for forest extraction in the context of biotic resource use exist (de Baan et al. 2012; Finkbeiner et al. 2014; Michelsen et al. 2014).

The discussion shows that conceptually differing approaches and models exist for characterizing contributions of resource utilization in products or production processes to the decreasing availability of resources (see i.a. Guinée et al. 2002; Lindeijer et al. 2002). These models can be used to enhance the evaluation of resource efficiency, but some shortcomings remain. In the next section, the shortcomings of the different methods are discussed in the context of their applicability and reliability for addressing resource availability.

---

[15] Threat levels measures the risk of extinction and thus biodiversity loss over a specific time and indicate how severe the risk of extinction for specific species is. Three levels exist to classify threatened species: critically endangered, endangered, and vulnerable (International Union for Conservation of Nature and Natural Resources 2014).

4 Shortcomings of LCIA Methods

In this section, available approaches and methods are evaluated with regard to their significance for the evaluation of resource efficiency. As outlined earlier, for a resource efficient development, the security of supply needs to be ensured and environmental impacts need to be decreased.

While the assessment of environmental pollution associated with the resource extraction process and use is common practice, the extent to which current LCIA methods are capable of addressing the impacts of natural resource extraction on the quantity or quality of the resource (and for biotic resources on the overall environment) is widely debated and no common agreement or methodology exists. This is mainly due to:

- The different perceptions of the relevant concepts of depletion and the controversial discussion of abiotic resource depletion as an environmental problem (see, e.g., Finnveden 2005; Guinée and Heijungs 1995; Steen 2006; Udo de Haes et al. 2002; UNEP 2010a; Weidema et al. 2005)
- The complexity of the assessment of biotic resource extraction as well as missing data to determine CFs for species

In the context of a comprehensive evaluation of resource use for efficient and sustainable production, not all described indicators are effective, and existing models are not sufficient. The shortcomings of existing methods and approaches for assessing abiotic and biotic resources use are outlined in the following sections.

4.1 Abiotic Resources

The assessment of abiotic resource depletion is shaped by a lack of consensus on methodologies and results of impact assessments (see, e.g., Berger and Finkbeiner 2011; Finnveden et al. 2009; Hauschild et al. 2013; Heijungs et al. 1997; Klinglmair et al. 2013; Lindeijer et al. 2002). For using LCIA methods for the assessment of resource efficiency, the link to the availability of resources should be clear and the significance of the indicators to describe resource security needs to be given.

Commonly, existing methods are classified according to the underlying concepts to model resource depletion. The ILCD Handbook distinguishes methods based on the inherent property of the material (category 1), the decreasing availability of stocks in nature (category 2), and future consequence of resource extraction (category 3) (see also Sect. 3.1).[16] In the following, on the basis of these categories, methodologies are evaluated with regard to their usefulness for addressing resource availability.

[16] In total, four different categories are outlined in the ILCD Handbook. However, as category 3 focuses on water, it is not included here (European Commission 2010c, 2011b).

4.1.1 Assessment Based on the Inherent Property of Materials

Category 1 methods use an inherent property of a resource, such as entropy or exergy, as the basis for the characterization (Bösch et al. 2007; Dewulf et al. 2007, 2008; European Commission 2010c; Finnveden and Östlund 1997; Valero et al. 2013). Exergy analysis is often perceived as a useful tool to asses natural resources as it takes into account variables such as composition ore grade and the state of technology (Valero and Valero 2009; Valero et al. 2009). However, this method does not account for availability loss as the exergy values of different abiotic resources will always stay the same. Furthermore, the impact pathway does not describe the depletion process, but the use of inherent properties has thus no relation to the reasons why mankind is worried about the use of resources (European Commission 2010b; Heijungs et al. 1997; Lindeijer et al. 2002; Steen 2006). Methods that use the inherent property of the material as the basis for the characterization have only a low relevance with regard to expressing resource depletion, as these properties do not have a direct link to the increasing scarcity of resources (European Commission 2010b; Hauschild et al. 2013; Vieira et al. 2011).

4.1.2 Assessment Based on Reserves and/or Annual Extraction Rates

The chemical and physical basis of abiotic resources is quantifiable and availability in the Earth's crust can generally be determined (Rankin 2011). In ADP method, the extraction rate of a resource is divided by the reserve squared (CML 2013). The assessment of stocks and extraction rates provides information about the geologic availability and the static lifetime of stocks (Guinée and Heijungs 1995; Guinée 1995; Guinée et al. 2002; Heijungs et al. 1997). However, the definition of the recoverable stock is a problem as, depending on the definition of this number, results will change dramatically. It is difficult to fix convincing boundaries for the determination of reserve numbers as the stock size very much depends on the required effort of extraction (Goedkoop and Spriensma 2001) (see additional discussion provided in Box 5.1). The EDIP 97 approach is based on economic reserves and extraction rates, but does not reflect the current importance of a resource, as the global annual production drops out of the equation during the weighting of the method (European Commission 2011b; Klinglmair et al. 2013).[17]

[17] The amount of the resource extracted is divided by the 2004 global production of the resource and weighted according to the quantity of the resources in economically exploitable reserves (European Commission 2011b).

4.1.3 Assessment Based on Future Consequences of Resource Extraction

Category 4 methods attempt to assess how the extraction of high concentration of resources today will affect future extraction of resources. Several methods and indicators have been proposed in this regard (Goedkoop and Spriensma 2001; Müller-Wenk 1999; Steen 1999, 2006; Weidema et al. 2005). These approaches aim at addressing the damage of resource use and are based on decreasing ore grades and additional energy requirements or costs associated with future resource extraction or with future willingness to pay to repair the damage. Steen (1999) defines the estimated costs to extract and produce resources, considering different technical scenarios and assuming a very long time perspective. However, he estimates the costs of such an operation with present day technology and the present day extraction with the energy requirements that would be needed in the future (Goedkoop and Spriensma 2001; Vieira et al. 2011). The EPS method has been criticized for the many assumptions made and the uncertainties associated with this methods are high (European Commission 2011b).

In the ReCiPe method, the damage to a resource is defined as the additional costs society has to pay as a result of an extraction (Goedkoop et al. 2008). The method uses a monetization of surplus energy demand for characterizing future efforts for resource extraction (Klinglmair et al. 2013). Similarly, the LIME method aims at quantifying costs of the overuse of a resource (European Commission 2010b; Itsubo and Inaba 2012). The Eco-indicator 99 and IMPACT 2002+ assume that extraction at present will require more energy-intensive extraction in the future (Goedkoop and Spriensma 2001; Jolliet et al. 2003). Hereby, the energy required for resource extraction is assumed to be inversely proportional to the ore grade. A new model developed by Vieira et al. (2012) takes a similar approach, modeling the decrease in ore grade due to an increase in metal extraction based on cumulative ore grade-tonnage relationships and displaying increasing costs. However, modeling future consequences involves a high degree of uncertainty, and estimates of future energy consumption or costs are not reliable (Lindeijer et al. 2002). Using a cost-oriented problem definition of resource depletion has a direct relation to the present social context but can hardly be calculated in the long term (Steen 2006).

The ADP or EDIP 97 methods are midpoint approaches, while, for example, Eco-indicator 99 or IMPACT2002+ include characterization factors at an endpoint level. Endpoint approaches, such as surplus energy or surplus costs, are damage oriented and depend on the societal valuation of resources rather than the physical availability. As endpoint approaches address the end of the cause-effect chain, more uncertainties are involved (Udo de Haes and Lindeijer 2002). Methods for assessing resource availability at the endpoint level are not well defined and considered too immature to be used for modeling geophysical dimension of abiotic resource availability (European Commission 2011b; Hauschild et al. 2013). Quantifying the actual damage of abiotic resource scarcity is challenging as effects occur well into the future, and future circumstances need to be considered to determine actual damages. Midpoint methods are problem oriented and refer to the decreasing availability of resources (e.g., based on stocks and/or extraction) and not the potential effects on society. Characterization factors on a midpoint level have higher acceptance in decision making and can be regarded as relevant (Udo de Haes and Lindeijer 2002).

Despite existing shortcomings, the ADP method is recommended by the ILCD Handbook,[18] in the Product Environmental Footprint (PEF), and also in the context of the Seventh Framework Project of the European Commission as current best practice for assessing resource depletion (on a midpoint level) (Dong et al. 2013; European Commission 2011b, 2013a; Guinée et al. 2002; Hauschild et al. 2013). In Box 5.1 the ADP methodology and existing shortcomings are described in more detail.

Box 5.1: Measuring Geologic Depletion of Abiotic Resources by Means of ADP

The ADP model defines the decrease of the resource itself as the key problem (van Oers et al. 2002). The characterization factors of this method are a function of the yearly extraction of a resource and the stock of the resource. This factor is derived for each element and is a relative measure with the depletion of "antimony" (Sb) as a reference for elements and MJ for fossil fuels (van Oers et al. 2002).

To evaluate the effect of extraction on the available resource stocks, the reserves are taken into account more than once in the ADP method, by putting the square of the reserve number in the denominator (Eq. 5.2). This is done to put emphasis on the share of the stock that is extracted (Guinée 1995).

$$\mathrm{ADP}_{\text{i, ultimate reserves}} = \frac{\text{extraction rate i}}{(\text{ultimate reserves i})^2} \times \frac{(\text{ultimate reserves antimony})^2}{\text{extraction rate antimony}} \tag{5.2}$$

In this context, Guinée (1995) proposes to use reserves that can ultimately be technically extracted as a reference figure. However, Guinée argues that data on this type of reserves are not exactly known. Thus, for providing a realistic picture of resource depletion, Guinée (1995) proposed to use the *ultimate*

(continued)

[18] International Reference Life Cycle Data System (ILCD). This consists primarily of the ILCD Handbook and the ILCD Data Network.

Box 5.1 (continued)
reserves[19] as a reference. The figure is often criticized as *ultimate reserves* cannot be extracted completely (European Commission 2011b). Thus, as an alternative, it was proposed to use *reserves*[20] or *reserve base*[21] as a reference (European Commission 2011b, 2013a). However, *reserves* or *reserve base* has a strong economic link and provide limited information with regard to the availability of geological stocks. *Economic reserves* and use (extraction) are codependent, as the search for new deposits depends on the probability of exploration and use of resources (Steen 2006). Those reserves are affected by many factors that can change in a very short time (e.g., available technologies, resource prices). The *economic reserves* and *reserve base* of most resources have increased over the past, even though the actual depletion problem (referring to the geologic availability of resources) must necessarily have increased (Guinée 1995). Thus, the assessment of *reserves* or *reserve base* is ephemeral (see, e.g., Kesler 2007) and not a good basis for the assessment of abiotic resource depletion.

As discussed in Box 5.1, the definition of geologic resource stocks as a basis for calculating the ADP is quite controversial. In addition to the reserves in the Earth's crust, minerals that accumulate in society (in-use stock and waste flows) need to be considered, too. Availability of such anthropogenic stocks can have a meaningful effect on the overall availability of certain materials and give an indication about the sustainability of their use. Urban mining (recycling) can be seen as an important measure for (future) resource supply and should be considered for the evaluation of abiotic resource availability (Klinglmair et al. 2013; Müller-Wenk 1999; Schneider et al. 2011).

While the impacts of the extraction are clearly linked to the environment, public discussion of resource depletion as an environmental problem is quite controversial (Finnveden 1996, 2005; Guinée 1995; Sala 2012; Schneider et al. 2013; Steen 1999; Weidema et al. 2005). Resource availability is often viewed as an economic or

[19] The quantity of resources that is ultimately available in the Earth's crust. Estimated by multiplying the average natural concentration of the resources in the Earth's crust by the mass or volume of the crust (Guinée 1995). The definition includes nonconventional and low-grade materials and common rocks.

[20] "Reserves" are stocks that are known and profitable to be exploited at current prices, state of technology, etc. (Tilton and Lagos 2007; USGS 2014).

[21] The term "reserve base" refers to the part of a resource that meets specific physical and chemical criteria, related to current mining and production practices (UNEP 2011; USGS 2014). The reserves base was used as an estimate of the size of those parts of resources that had reasonable potential for becoming economic within planning horizons. However, these estimates were based on expert opinion rather than on actual data. The USGS discontinued reporting of estimates of the reserve base in 2010.

societal problem, beyond an environmental perception. The question of what it is that needs to be protected with regard to abiotic resources and which impacts are relevant is not that straightforward (Hauschild et al. 2013). Some people believe that the current rate of dissipating abiotic resources is a significant impoverishment of nature, to the detriment of future generations (Lindeijer et al. 2002). Others argue that it is not clear to what extent biogeochemical cycles are affected and if potentially environmental changes occur due to resource depletion (and not extraction) (Sala 2012) and that abiotic resource availability has socioeconomic rather than environmental relevance. As abiotic resources have a functional value, available models already reflect an economic orientation of the evaluation of resource availability and are concerned with the decrease of resource availability for human use (Klinglmair et al. 2013) rather than the environmental impact of resource depletion itself. Thus, an extension of existing impact assessment practice toward a wider interpretation and evaluation of the problem is necessary to account for the availability of resources for human purposes and to increase the efficiency of resource use. This is taken up and discussed further in Sect. 5.1.

4.2 Biotic Resources

While impacts associated with the growth and extraction of biotic resources like land use, water consumption, or climate change are addressed in the respective impact categories, the depletion of biotic resources is not properly addressed so far (Finkbeiner et al. 2014). Current approaches for the assessment of biotic resources are insufficient as they do not depict depletion of biotic resources in the necessary extent. Due to the complexity of analyzing biodiversity, developing impact assessment methods is challenging. For evaluating the degree of depletion of biotic resources, the impact assessment method should reflect whether a certain species is already threatened or not. The extraction of threatened species can lead more likely to extinction than extraction of nonthreatened species. Furthermore, as extraction of one species can lead to the depletion of another species (Langlois et al. 2014), interdependencies between species have to be considered.

Most of the methods and indicators introduced in Sect. 3.2 have been developed for the fishery sector. In the following, these methods are analyzed regarding their applicability to address resource availability and their capability to adequately measure impacts of biotic resource extraction.

NPP-based indicators are used by several method developers to evaluate fishery (Hornborg et al. 2012; Libralato et al. 2008; Papatryphon et al. 2004a, b; Pelletier et al. 2006; Vázquez-Rowe et al. 2012b). Even though NPP-based indicators are adequate to measure the ecological efficiency and therefore allow a comparison of different aquatic as well as aquatic and terrestrial products, no indication is given about the effects on resource depletion. NPP can only account for the overall amount of used biomass but does not consider the species stock or the replenishment rate. The same is true for the method from Steen (1999), where fish and wood are only evaluated by summing up their overall amount in kilogram. These indicators simply allow a conclusion regarding resource use in general, but not if these used resources are actually depleted in the course of production.

The MSY, used for accounting for the renewability rates of species (Emanuelsson et al. 2014; Langlois et al. 2012, 2014), seems to be the current best available practice for determining potential biotic resource depletion. The MSY approach is based on the harvesting rate, the carrying capacity which determines the degree of replenishment, and the species stock and can thus also be seen as the implementation of the proposed method for determining biotic resource depletion by Lindeijer et al. (2002). As the MSY method has been used in legislation linked to fishing for many years, data for several species is available (Langlois et al. 2014; Ricard et al. 2012). Nevertheless, applying this indicator is still associated with several challenges. When determining MSY, interconnections and trophic interactions of species in multispecies systems are often neglected due to lack of data and knowledge. Thus impacts on non-harvested species,[22] which are proven to be significant, are not taken into account (Beddington and May 1980; Ghosh and Kar 2014; Legović and Geček 2010; Reynolds 2008). Furthermore, for most species, stock data is missing and is therefore only estimated within the background models. This leads to high uncertainties as these models can, for example, not adequately include biological structures or age-specific patterns. Additionally, misreports of fish catches, especially for discards, limit reliable stock estimation (Hilborn 2011; Reynolds 2008). Even though the MSY has its shortcomings, the field of fishery is the most advanced so far when it comes to determining biotic resource depletion. However, missing in the current literature is guidance for the interpretation of the results considering the challenges of the MSY method. The concept of maximum sustainable yield could also be applied to forests (see Bennett 1999), but has not been implemented in any case study so far. Approaches for evaluating forests and other biotic resources in the context of their potential depletion due to extraction are still missing.

Similar to abiotic resources, availability of biotic resources can also be constrained by additional aspects, which need to be considered for a comprehensive assessment of resource availability for increasing resource efficiency (Finkbeiner et al. 2014). Factors like natural disasters (e.g., forest fires or pest infestation) and logistic constraints (e.g., storage stability) may affect biotic resource availability (Finkbeiner et al. 2014; VDI 2013).

Next to the shortcomings of accounting for the depletion and availability of biotic resources, effects of extraction on the natural environment and its functions cannot be assessed adequately so far. Accounting for loss of biodiversity or impacts on LSFs in LCA is challenging as more research on interdependencies in and between ecosystems and species as well as sufficient inventory data is needed

[22] The term "non-harvested" species refers to species which can be depleted without being extracted themselves simply due to the extraction of species they rely on for a food source within their food web.

(Finkbeiner et al. 2014). So far, no sufficient and applicable methods have been proposed for evaluating effects of biotic resource extraction and depletion on biodiversity. Langlois and colleagues (Langlois et al. 2012, 2014) propose to use a NPP-based indicator to account for biodiversity loss. NPP-based indicators have been used before for measuring biodiversity in different frameworks as a correlation between decreasing NPP values and when loss of biodiversity exists (Catovsky et al. 2002; Costanza et al. 2007; Itsubo and Inaba 2012). However, NPP-based indicators are not comprehensive enough to adequately assess complex impacts on biodiversity, and LSFs as interdependencies of species and ecosystems cannot be assessed properly. Furthermore, approaches for assessing impacts on animals and plants due to biotic resource extraction and depletion and relating effects on LSFs (e.g., pollination ratio of bees) do not exist for the time being. Modeling these complex systems and all possible interconnections is a challenge (see Finkbeiner et al. 2014).

So far biotic resource depletion and availability have only been discussed for wild animals and plants but not for agri-, silvi-, and aquacultural products, as man-made products are considered as nondepletable resources. Even though this might be true for agricultural products like crops, depletion can occur when plants have long growth periods and low yields like shea trees. Furthermore, economic constrains can lead to decreased availability for biotic resources and can affect man-made products as well as wild biotic resources (Finkbeiner et al. 2014). Furthermore, impacts on biodiversity and LSFs due to extraction of man-made biotic resources can occur. For example, efficiency of crop pollination depends on the amount of natural habitat cover in the surrounding area (Vaissière et al. 2011).

The lack of consensus on the evaluation of natural resource depletion and the lack of commonly applied and available methods to address the effects of abiotic and biotic resource extraction on their availability and the natural environment create inconsistency and limit decision-making support (see also Finnveden et al. 2009; Hauschild et al. 2013; Klinglmair et al. 2013; Lindeijer et al. 2002). Further research is needed to enhance and broaden the assessment of abiotic and biotic resource use to deliver comprehensive measures for the assessment of resource efficiency and also to account for the problem in the context of sustainable development.

5 Research Needs and Methodological Developments

The assessment of abiotic and biotic resource use in LCA has still several shortcomings as highlighted in the previous chapter. So far, their availability in nature has been the focal point of the assessment of resources in LCA. The AoP "natural resources" relates to the use of resources with implications for their present, but mainly future availability (see, e.g., Lindeijer et al. 2002), and the availability of natural resource stocks is identified as the main concern. However, resource availability has a clear link to technological progress and human well-being today. The general concern with regard to resources is a shortage in supply relative to the interests and needs of humans and not only the natural availability as such (see also DeYoung Jr. et al. 1987; Mancini et al. 2013). Acknowledging the definition of natural resources according to their usefulness for human purposes, the United Nations Environment Programme specified the "resource provision capability for human welfare" as the correct description of the AoP addressed in LCA (see Lindeijer et al. 2002; UNEP 2010a; Udo de Haes et al. 2002). This definition already goes beyond a direct link to environmental consequences of resource use. In fact, "resource provision capability for human welfare" is already aptly describing the general concern over the access to resources and the availability for human use. In the context of achieving sustainable development, the amount of material available to society (now or in the future) is influenced by physical realities, politics, economic circumstances, and social or environmental concerns (see also Schneider 2014). Consequently, as already pointed out in the previous section, for the assessment of resource efficiency and for achieving sustainability goals, a more comprehensive approach for the evaluation of resource availability is needed. Any kind of scarcity is relevant and can have consequences on the efficiency of resource use and on the implementation of products. Shortages in resource supply can negatively affect the ability to maintain and expand the man-made environment and impede sustainable development. As shown in Fig. 5.4, next to physical scarcity, which is a long-term concern and caused by the depletion of resources, short-term concerns, caused by constraints in the supply chain (e.g., socioeconomic constraints), can affect the accessibility and thus the availability of resource as well (see, e.g., European Commission 2010a; Graedel et al. 2012; Schneider et al. 2013; Schneider 2014). For limiting the risks linked to the security of resource supply, a more comprehensive assessment of resource availability would be beneficial. Resource efficiency, as a measure to achieve sustainable development, should encompass all relevant dimensions of availability.

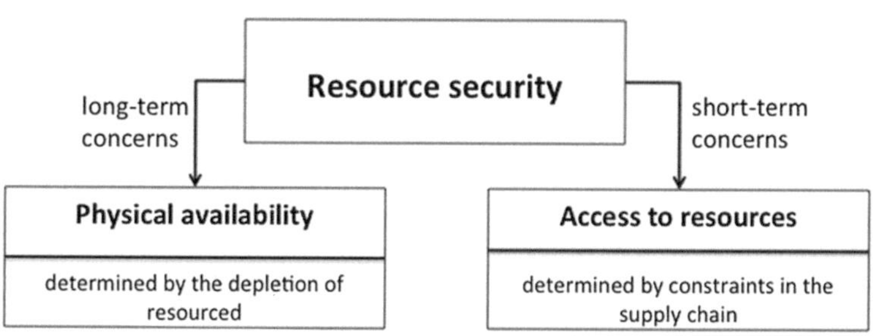

Fig. 5.4 Dimensions of resource security

5.1 Abiotic Resources

Unlike biotic resources that have a clear link to the environment, abiotic resource availability has mainly a functional value, referring to the use for human purposes. It is debatable whether abiotic resource availability is an environmental or economic issue and whether an analysis of the geologic availability, as currently conducted in LCA, is sufficient to capture the whole implications of current resource use on their future availability. Minerals have no intrinsic values in themselves while locked up in ore bodies buried in the earth and are considered to be "outside the biosphere," as a reduction of natural stocks has no direct influence on ecosystems (Petrie 2007; Udo de Haes et al. 2002) (see also Sect. 3.2). The geologic abundance of abiotic resources is only one of several factors affecting resource availability. The consideration of additional aspects of resource availability, like the availability of stocks in the anthroposphere or the scarcity of resources caused by economic factors, is currently not considered sufficiently in most product assessment.

Stewart and Weidema (2005a) point out that not the extraction of minerals from the environment and natural stocks should be of concern, but rather the dissipative[23] use and disposal of materials. Abiotic resource stocks are not "used up" but rather transferred from the Earth's crust into the anthroposphere. Regarding the functionality of metallic minerals (see also Sect. 1.1), the loss of "potential functions of resources in the future due to the use in products and product systems at present" (van Oers et al. 2002) needs to be assessed, beyond their natural availability. From a functional point of views, it is irrelevant whether the abiotic resource is available in the environment or economy if the function at present attached in economic goods is still available for future applications. Materials like copper or aluminum can be recycled and thus still fulfill a function within the economy. However, if the material is dissipated and is not present in sufficient quality anymore, its potential functions will be lost for mankind. Thus, van Oers et al. (2002) suggested that there is a need to develop a method that acknowledges that a resource is only depleted when it is dissipated rather than extracted (van Oers et al. 2002). The differentiation between dissipative and non-dissipative use is needed for assessing potential resource depletion (see, i.a., Goedkoop and Spriensma 2001). An extension of existing models for the evaluation of mineral resource depletion by acknowledging metal stocks in the anthroposphere and including these stocks into the assessment of physical availability of material stocks is needed. A first approach to capture anthropogenic resource stocks in addition to geologic stocks was published by Schneider et al. (2011) (the anthropogenic stock extended abiotic depletion potential, AADP); however, several shortcomings remain.

The availability of abiotic resources for products and product systems is not only affected by the general quantity or physical availability of resources. Mineral

[23] Dissipation refers to the state where elements become so dilute or change their chemical form so they can no longer fulfill the required function (UNEP 2010a).

availability for products is also related to the accessibility of resources at the time of production and can also be constrained by, for example, economic and social aspects. Thus, approaches to address resource availability should not be limited the assessment of physical depletion.

The assessment of economic aspects related to resource availability has gained some attention over the past years. Various papers and working groups are dealing with the determination of risks associated with resource supply (Angerer et al. 2009a, b; Defra 2012; Erdmann and Behrendt 2010; European Commission 2010a; Graedel et al. 2012; Nassar et al. 2012; National Research Council 2008; Nuss et al. 2014; Rosenau-Tornow et al. 2009; Schneider et al. 2013; VDI 2013). Several indicators were introduced assessing the supply risk considering economic aspects and vulnerability of countries or companies to supply disruptions. Supply risks addressed refer to interruptions in the supply chain due to market imbalances, political risks, etc. However, the definition and determination of supply risk is complex, and existing methodologies are often immature and lack transparency. Recent publications aim to integrate economic aspects into the evaluation of resource availability on a product level. Models based on existing LCA methodology and terminology and focusing specifically on indicators for evaluation on a product level enable an easy implementation in current resource assessment practice and can eventually be used to improve the evaluation of resource efficiency (see Schneider et al. 2013; Schneider 2014; VDI 2013).

5.2 *Biotic Resources*

As the extraction of biotic resources can lead to depletion of resources (covered by the AoP "natural resources") as well as loss of biodiversity (addressed in the AoP "natural environment"), impact assessment methods have to be developed and implemented for both AoPs.

As shown in Sect. 3.2 for the biotic resource "fish," several approaches exist for the assessment of potential depletion. The MSY approach and the approaches suggested by Lindeijer et al. (2002) and Guinée et al. (1993) seem promising to adequately evaluate the depletion for the AoP natural resources. However, it has to be clarified to what extent the models are applicable to other biotic resources like forests and terrestrial animals, and if necessary background data is available. Additionally, factors like renewability rate and thread level need to be included within impact assessment methods as well.

Furthermore, aspects like political stability as well as additional factors, for example, abiotic factors influencing the availability of biotic resources, should be analyzed in more detail and transferred into characterization models for a more comprehensive assessment of biotic resources.

Evaluating impacts for the AoP natural environment due to biotic resource extraction is challenging as very few interdependencies within ecosystem elements and between ecosystems are established. As impacts are highly dependent on local and regional factors, specific extraction sites have to be known to adequately address impacts on biodiversity and LSFs. Thus, one aim could be to develop a reliable screening indicator which is able to detect potential impacts on the environment due to biotic resource extraction. This indicator could be based on plant and animal population data of the surrounding ecosystem, assuming the potential loss of biodiversity equals the biodiversity of the ecosystem surrounding the extraction site. Thus, the indicator might be more adequate for systems resembling ecosphere conditions, for example, shea trees, than for systems which are highly influenced by human activities, for example, large agricultural areas. Furthermore, existing land use methods for forestry could be used as a starting point to address impacts on the environment due to extraction and depletion of forests.

The assessment of availability of biotic materials from man-controlled cultures has not been considered so far, even though the importance of man-made biotic resources outside the food and feed sector is rapidly increasing, especially as energy sources but also as a substitute material in industrial production systems. Moreover, resources from nature as well as from man-controlled processes will mainly provide a similar direct functional value for humans. Services like feed, food, shelter, etc., do not necessarily depend on the origin of the biotic resource itself, for example, wild swine meat will have the same nutritional value as swine meat from agricultural processes. The indirect impacts on the environment deriving from biotic resource extractions on the other hand are significantly different and have to be considered in the development of an adequate impact assessment method, respectively.

Even though for most agri-, silvi-, and aquacultural products depletion is not an actual issue (exceptions might be plants with long growth periods and low yields like shea plants), availability aspects, for example, due to economic factors, could be of relevance as well. As impacts of biotic resource extraction are different from the impacts arising from extraction of man-made culture (Lindeijer et al. 2002), different impact assessments are needed to reflect the diverse characteristics of both biotic resource types (even though some aspects might be similar).

As biotic resources have various uses – in the food sector, as animal feed, as energy source, or for industrial production processes – availability aspects should be evaluated in the context of sustainable development to avoid undesired trade-offs. Increasing availability of biotic resources due to converting natural areas into agricultural ones can lead to more prosperity within the region (and thus influencing the social dimension of sustainability) but can also affect life-support functions of the local ecosystem resulting in damage of the ecosystem (influencing the environmental dimension of sustainability). Therefore, trade-offs need to be identified and properly assessed to determine the point with the lowest environmental impacts and highest prosperity.

For man-made biotic resources, impacts caused by extraction should be evaluated as well. As the main reason for biotic resource production is the extraction itself and impacts of harvesting techniques are considered in other impact categories, for example, land use or eutrophication, impacts on biodiversity and LSFs due to actual resource extraction might be only of interest when the analyzed system is less influenced by human activities. Even though, there might not be effects on the surrounding ecosystem due to extraction of maize and excessive removal of, for example, long-grown trees, even out of managed systems can influence life-support functions (e.g., Cardinale et al. 2012; Lenzen et al. 2012). Furthermore, when considering man-made biotic resource availability, aspects are also of relevance for the AoP natural environment, as studies show that diversity within crop species can influence the ecosystem quality (e.g., Hajjar et al. 2008).

For a comprehensive and consistent assessment of resource availability in the context of innovative and complex products, methods need to be developed that support a comparison of biotic and abiotic resources. However, as biotic resources are quite inhomogeneous in their characteristics, finding an indicator set to properly cover all related aspects (e.g., food web features) seems challenging, especially when considering additional aspects of resource availability.

To increase resource efficiency and to support sustainable development, the assessment of resource use needs to include additional aspects, beyond current practice. As resource security is necessary for sustainable production and consumption, the evaluation of resource use should be enhanced considering resource functionality and toward the assessment of all dimensions of sustainability.

6 Toward Life Cycle Sustainability Assessment

The current evaluation of resource efficiency lacks significance in the context of sustainable development. The use of LCIA methods and indicators as a basis for the assessment of resource efficiency, beyond mass-based indicators, enables improved decision making and increases the significance of results. However, for decision-making support on a product level and in the context of sustainable development, a more holistic perspective needs to be applied.

While the environmental dimension is widely discussed and several methods exist, models for the social and economic dimension of resource availability are still missing. The assessment of resource availability for products solely in the context of LCA needs to be challenged. Consideration of resource availability as an environmental problem does not cover the whole dimension this topic induces. As outlined in the previous section, resource depletion is only one of several factors that threaten availability. Criteria affecting economic systems as well as potential social and environmental constraints to resource provision need to be assessed, complementary to existing environmental LCA models, to sustain industrial production and to increase resilience toward supply disruption (see also Graedel and Erdmann 2012; UNEP 2010a) (see Fig. 5.5). A comprehensive assessment of sustainability, enabling product development and implementation to be in line with considerations of inter- and intragenerational equity, cannot be achieved so far. In the LCA community, much attention has recently been devoted to the development of life cycle sustainability assessment (LCSA) (see, e.g., Finkbeiner

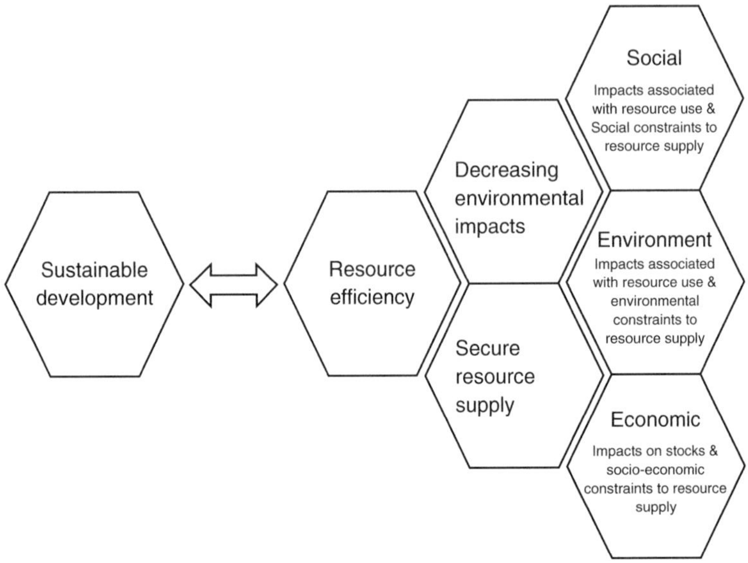

Fig. 5.5 Implementation of resource efficiency in the context of sustainable development

2011; Jørgensen et al. 2013; Klöpffer 2008; Valdivia et al. 2011), which could also be applied to accomplish a comprehensive assessment of resource use.

So far, resource availability has only been addressed under the frame of environmental assessment. Considering the fact that scarcity of resources can affect human productivity, a holistic and realistic assessment of resources use has to go beyond the analysis of mere availability of resources in the natural environment or the environmental impacts of their extraction and consider the complexity of sustainable development. Sound material choices and informed product development cannot be achieved without considering both the short-term and long-term availability of mineral resources. A comprehensive evaluation of resource efficiency under consideration of the triple bottom line of sustainability is needed. An analysis toward LCSA, including also social and economic information, is essential to find more sustainable means of resource use.

Even though sustainability is nowadays an accepted concept, the challenge of including all three dimensions into the assessment of resource availability remains. To bring the comprehensive assessment of resource availability in the context of increasing the efficiency of resource use into practice, operational approaches and tools are required. Results need to be presented in a comprehensive yet simple way and need to be in line with the goals of resource efficiency. While existing methods and indicators in LCA deliver a first input for the assessment of resource efficiency, additional indicators are necessary to cover the different dimensions of resource security and to offer guidance for sustainable decision making.

References

Achzet B, Reller A, Zepf V, Rennie C, Ashfield M, Simmons J (2011) Materials criticality to the energy industry. An introduction. University of Augsburg, BP, ON Communication, Augsburg

Angerer G, Erdmann L, Marscheider-Weidemann F, Scharp M, Lüllmann A, Handke V, Marwerde M (2009a) Rohstoffe für Zukunftstechnologien. ISI-Schriftenreihe "Innovationspotenziale". Fraunhofer IRB Verlag, Stuttgart

Angerer G, Marscheider-Weidemann F, Wendl M, Witschel M (2009b) Lithium für Zukunftstechnologien. Fraunhofer ISI, Karlsruhe

Aubin J, Papatryphon E, Van der Werf HMG, Petit J, Morvan YM (2006) Characterisation of the environmental impact of a turbot (Scophthalmus maximus) re-circulating production system using life cycle assessment. Aquaculture 261(4):1259–1268. doi:10.1016/j.aquaculture.2006.09.008

Auty RM (1993) Sustaining development in mineral economics RTZ. Routledge, London

Avadí A, Fréon P (2013) Life cycle assessment of fisheries: a review for fisheries scientists and managers. Fish Res 143:21–38. doi:10.1016/j.fishres.2013.01.006

Azapagic A (2004) Developing a framework for sustainable development indicators for the mining and minerals industry. J Clean Prod 12(6):639–662

Bardi U (2011) Revisiting the limits to growth. Springer, New York

Bardi U (2013) Der geplünderte Planet – Die Zukunft der Menschen im Zeitalter schwindender Ressourcen. Oekom, München

Beddington JR, May RM (1980) Maximum sustainable yields in systems subject to harvesting at more than one trophic level. Math Biosci 51(3–4):261–281. doi:10.1016/0025-5564(80)90103-0

Behrens A, Giljum S, Kovanda J, Niza S (2007) The material basis of the global economy – worldwide patterns of natural resource extraction and their implications for sustainable resource use policies. Ecol Econ 64(2):444–453

Bennett E (1999) Hunting for sustainability in tropical forests. Columbia University Press, New York

Berger M, Finkbeiner M (2011) Correlation analysis of life cycle impact assessment indicators measuring resource use. Int J Life Cycle Assess 16(1):74–81

Berger M, Warsen J, Krinke S, Bach V, Finkbeiner M (2012) Water footprint of European cars: potential impacts of water consumption along automobile life cycles. Environ Sci Technol 46(7):4091–4099

Berger M, Pfister S, Motoshita M (2016) Water footprinting in life cycle assessment – how to count the drops and assess the impacts? Chapter 3 "Special Types of Life Cycle Assessment" (Finkbeiner M ed). In: LCA compendium – the complete world of life cycle assessment (Klöpffer W, Curran MA, series eds). Springer, Dordrecht, pp 73–114

BIO Intelligence Service (2012) Assessment of resource efficiency indicators and targets. Final report prepared for the European Commission, DG Environment. Institute for Social Ecology (SEC) and Sustainable Europe Research Institute (SERI)

Bösch ME, Hellweg S, Huijbregts M, Frischknecht R (2007) Applying cumulative exergy demand (CExD) indicators to the ecoinvent database. Int J Life Cycle Assess 12(3):181–190

Brentrup F, Küsters J, Lammel J, Kuhlmann H (2002) Impact assessment of abiotic resource consumption. Int J LCA 7(5):301–307

Bringezu S, Schütz H (2010) Material use indicators for measuring resource productivity and environmental impacts. Paper within the framework of the "Material Efficiency and Resource Conservation" (MaRess) Project. Wuppertal Institute for Climate, Environment and Energy, Wuppertal

BUWAL (1998) Bewertung in Ökobilanzen mit der Methode der ökologischen Knappheit – Ökofaktoren 1997. Schriftenreihe Umwelt, Nr. 297 – Ökobilanzen. Federal Office for Environment, Forest and Landscape, Bern

Cardinale BJ, Duffy JE, Gonzalez A, Hooper DU, Perrings C, Venail P, Narwani A, Mace GM, Tilman D, Wardle DA, Kinzig AP, Daily GC, Loreau M, Grace JB, Larigauderie A, Srivastava DS, Naeem S (2012) Biodiversity loss and its impact on humanity. Nature 486(7401):59–67. doi:10.1038/nature11148

Catovsky S, Bradford MA, Hector A (2002) Biodiversity and ecosystem productivity: implications for carbon storage. Oikos 97(3):443–448

CML (2013) CML – IA characterisation factors. Universiteit Leiden, Institute of Environmental Sciences (CML), Leiden

Costanza R, Fisher B, Mulder K, Liu S, Christopher T (2007) Biodiversity and ecosystem services: a multi-scale empirical study of the relationship between species richness and net primary production. Ecol Econ 61(2–3):478–491. doi:10.1016/j.ecolecon.2006.03.021

Curran M, de Baan L, de Schryver AM, van Zelm R, Hellweg S, Koellner T, Sonnemann G, Huijbregts MAJ (2011) Toward meaningful end points of biodiversity in life cycle assessment. Environ Sci Technol 45(1):70–79

de Baan L, Alkemade R, Koellner T (2012) Land use impacts on biodiversity in LCA: a global approach. Int J Life Cycle Assess 18(6):1216–1230. doi:10.1007/s11367-012-0412-0

de Haes HA U, Finnveden G, Goedkoop M, Hauschild M, Hertwich EG, Hofstetter P, Jolliet O, Klöpffer W, Krewitt W, Lindeijer EW, Mueller-Wenk R, Olson SI, Pennington DW, Potting J, Steen B (2002) Life cycle impact assessment: striving towards best practice. Published by the Society of Environmental Toxicology and Chemistry (SETAC), Pensacola

de Udo Haes H, Lindeijer E (2002) The conceptual structure of life cycle impact assessment. In: de Udo Haes HA, Finnveden G, Goedkoop M et al (eds) Life-cycle impact assessment: striving towards best practice. Published by the Society of Environmental Toxicology and Chemistry (SETAC), Pensacola, pp 209–226

Defra (2012) A review of national resource strategies and research. Department for Environment, Food and Rural Affairs, London

Demurtas A, Sousanoglou A, Morton G, Humphris-Bach A, Cole A (2014) Resource Efficiency Scoreboard 2014 Highlights, European Union, Belgium

Dewulf J, Bösch ME, de Meester B, van der Vorst G, van Langenhove H, Hellweg S, Huijbregts MAJ (2007) Cumulative exergy extraction from the natural environment (CEENE): a comprehensive life cycle impact assessment method for resource accounting. Environ Sci Technol 41(24):8477–8483

Dewulf J, Van Langenhove H, Muys B, Bruers S, Bakshi BR, Grubb GF, Paulus DM, Sciubba E (2008) Exergy: its potential and limitations in environmental science and technology. Environ Sci Technol 42(7):2221–2232

DeYoung JH Jr, Barton PB, Boulègue J, Dahmen P, Gocht WRA, Kürsten MOC, Phillips WGB, Price RA, Sheldon RP, Skinner BJ, Vogt JH, Alten EGL, Weisser JD, Williams N (1987) Assessment of non-energy mineral resources. In: McLaren DJ, Skinner BJ (eds) Resources and world development. Dahlem Workshop Reports. Wiley, Dahlem, pp 508–523

Doka G, Hillier W, Kaila S, Köllner T, Kreißig J, Muys B, Quijano JG, Salpakivi-Salomaa P, Schweinle J, Swan G, Wessman H (2006) The assessment of environmental impacts caused by land use in the life cycle assessment of forestry and forest products – guidelines, hints and recommendations – final report of Working Group 2 "Land use" of COST Action E9. www.doka.ch/COSTE9LandUseDoka.pdf

Dong Y, Laurent A, Hauschild MZ (2013) Recommended assessment framework, characterisation models and factors for environmental impacts and resource use. Report prepared within the Seventh Framework Project of the European Union – development and application of a standardized methodology for the prospective sustainability assessment of technologies

Dreyer LC, Hauschild MZ, Shierbeck J (2006) A framework for social life cycle impact assessment. Int J Life Cycle Assess 11(2):88–97

Efole Ewoukem T, Aubin J, Mikolasek O, Corson MS, Tomedi Eyango M, Tchoumboue J, van der Werf HMG, Ombredane D (2012) Environmental impacts of farms integrating aquaculture and agriculture in Cameroon. J Clean Prod 28:208–214. doi:10.1016/j.jclepro.2011.11.039

Emanuelsson A, Ziegler F, Pihl L, Sköld M, Sonesson U (2014) Accounting for overfishing in life cycle assessment: new impact categories for biotic resource use. Int J Life Cycle Assess 19 (5):1156–1168. doi:10.1007/s11367-013-0684-z
Erdmann L, Behrendt S (2010) Kritische Rohstoffe für Deutschland. Institut für Zukunftsstudien und Technologiebewertung (IZT), Berlin
European Commission (2005) Thematic strategy on the sustainable use or natural resources. COM (2005) 670. Brussels
European Commission (2010a) Critical raw materials for the EU. Report of the ad-hoc working group on defining critical raw materials. European Commission, Brussels
European Commission (2010b) International reference life cycle data system (ILCD) handbook – general guide for life cycle assessment – detailed guidance. EUR 24708 EN. Publications Office of the European Union, Luxembourg
European Commission (2010c) International Reference Life Cycle Data System (ILCD) Handbook – Framework and Requirements for Life Cycle Impact Assessment Models and Indicators. EUR 24709 EN. Luxembourg
European Commission (2011a) Analysis associated with the Roadmap to a Resource Efficient Europe. SEC(2011) 1067 final, Brussels
European Commission (2011b) International Reference Life Cycle Data System (ILCD) handbook – recommendations for life cycle impact assessment in the European context. Publications Office of the European Union, Luxemburg
European Commission (2011c) Roadmap to a resource efficient Europe. COM (2011) 571 final. Brussels
European Commission (2012) Options for Resource Efficiency Indicators. European Commission, Directorate-general Environment, Brussels
European Commission (2013a) Commission Recommendation of 9 April 2013 on the use of common methods to measure and communicate the life cycle environmental performance of products and organisations
European Commission (2013b) Science for environment policy in-depth report: resource efficiency indicators. Science Communication Unit, University of the West of England, Bristol
FAO (2006) The state of world fisheries and aquaculture – opportunities and challenges 2006. Food and Agriculture Organization of the United Nations, Rome
FAO (2012) Capture fisheries. Food and Agriculture Organization of the United Nations, Rome
FAO (2014a) State of the World's forests – enhancing the socioeconomic benefits from forests. The Food and Agriculture Organization of the United Nations (FAO), Rome
FAO (2014b) The state of world fisheries and aquaculture – opportunities and challenges 2014. Food and Agriculture Organization of the United Nations, Rome
Finkbeiner M (ed) (2011) Toward life cycle sustainability management. Springer Science + Business Media, Berlin
Finkbeiner M, Ackermann R, Bach V, Berger M, Brankatschk G, Chang Y-J, Grinberg M, Lehmann A, Martínez-Blanco J, Minkov N, Neugebauer S, Scheumann R, Schneider L, Wolf K (2014) Challenges in life cycle assessment: An overview of current gaps and research needs". Chapter 7 "Background and Future Prospects in Life Cycle Assessment" (Klöpffer W ed). In: LCA Compendium – The Complete World of Life Cycle Assessment (Klöpffer W, Curran MA, series eds). Springer, Dordrecht, pp 207–258
Finnveden G (1996) Resources and related impact categories. In: de Udo Haes HA (ed) Towards a methodology for life-cycle impact assessment. Society of Environmental Toxicology and Chemistry (SETAC), Brussels, pp 39–48
Finnveden G (2005) The resource debate needs to continue. Int J Life Cycle Assess 10(5):372
Finnveden G, Östlund P (1997) Exergies of natural resources in life cycle assessment and other applications. Energy 22(9):923–931
Finnveden G, Hauschild MZ, Ekvall T, Guinée J, Heijungs R, Hellweg S, Koehler A, Pennington D, Suh S (2009) Recent developments in life cycle assessment. J Environ Manag 91(1):1–21

Frischknecht R, Steiner R, Jungbluth N (2009) The ecological scarcity method: Eco-factors 2006 – a method for impact assessment in LCA. Environmental studies no. 0906. Federal Office for the Environment (FOEN), Bern

Ghosh B, Kar TK (2014) Sustainable use of prey species in a prey–predator system: jointly determined ecological thresholds and economic trade-offs. Ecol Model 272:49–58. doi:10.1016/j.ecolmodel.2013.09.013

Giljum S, Polzin C (2009) Resource efficiency for sustainable growth: global trends and European policy scenarios. Background paper. Sustainable Europe Research Institute (SERI), Vienna

Giljum S, Behrens A, Hinterberger F, Lutz C, Meyer B (2008) Modelling scenarios towards a sustainable use of natural resources in Europe. Environ Sci Pol 11(3):204–216

Giurco D, Cooper C (2012) Mining and sustainability: asking the right questions. Miner Eng 29:3–12

Goedkoop M, Spriensma R (2001) The eco-indicator 99 – a damage-oriented method for life cycle impact assessment, vol 2. PRé Consultants B.V, Amersfoort

Goedkoop M, Heijungs R, Huijbregts M, Schryver AD, Struijs J, Zelm Rv (2008) ReCiPe 2008: a life cycle impact assessment method which comprises harmonised category indicators at the midpoint and the endpoint level. Pré Consultants, CML, RUN, RIVM, Amersfoort

Gontier M, Balfors B, Mörtberg U (2006) Biodiversity in environmental assessment – current practice and tools for prediction. Environ Impact Assess Rev 26(3):268–286

Gordon RB, Bertram M, Graedel TE (2006) Metal stocks and sustainability. Proc Natl Acad Sci U S A 103(5):1209–1214

Graedel TE, Erdmann L (2012) Will material scarcity impede routine industrial use? MRS Bull 37(4):325–331

Graedel TE, Barr R, Chandler C, Chase T, Choi J, Christofferson L, Friedlander E, Henly C, Jun C, Nassar NT, Schechner D, Warrne S, Yang M-y, Zhu C (2012) Methodology of metal criticality determination. Environ Sci Technol 46(2):1063–1070

Guinée JB (1995) Development of a methodology for the environmental life-cycle assessment of products. Leiden University, Leiden

Guinée J, Heijungs R (1995) A proposal for the definition of resource equivalency factors for use in product LCA. Environ Toxicol Chem 14(5):917–925

Guinée JB, Heijungs R, Haes HAU, Huppes G (1993) Quantitative life cycle assessment of products – 2. Classification, valuation and improvement analysis. J Clean Prod 1(2):81–91

Guinée JB, Bruijn H, van Duin R, Gorree M, Heijungs R, Huijbregts M, Huppes G, Kleihn R, de Koning A, van Oers L, Sleeswijk A, Suh S, Udo de Haes HA (eds) (2002) Handbook on life cycle assessment – an operational guide to the ISO standards. Kluwer Academic Publishers, Dordrecht

Hagelüken C, Meskers CEM (2010) Complex life cycles of precious and special metals. In: Graedel TE, Voet EVD (eds) Linkages of sustainability. MIT Press, Cambridge, pp 163–197

Hajjar R, Jarvis DI, Gemmill-Herren B (2008) The utility of crop genetic diversity in maintaining ecosystem services. Agric Ecosyst Environ 123(4):261–270. doi:10.1016/j.agee.2007.08.003

Hansen SB, Olsen SI, Ujang Z (2013) Carbon balance impacts of land use changes related to the life cycle of Malaysian palm oil-derived biodiesel. Int J Life Cycle Assess 19(3):558–566. doi:10.1007/s11367-013-0672-3

Hauschild M, Wenzel H (1998) Environmental assessment of products, vol 2, Scientific backgrounds. Chapman & Hall, New York

Hauschild M, Goedkoop M, Guinée J, Heijungs R, Huijbregts M, Jolliet O, Margni M, de Schryver A, Humbert S, Laurent A, Sala S, Pant R (2013) Identifying best existing practice for characterization modelling in life cycle impact assessment. Int J Life Cycle Assess 18 (3):683–697

Heijungs R, Guinée J, Huppes G (1997) Impact categories for natural resources and land use. CML Report 138. Leiden University, Centre of Environmental Science (CML), Leiden

Helin T, Holma A, Soimakallio S (2014) Is land use impact assessment in LCA applicable for forest biomass value chains? Findings from comparison of use of Scandinavian wood, agro-biomass and peat for energy. Int J Life Cycle Assess 19(4):770–785. doi:10.1007/s11367-014-0706-5

Henzen C (2008) The impact of land use on biodiversity in the framework of life cycle assessment. University of Basel, Basel

Hilborn R (2011) Future directions in ecosystem based fisheries management: a personal perspective. Fish Res 108(2–3):235–239. doi:10.1016/j.fishres.2010.12.030

Hornborg S, Nilsson P, Valentinsson D, Ziegler F (2012) Integrated environmental assessment of fisheries management: Swedish Nephrops trawl fisheries evaluated using a life cycle approach. Mar Policy 36(6):1193–1201. doi:10.1016/j.marpol.2012.02.017

Humbert S, De Schryver A, Bengoa X, Margni M, Joliet O (2012) IMPACT 2002+: user guide. Version Q2.21. Quantis, CIRAIG, Center for Risk Science and Communication, Ann Arbor

International Union for Conservation of Nature and Natural Resources (2014) The IUCN (International Union for Conservation of Nature) red list of threatened species. http://www.iucnredlist.org/

ISO 14040 (2006) Environmental management – life cycle assessment – principles and framework. European Committee for Standardisation, Brussels

ISO 14044 (2006) Environmental management – life cycle assessment – requirements and guidelines. European Committee for Standardisation, Brussels

ISO 14045 (2012) Environmental management – eco-efficiency assessment of product systems – principles, requirements and guidelines. European Committee for Standardisation, Brussels

Itsubo N, Inaba A (2012) LIME 2 – life-cycle impact assessment method based on endpoint modeling – summary. Life cycle assessment society of Japan

Jerbi MA, Aubin J, Garnaoui K, Achour L, Kacem A (2012) Life cycle assessment (LCA) of two rearing techniques of sea bass (*Dicentrarchus labrax*). Aquac Eng 46:1–9. doi:10.1016/j.aquaeng.2011.10.001

Jolliet O, Margni M, Charles R, Humbert S, Payet J, REbitzer G, Rosenbaum R (2003) IMPACT 2002+: a new life cycle impact assessment methodology. Int J Life Cycle Assess 8(6):324–330

Jolliet O, Müller-Wenk R, Bare J, Brent A, Goedkoop M, Heijungs R, Itsubo N, Pena C, Pennington D, Potting J, Rebitzer G, Stewart M, Haes HU, Weidema B (2004) The LCIA midpoint-damage framework of the UNEP/SETAC life cycle initiative. Int J Life Cycle Assess 9(6):394–404

Jørgensen A, Herrmann IT, Bjorn A (2013) Analysis of the link between a definition of sustainability and the life cycle methodologies. Int J Life Cycle Assess 18(8):1440–1449

Kesler SE (1994) Mineral resources, economics and the environment. Macmillan College Publishing Company, New York

Kesler SE (2007) Mineral supply and demand into the 21st century. U. S Geological Survey, Reston

Kleijn R (2012) Materials and energy: a story of linkages. Leiden University, Leiden

Klinglmair M, Sala S, Brandao M (2013) Assessing resource depletion in LCA: a review of methods and methodological issues. Int J Life Cycle Assess 19(3):580–592

Klöpffer W (2008) Life cycle sustainability assessment of products. Int J Life Cycle Assess 13:89–95

Klöpffer W, Renner I (1994) Methodik der Wirkungsbilanz im Rahmen von Produkt-Ökobilanzen unter Berücksichtigung nicht oder nur schwer quantifizierbarer Umwelt-Kategorien. Imweltbundesamt, Berlin

Koellner T, Baan L, Beck T, Brandão M, Civit B, Margni M, Canals LM, Saad R, Souza DM, Müller-Wenk R (2013) UNEP-SETAC guideline on global land use impact assessment on biodiversity and ecosystem services in LCA. Int J Life Cycle Assess 18:1188–1202. doi:10.1007/s11367-013-0579-z

Langlois J, Fréon P, Delgenes JP, Steyer JP, Hélias A (2012) Biotic resources extraction impact assessment in LCA of fisheries. Paper presented at the LCA Food, Saint-Malo, October 2–4, 2012

Langlois J, Fréon P, Delgenes J-P, Steyer J-P, Hélias A (2014) New methods for impact assessment of biotic-resource depletion in life cycle assessment of fisheries: theory and application. J Clean Prod 73:63–71. doi:10.1016/j.jclepro.2014.01.087

Legović T, Geček S (2010) Impact of maximum sustainable yield on independent populations. Ecol Model 221(17):2108–2111. doi:10.1016/j.ecolmodel.2010.05.015

Lenzen M, Moran D, Kanemoto K, Foran B, Lobefaro L, Geschke A (2012) International trade drives biodiversity threats in developing nations. Nature 486(7401):109–112

Libralato S, Coll M, Tudela S, Palomera I, Pranovi F (2008) Novel index for quantification of ecosystem effects of fishing as removal of secondary production. Mar Ecol Prog Ser 355:107–129. doi:10.3354/meps07224

Lindeijer EW, Müller-Wenk R, Steen B et al (2002) Impact assessment of resources and land use. In: de Udo Haes HA, Finnveden G, Goedkoop M (eds) Life-cycle impact assessment: striving towards best practice. Published by the Society of Environmental Toxicology and Chemistry (SETAC), Pensacola, pp 11–64

Mancini L, De Camillis C, Pennington D (eds) (2013) Security of supply and scarcity of raw materials. Publications Office of the European Union, Luxemburg

Meadows DH, Meadows DL, Randers J, Behrens WW (1972) The limits to growth. A report of the club of Rome. Universe Books, New York

Meadows DH, Randers J, Meadows DL (2004) Limits to growth – the 30-year update. Chelsea Green Publishing Company, Vermont

Michelsen O, McDevitt JE, Coelho CRV (2014) A comparison of three methods to assess land use impacts on biodiversity in a case study of forestry plantations in New Zealand. Int J Life Cycle Assess 19(6):1214–1225. doi:10.1007/s11367-014-0742-1

Mikesell RF (1994) Sustainable development and mineral resources. Resour Policy 20(2):83–86

Milà i Canals L (2003) Contribution to LCA methodology for agricultural systems. Site-dependency and soil degradation impact assessment. Universitat Autónoma de Barcelona, Bellaterra

Milà i Canals L (2007) Land use in LCA: a new subject area and call for papers. Int J Life Cycle Assess 12:1 (1 Editorial)

Milà i Canals L, de Baan, L (2015) Land use. Chapter 11 "Life Cycle Impact Assessment" (Hauschild M, Huijbregts MAJ eds). In: LCA compendium – the complete world of life cycle assessment (Klöpffer W, Curran MA, series eds). Springer, Dordrecht, pp 197–222

Milà i Canals L, Bauer C, Depestele J, Dubreuil A, Freiermuth Knuchel R, Gaillard G, Michelsen O, Müller-Wenk R, Rydgren B (2006) Key elements in a framework for land use impact assessment within LCA (11 pp). Int J Life Cycle Assess 12(1):5–15. doi:10.1065/lca2006.05.250

Milà i Canals L, Bauer C, Depestele J, Dubeuil A, Knuchel R, Gaillard G, Michelsen O, Mueller-Wenk R, Rydgren B (2007) Key elements in a framework for land use impact assessment within LCA. Int J Life Cycle Assess 12(1):5–15

Millennium Ecosystem Assessment (2005) Ecosystems and human well-being biodiversity synthesis. World Resources Institute, Washington, DC

Müller-Wenk R (1999) Depletion of abiotic resources weighted on base of 'virtual' impacts of lower grade deposits used in the future. IWÖ-Diskussionsbeitrag nr. 57. St. Gallen

Müller-Wenk R, Brandão M (2010) Climatic impact of land use in LCA—carbon transfers between vegetation/soil and air. Int J Life Cycle Assess 15(2):172–182. doi:10.1007/s11367-009-0144-y

Nassar NT, Barr R, Browning M, Diao Z, Fiedlander E, Harper EB, Henly C, Kavlak G, Kwatra S, Jun C, Warren S, Yang M-Y, Graedel TE (2012) Criticality of the geological copper family. Environ Sci Technol 46(2):1071–1078

National Research Council (2008) Minerals, critical minerals, and the U.S. Economy. National Academies Press, Washington, DC

Núñez M, Antón A, Muñoz P, Rieradevall J (2012) Inclusion of soil erosion impacts in life cycle assessment on a global scale: application to energy crops in Spain. Int J Life Cycle Assess 18 (4):755–767. doi:10.1007/s11367-012-0525-5

Nuss P, Harper EM, Nassar NT, Reck BK, Graedel TE (2014) Criticality of iron and its principal alloying elements. Environ Sci Technol 48(7):4171–4177

Olson DM, Dinerstein E, Wikramanayake ED, Burgess ND, Powell GVN, Underwood EC, D'amico JA, Itoua I, Strand HE, Morrison JC, Loucks CJ, Allnutt TF, Ricketts TH, Kura Y, Lamoreux JF, Wettengel WW, Hedao P, Kassem KR (2011) Terrestrial ecoregions of the world: a new map of life on earth – a new global map of terrestrial ecoregions provides an innovative tool for conserving biodiversity. Bioscience 51(11):933–938

Papatryphon E, Petit J, Kaushik SJ, Werf HM (2004a) Environmental impact assessment of salmonid feeds using Life Cycle Assessment (LCA). Ambio J Hum Environ 33(6):316–323

Papatryphon E, Petit J, Werf HMGvd (2004b) The development of life cycle assessment for the evaluation of rainbow trout farming in France. Paper presented at the proceedings of the 4th international conference on life cycle assessment in the agri-food sector, 6–8 October 2004

PE International (2013) GaBi 6 software-system and database for life cycle engineering. Echterdingen, Stuttgart

Pelletier NL, Ayer NW, Tyedmers PH, Kruse SA, Flysjo A, Robillard G, Ziegler F, Scholz AJ, Sonesson U (2006) Impact categories for life cycle assessment research of seafood production systems: Review and prospectus. Int J Life Cycle Assess 12(6):414–421. doi:10.1065/lca2006.09.275

Pennington DW, Potting J, Finnveden G, Lindeijer E, Jolliet O, Rydberg T, Rebitzer G (2004) Life cycle assessment part 2: current impact assessment practice. Environ Int 30(5):721–739

Petrie J (2007) New models of sustainability for the resources sector – a focus on minerals and metals. Trans IChemE B Process Saf Environ Prot 85(B1):88–98

Pfister S (2015) Water use. Chapter 12 "Life Cycle Impact Assessment" (Hauschild M, Huijbregts MAJ eds). In: LCA compendium – the complete world of life cycle assessment (Klöpffer W, Curran MA, series eds), pp 223–245

Radetzki M (2002) Is resource depletion a threat to human progress? Oil and other critical exhaustible materials. Paper presented at the energex' 2002, Cracow, 19–24 May

Rankin WJ (2011) Minerals, metals and sustainability: meeting future material needs. CSIRO Publishing, Collingwood

Reuter MA, Heiskanen K, Boin U, Schaik A, Verhoef E, Yang Y, Georgalli G (2005) The metrics of material and metal ecology, vol 16, Developments in Mineral Processing. Elsevier, Amsterdam

Reynolds JD (2008) Conservation of Exploited Species. Conservation biology, vol 6. Cambridge University Press, Cambridge

Ricard D, Minto C, Jensen OP, Baum JK (2012) Examining the knowledge base and status of commercially exploited marine species with the RAM legacy stock assessment database. Fish Fish 13(4):380–398. doi:10.1111/j.1467-2979.2011.00435.x

Ritthoff M, Rohn H, Liedtke C (2002) Calculating MIPS – resource productivity of products and services. Wuppertal Spezial 27e. Wuppertal Institute for Climate, Environment and Energy, Wuppertal

Rosenau-Tornow D, Buchholz P, Riemann A, Wagner M (2009) Assessing the long-term supply risks for mineral raw materials – a combined evaluation of past and future trends. Resour Policy 34(4):161–175

Sala S (2012) Assessing resource depletion in LCA: a review of methods and methodological issues. In: Mancini L, De Camillis C, Pennington D (eds) Security of supply and scarcity or raw materials. Towards a methodological framework for sustainability assessment. European Commission, Joint Research Center, Institute for Environment and Sustainability. Publication Office of the European Union, Luxemburg, pp 24–27

Sala S, Farioli F, Zamagni A (2013) Progress in sustainability science: lessons learnt from current methodologies for sustainability assessment: Part 1. Int J Life Cycle Assess 18(9):1653–1672

Schenck RC (2001) Land use and biodiversity indicators for life cycle impact assessment. Int J Life Cycle Assess 6(2):114–117

Schneider L (2014) A comprehensive approach to model abiotic resource provision capability in the context of sustainable development. Doctoral thesis, Technical University Berlin, Berlin

Schneider L, Berger M, Finkbeiner M (2011) The anthropogenic stock extended abiotic depletion potential (AADP) as a new parameterisation to model the depletion of abiotic resources. Int J Life Cycle Assess 16(9):929–936

Schneider L, Berger M, Schüler-Hainsch E, Knöfel S, Ruhland K, Mosig J, Bach V, Finkbeiner M (2013) The economic resource scarcity potential (ESP) for evaluating resource use based on life cycle assessment. Int J Life Cycle Assess 19(3):601–610

Science and Technology Committee (2011) Strategically important metals. Fifth report of session 2010–12. House of Commons, London

Skinner (1976) A second iron age ahead? Am Sci 64(3):158–169

Souza D, Flynn D, de Clerck F, Rosenbaum R, de Melo LH, Koellner T (2013) Land use impacts on biodiversity in LCA: proposal of characterization factors based on functional diversity. Int J Life Cycle Assess. doi:10.1007/s11367-013-0578-0

Steen B (1999) A systematic approach to environmental priority strategies in product development (EPS). Version 2000 – models and data. Report 1999:5. Chalmers University of Technology, Centre for Environmental Assessment of Products and material Systems, Gothenburg

Steen BA (2006) Abiotic resource depletion – different perceptions of the problem with mineral deposits. Int J Life Cycle Assess 11(1):49–54

Steffan-Dewenter I, Kessler M, Barkmann J, Bos MM, Buchori D, Erasmi S, Faust H, Gerold G, Glenk K, Gradstein SR, Guhardja E, Harteveld M, Hertel D, Höhn P, Kappas M, Köhler S, Leuschner C, Maertens M, Marggraf R, Migge-Kleian S, Mogea J, Pitopang R, Schaefer M, Schwarze S, Sporn SG, Steingrebe A, Tjitrosoedirdjo SS, Tjitrosoemito S, Twele A, Weber R, Woltmann L, Zeller M, Tscharntke T (2007) Tradeoffs between income, biodiversity, and ecosystem functioning during tropical rainforest conversion and agroforestry intensification. Proc Natl Acad Sci U S A 104:4973–4978

Steinbach V, Wellmer F-W (2010) Consumption and use of non-renewable mineral and energy raw materials from an economic geology point of view. Sustainability 2(5):1408–1430

Stewart M, Weidema B (2005) A consistent framework for assessing the impacts from resource use, a focus on resource functionality. Int J Life Cycle Assess 10(4):240–247

Swart P, Alvarenga RAF, Dewulf J (2015) Abiotic resource use. Chapter 13 "Life Cycle Impact Assessment" (Hauschild M, Huijbregts MAJ eds). In: LCA Compendium – the complete world of life cycle assessment (Klöpffer W, Curran MA, series eds), pp 247–269

Tilton JE (2003) On borrowed time? Assessing the threat of mineral depletion. Resources for the Future, Washington, DC

Tilton JE, Lagos G (2007) Assessing the long-run availability of copper. Resour Policy 32(1–2):19–23

Turner G (2008) A comparison of the limits to growth with thirty years of reality. CSIRO Sustainable Ecosystems, Canberra

Udo de Haes HA U, Finnveden G, Goedkoop M, Hauschild M, Hertwich EG, Hofstetter P, Jolliet O, Kl-pffer W, Krewitt W, Lindeijer EW, Mueller-Wenk R, Olson SI, Pennington DW, Potting J, Steen B (2002) Life cycle impact assessment: striving towards best practice. Published by the Society of Environmental Toxicology and Chemistry (SETAC), Pensacola

UNEP (2010a) Assessing the environmental impacts of consumption and production; priority products and materials. A report of the Working Group on the Environmental impacts of products and materials to the International Panel for Sustainable Resource Management. Hertwich E, van der Voet E, Suh S, Tukker A, Huijbregts M, Kazmierczyk P, Lenzen M, McNeely J, Moriguchi Y, Paris

UNEP (2010b) Metal stocks in society. United Nations Environment Programme, International Panel for Sustainable Resource Management, Paris

UNEP (2010c) United Nations Environment Programme (UNEP): connecting the dots – Biodiversity, adaptation, food security and livelihoods United Nations Environment Programme, Nairobi

UNEP (2011) Decoupling natural resource use and environmental impacts from economic growth. A report of the working group on decoupling to the International Resource Panel. United Nations Environment Programme, Paris

United Nations (2005) Handbook of national accounting: integrated environmental and economic accounting 2003, vol 61, Studies in Methods. United Nations, European Commission, International Monetary Fund, Organisation for Economic Co-operation and Development, World Bank, New York

USGS (2014) Mineral commodity summaries. U.S. Geological Survey, Department of the Interior, Reston

Vaissière B, Freitas B, Gemmill-Herren B (2011) Protocol to detect and assess pollination deficits in crops: a handbook for its use. FAO, Rome

Valdivia S, Ugaya CML, Sonnemann G, Hildenbrand J (2011) Towards a life cycle sustainability assessment – making informed choices on products, UNEP/SETAC Life Cycle Initiative, Paris. ISBN: 978-92-807-3175-0

Valero A, Valero A (2009) The crepuscular planet. Part II: a model for the exhausted continental crust. Paper presented at the ECOS, Paraná, 31 August – 3 September

Valero A, Valero A, Amaya M (2009) Inventory of the exergy resource on earth including its mineral capital. Energy 35(2):989–995

Valero A, Valero A, Domínguez A (2013) Exergy replacement cost of mineral resources. J Environ Account Manag 1(2):147–158

van der Voet E (2005) Policy review on decoupling: development of indicators to assess decoupling of economic development and environmental pressure in the EU-25 and AC-3 countries. CML report 166. Leiden University, Leiden

van Oers L, deKoning A, Guinée J, Huppes G (2002) Abiotic resource depletion in LCA. Road and Hydraulic Engineering Institute, Leiden

van Zelm R, Muchada PAN, van der Velde M, Kindermann G, Obersteiner M, Huijbregts MAJ (2014) Impacts of biogenic CO_2 emissions on human health and terrestrial ecosystems: the case of increased wood extraction for bioenergy production on a global scale. GCB Bioenergy: n/a–n/a. doi:10.1111/gcbb.12153

Vázquez-Rowe I, Hospido A, Moreira MT, Feijoo G (2012a) Best practices in life cycle assessment implementation in fisheries. Improving and broadening environmental assessment for seafood production systems. Trends Food Sci Technol 28(2):116–131. doi:10.1016/j.tifs.2012.07.003

Vázquez-Rowe I, Moreira MT, Feijoo G (2012b) Inclusion of discard assessment indicators in fisheries life cycle assessment studies. Expanding the use of fishery-specific impact categories. Int J Life Cycle Assess 17(5):535–549. doi:10.1007/s11367-012-0395-x

VDI (2013) http://www.vdi.de/technik/fachthemen/energie-und-umwelt/fachbereiche/ressourcenmanagement/themen/richtlinienwerk-zur-ressourceneffizienz-zre/

Verhoef EV (2004) The ecology of metals. Delft University of Technology, Delft

Vieira M, Ponsionen TC, Goedkoop M, Huijbregts MAJ (2011) Mineral and fossil resources. Development and application of environmental life cycle impact assessment methods for improved sustainability characterisation of technologies (LC-IMPACT). Seventh Framework Programme, European Union, Brussels

Vieira MDM, Goedkoop MJ, Storm P, Huijbregts MAJ (2012) Ore grade decrease as life cycle impact indicator for metal scarcity: the case of copper. Environ Sci Technol 46 (23):12772–12778

Vogtländer JG, Velden NM, Lugt P (2013) Carbon sequestration in LCA, a proposal for a new approach based on the global carbon cycle; cases on wood and on bamboo. Int J Life Cycle Assess 19(1):13–23. doi:10.1007/s11367-013-0629-6

Wäger PA, Widmer R, Stamp A (2011) Scarce technology metals – applications, criticalities and intervention options. Federal Office for the Environment (FOEN), Bern

Watson RT, Rosswall T, Steiner A, Töpfer K, Arico S, Bridgewater P (2005) Ecosystems and human well-being: biodiversity synthesis. In: Watson RT, Rosswall T, Steiner A, Töpfer K, Arico S, Bridgewater P (eds) Ecosystems. World Resources Institute, Washington. doi:10.1196/annals.1439.003

Weidema B, Finnveden G, Stewart M (2005) Impacts from resource use – a common position paper. Int J Life Cycle Assess 10(6):382

Wellmer F-W, Becker-Platen JD (2002) Sustainable development and the exploitation of mineral and energy resources: a review. Int J Earth Sci 91(5):723–745

Weterings R, Bastein T, Tukker A, Rademaker M, Ridder M (2013) Resources for our future – key issues and best practices in resource efficiency. The Hague Centre for Strategic Studies (HCSS) and TNO, Amsterdam

Wuppertal Institute for Climate EaE (2013) Economy-wide indicators. Topics Online "Economy-wide Material Flow Analysis and Indicators". Wuppertal Institute for Climate, Environment and Energy, Wuppertal

Yellishetty M, Ranjith PG, Tharumarajah A, Bhosale S (2009) Life cycle assessment in the minerals and metals sector: a critical review of selected issues and challenges. Int J Life Cycle Assess 14(3):257–267

Yellishetty M, Mudd GM, Ranjith PG (2011) The steel industry, abiotic resource depletion and life cycle assessment: a real or perceived issue? J Clean Prod 19(1):78–90

Ziegler F, Valentinsson D (2008) Environmental life cycle assessment of Norway lobster (*Nephrops norvegicus*) caught along the Swedish west coast by creels and conventional trawls—LCA methodology with case study. Int J Life Cycle Assess 13(6):487–497. doi:10.1007/s11367-008-0024-x

Ziegler F, Nilsson P, Mattsson B, Walther Y (2003) Life Cycle assessment of frozen cod fillets including fishery-specific environmental impacts. Int J Life Cycle Assess 8(1):39–47. doi:10.1007/bf02978747

Ziegler F, Emanuelsson A, Eichelsheim JL, Flysjö A, Ndiaye V, Thrane M (2011) Extended life cycle assessment of southern pink shrimp products originating in *Senegalese Artisanal* and industrial fisheries for export to Europe. J Ind Ecol 15(4):527–538. doi:10.1111/j.1530-9290.2011.00344.x

Chapter 6
Input–Output and Hybrid LCA

Shinichiro Nakamura and Keisuke Nansai

Abstract Known as hybrid LCA, integrated use of economic input–output (IO) analysis and process-based LCA (PLCA) has become a major tool of LCA inventory analysis. Proceeding from the basics of IO, this chapter discusses the issues of monetary versus physical data, multiregional extension, end-of-life phase with waste management and recycling, cost and price (with implications for life cycle costing), technology choices, and substitution. Besides the strengths of hybrid LCA, several often-cited "weaknesses" are also addressed.

Keywords CES functions • CGE • Database • Computable general equilibrium (CGE) • LCC • Life cycle assessment (LCA) • Life cycle costing (LCC) • Linear programming (LP) • LP • Monetary and physical tables • MRIO • Multiregional input–output (MRIO) • Productive conditions • Recycling • Supply and use tables • Waste input–output

Acronyms

CGE	Computable general equilibrium
EEIO	Environmentally extended IO
EoL	End of life
EPR	Extended producer responsibility
EXIOPOL	Externality data and input–output tools for policy analysis
GDP	Gross domestic product
GE	General equilibrium model
ICP	International comparison program
IOT	IO table

S. Nakamura (✉)
Faculty of Political Science and Economics, Waseda University, Tokyo 169-8050, Japan
e-mail: nakashin@waseda.jp

K. Nansai
National Institute for Environmental Studies, Center for Material Cycles and Waste Management Research, Ibaraki 305-8506, Japan

M. Finkbeiner (ed.), *Special Types of Life Cycle Assessment*, LCA Compendium – The Complete World of Life Cycle Assessment,
DOI 10.1007/978-94-017-7610-3_6

MIOT	Monetary IO table
MRIO	Multiregional input–output table
PIOT	Physical IOT
PLCA	Process-based LCA
PPP	Purchasing power parity
ROW	Rest of the world
WIO-MFA	Waste input–output material flow analysis

1 Introduction: Economic Input–Output Analysis for Life Cycle Assessment

Integrating economic input–output analysis (IO) and process-based LCA (PLCA) has become a major tool of LCA inventory analysis (Suh et al. 2004). This is termed "hybrid" LCA to refer to the fact that it is an integration of PLCA, the standard or classical form of LCA, and IO, with the aim of combining the strengths of each method (Wiedmann 2009). With its roots in the energy analysis literature of the 1970s (Wright 1974; Bullard and Herendeen 1975), the methodological framework is not in itself new.

This chapter aims to provide a concise yet comprehensive account of hybrid LCA, with an emphasis on methodological aspects. Reflecting its increased use in LCA, the literature on hybrid LCA abounds with reviews, among others, in Minx et al. (2008), Suh (2009), Williams et al. (2009). Our review is comprehensive in regard to the description of the static IO model, starting from the very basic level of a one-sector IO. It is further distinguished by its consideration of several topics not fully discussed in the literature, including the end-of-life (EoL) phase, the cost and price IO model, and extension from a square to a rectangular technology matrix to accommodate the presence and choice of alternative technologies. There is also a discussion of computable general equilibrium (CGE) models, the relevance of which for LCA is under increasing discussion in connection with consequential LCA.

2 Basics of Input–Output Analysis for Life Cycle Assessment

IO is concerned with quantitatively capturing the interdependences among different sectors of the economy via the flow of input and output flows at high levels of sectoral resolution. Interdependences emerge because sectors require each other's outputs as inputs. Central to IO is the representation of these interdependences in terms of a matrix of technical coefficients that closely resembles the concept of technology matrix in LCA (Heijungs and Suh 2002; Suh et al. 2004).

In LCA, a unit process of production (henceforth, called "unit process" for simplicity) is the basic entity of production (or any sort of conversion process, to be general) and is represented by the flow of inputs entering it and outputs leaving it. The outputs of a unit process are made up of products, by-products, wastes, and emissions. A unit process can be classified by its major product, the production of which is the primary purpose of its operation. In the following, the term a "sector" (of production) is used synonymously with "unit process." Accordingly, a sector can be associated with its primary product.

We start with a simple case where each unit process is characterized by a single product, while in reality joint production is usually the case. The case of joint production including by-products and wastes is discussed in Sect. 3.4. Issues associated with the use and make matrices that are frequently employed in the compilation of IO data (IO table) to cope with the presence of joint products are discussed in Sect. 3.6.

First we consider the simplest case of an economy consisting of a single producing sector. While this case may appear odd, given the multi-sectoral nature of IO, it will help familiarize readers with the basic concepts of IO without burdening them with mathematics: simple arithmetic suffices. The multi-sectoral nature of IO is then considered for the case of an economy with two producing sectors. This involves systems of simultaneous equations. Rewriting these in terms of matrix algebra yields the basic results of IO, which can be applied to the general case of n producing sectors.

2.1 Input–Output Model with One Sector

Consider an economy with a single sector producing a single product. Take as an example the production of grain, with water, soil, nutrients, sunlight, and CO_2 all in abundant supply. Under this condition, seeds are the sole input of concern.

2.1.1 Balance Equations

Write x_1 for the amount of product 1 produced (in this example, grain) and x_{11} for the amount of product 1 that is inputted for production (seeds). The term x_{11} refers to the input consumed in the course of production to produce the output, and is called the intermediate input: it is used to meet intermediate demand. Writing y_1 for the surplus of production over input yields the following balance:

$$\underbrace{x_1}_{\text{output}} = \underbrace{x_{11}}_{\text{intermediate demand}} + \underbrace{y_1}_{\text{final demand}} \tag{6.1}$$

If this sector is productive, $y_1 > 0$. Positive surplus of output over input is a necessary condition for an economy to be productive; an economy without surplus would collapse. Because this sector is the sole sector of production in this economy, this condition must hold for this economy to be productive. The term y_1 is called final demand, because it is not consumed during production but remains available for final use. Satisfaction of this demand is the driving force of an economy (see Sect. 2.3.1 for further details). In the present example of grain, $y_1 > 0$ implies the presence of grain available for household consumption.

2.1.2 IO Coefficient and IO Model

The following simple relationship of proportionality is assumed between the amount of input, x_{11}, and the amount of output, x_1:

$$x_{11} = a_{11}x_1 \tag{6.2}$$

a_{11} refers to the amount of input 1 used to produce one unit of product 1, and is called the input coefficient. The proportionality between input and output in (6.2) corresponds to the technology characterized by constant returns to scale, that is, changes in the level of output do not change the proportion of input to output. This assumption is implicit in LCA using constant technology coefficients.

Because no output is obtained from zero input, $a_{11} > 0$. Furthermore, $a_{11} < 1$ must hold for the process to be "productive" as mentioned above. Otherwise, all the output would be consumed for intermediate use, leaving none left for final use, satisfaction of which is the very purpose of production. It follows that a_{11} has to satisfy the following condition:

$$0 < a_{11} < 1 \tag{6.3}$$

Substitution of (6.2) into (6.1) results in

$$x_1 = a_{11}x_1 + y_1 \tag{6.4}$$

Suppose, now, that one wants to know the amount of production x_1 required to satisfy a given amount of final demand when the state of technology is represented by input coefficient a_{11} and there are no immediate constraints on productive capacity. The answer is given by solving (6.4) for x_1:

$$x_1 = (1 - a_{11})^{-1}y_1 > y_1 \tag{6.5}$$

where the existence and positivity of the solution as well as the inequality follow from (6.3). Though simple, (6.5) encapsulates the essence of IO. The term $(1 - a_{11})^{-1}$ represents the Leontief inverse coefficient, a scalar version of the well-known Leontief inverse matrix, and gives the amount of product 1 that is directly and indirectly required to satisfy a unit of final demand for the product.

An alternative way to view this is to resort to the extension of the inverse coefficient as the sum of an infinite series:

$$(1 - a_{11})^{-1} = 1 + a_1 + a_{11}^2 + a_{11}^3 + \cdots \tag{6.6}$$

where a_{11} refers to the direct use required to produce one unit of product, or the input in the first "tier," a_{11}^2 to the input required to produce a_{11}, or the input in the second tier. Conversion of the series follows from (6.3).

2.2 *Input–Output with Two Sectors and More*

2.2.1 IO with Two Sectors

The presence of inter-sectoral dependence makes the case of two sectors fundamentally different from the above case of one sector. This interdependence arises when a sector used another sector's output as an input. Writing x_{ij} for the input of product i to sector j, the balance Eq. (6.1) is extended to

$$\begin{aligned} x_1 &= x_{11} + x_{12} + y_1 \\ x_2 &= x_{21} + x_{22} + y_2 \end{aligned} \tag{6.7}$$

Writing $a_{ij}, i \neq j$, for the coefficient that refers to the input of product i per unit of output of product j,

$$x_{ij} = a_{ij}x_j, i, j = 1, 2. \tag{6.8}$$

In accordance with (6.3), it is assumed that the following conditions are satisfied:

$$a_{ij} \geq 0, \tag{6.9}$$

$$1 - a_{ii} > 0 \tag{6.10}$$

Substitution of (6.8) into the above balance equations gives

$$\begin{aligned} x_1 &= a_{11}x_1 + a_{12}x_2 + y_1 \\ x_2 &= a_{21}x_1 + a_{22}x_2 + y_2 \end{aligned} \tag{6.11}$$

or

$$\begin{aligned} (1 - a_{11})x_1 - a_{12}x_2 &= y_1 \\ -a_{21}x_1 - (1 - a_{22})x_2 &+ y_2 \end{aligned} \tag{6.12}$$

The amounts of x_1 and x_2 that are required to satisfy the final demand are then given by

$$x_1 = \frac{1 - a_{22}}{d} y_1 + \frac{a_{12}}{d} y_2 = b_{11} y_1 + b_{12} y_2$$
$$x_2 = \frac{a_{21}}{d} y_1 + \frac{1 - a_{11}}{d} y_2 = b_{21} y_1 + b_{22} y_2 \quad (6.13)$$

where

$$d = (1 - a_{11})(1 - a_{22}) - a_{12} a_{21} \quad (6.14)$$

Nonnegativity of solution follows if (6.10) and

$$d = (1 - a_{11})(1 - a_{22}) - a_{12} a_{21} > 0 \quad (6.15)$$

are satisfied, which in input–output economics are together called Hawkins–Simon (HS) condition. This condition is the productive condition for the two-sector case.

Because negative production makes no sense, the HS condition must always be satisfied. As the number of sectors grows, its calculation becomes increasingly cumbersome, however. Fortunately, practitioners of IO can make use of the following useful theoretical result.

Theorem 1 Nikaido (1970) *The following conditions (I), (II), and (HS) are equivalent:*

(I) *For a certain* $f_i > 0, i = 1, 2$, (6.12) *has a nonnegative solution* $x_i \geq 0, i = 1, 2$.
(II) *For any* $f_i \geq 0, i = 1, 2$, (6.12) *has a nonnegative solution.*
(HS) $1 - a_{ii} > 0, i = 1, 2$ *and* $d > 0$.

The practical usefulness of this theorem follows from the fact that in real application of IO, it is usual practice to obtain the required input coefficients from the IO table of a real economy. In such tables, final demand and production volumes are positive, and hence the above condition (I) is automatically met. It follows from the theorem that the input coefficients thus obtained satisfy (HS), and the quantity model (6.12) always has nonnegative solutions for any nonnegative final demand. In short, for the input coefficients obtained from real IO data, one does not have to bother with (HS). The theorem does not hold, however, if certain a_{ij} are negative, which may be the case in both IO and LCA if by-products are involved (see Sect. 3.4 below).

2.2.2 IO with *n* Sectors

Using matrix notation and algebra, extension of IO from two sectors to an arbitrary number of sectors *n* is straightforward. Use of matrices is essential, for otherwise notation becomes too messy. It is worth pointing out that matrices are routinely used in PLCA as well (Minx et al. 2008). Furthermore, the structure of IO and matrix based PLCA is almost identical computationally (Suh et al. 2004).

For ease of elucidation, we first consider the case of $n=2$, with the understanding that once the basic formulae have been derived, they are applicable to any $n \geq 1$. In matrix notation, (6.11) can be written as

$$\underbrace{\begin{pmatrix} x_1 \\ x_2 \end{pmatrix}}_{x} = \underbrace{\begin{pmatrix} a_{11} & a_{12} \\ a_{21} & a_{22} \end{pmatrix}}_{A} \underbrace{\begin{pmatrix} x_1 \\ x_2 \end{pmatrix}}_{x} + \underbrace{\begin{pmatrix} y_1 \\ y_2 \end{pmatrix}}_{y} \tag{6.16}$$

or

$$x = Ax + y \tag{6.17}$$

and (6.12) as

$$(I - A)x = y \tag{6.18}$$

where

$$I = \begin{pmatrix} 1 & 0 \\ 0 & 1 \end{pmatrix} \tag{6.19}$$

and the solution

$$x = (I - A)^{-1} y \tag{6.20}$$

The HS condition for A in Theorem 1 now reads that all the principal minors of $I - A$ are positive, which is given by (6.10) and (6.15) for the case of $n=2$. Equation 6.20 is the fundamental equation in IO, and the inverse matrix on the right-hand side is called the Leontief inverse matrix.

2.3 Input–Output Tables and Data

In terms of the provision of data for analysis, IO differs considerably from PLCA by the fact that the underlying data of IO, IO table (IOT), is an essential component of every national accounting system, and is routinely compiled and published on regular bases in a large number of nations/regions. Its principal aim has been and is to record all the flows among production sectors of the economy including both goods and services that occurred in a given nation/region within a given year: the use in LCA is not its immediate aim. Accordingly, in most cases, publicly available IOTs would need some modifications and/or extensions before they can be applied to LCA. Prerequisite for it is a proper understanding of the basic structure of publicly available IO tables, to a brief description of which we now turn (see standard text books like Miller and Blair (2009) for details).

2.3.1 Final and Intermediate Demand in IO Accounts

In the input–output accounts, final demand or final use transactions consist of the transactions that make up the final expenditure components of GDP (gross domestic product): personal consumption expenditures, fixed investment, change in inventories, exports of goods and services, imports of goods and services, and government consumption expenditures and fixed investment (including investment by government enterprises) (Horowitz and Planting 2009). Intermediate consumption includes goods and services, such as energy, materials, and purchased services, which are entirely used up by producers in the course of production to produce output of goods and services during the accounting period. These inputs are sometimes referred to as current account expenditures. They do not include any capital account purchases that refer to durable goods that are used over a number of years (United Nations 2003; Bureau of Economic Analysis 2009).

2.3.2 IO Tables and National Accounts

The Basic Balance in an Economy

For any economy, the total supply of goods and services originating from both domestic and foreign sources must equal the total consumption of the economy (United Nations (2003)). Accordingly, for any product (goods and services) the following balance holds

$$\begin{aligned}\text{output} + \text{imports} = {} & \text{intermediate consumption} + \text{final consumption} \\ & + \text{gross capital formation} + \text{exports}\end{aligned} \tag{6.21}$$

or, using our notations,

$$\begin{aligned}x_i^d + x_i^m & = \sum_j \left((x_{ij}^d + x_{ij}^m) + (y_{iC}^d + y_{iC}^m) + \right. \\ & \left. (y_{iI}^d + y_{iI}^m) + (y_{iX}^d + y_{iX}^m) \right)\end{aligned} \tag{6.22}$$

where the superscript d refers to domestic flow and m to the imported flow, C to final consumption, I to gross capital formation, and X to export. x_{ij}^m refers to the intermediate consumption in sector j of i of domestic origin, y_{iC}^m to the final consumption of imported i, and y_{iX}^d to the export of domestic i. The export of imported i, y_{iX}^m will in many cases be zero.

The equality between supply and demand holds because of the inclusion in gross capital formation of any change in the level of inventory. A surplus of supply over demand is counted as an increase in the stock of inventory, and an excess of demand over supply as a decrease in the stock. When measured in monetary terms, we also have

$$\begin{aligned}\text{gross value added} &= \text{output} - \text{intermediate consumption}\\ &= \text{final consumption} + \text{gross capital formation}\\ &\quad + \text{exports} - \text{imports}\end{aligned} \tag{6.23}$$

Value added consists of the costs, such as compensation, profits, and depreciation, which are related to these inputs, or compensations for labor and capital services (Bureau of Economic Analyses 2009). Denoting the flow in monetary terms by attaching '*', this becomes

$$\begin{aligned}&\sum_j v_j^* = \sum_j x_j^{d*} - \sum_{i,j}\left(x_{ij}^{d*} + x_{ij}^{m*}\right) =\\ &\sum_{j=C,I}\sum_i\left(y_{ij}^{d*} + y_{ij}^{m*}\right) + \sum_i y_{iX}^{d*} - \sum_{j=C,I,X}\sum_i y_{ij}^{m*}\end{aligned} \tag{6.24}$$

where v_j^* refers to gross value added generated in sector j.

Gross Domestic Product (GDP)

In the presence of taxes and subsidies on goods and services, taxes have to be added to output and subsidies subtracted from output in order to arrive at a uniform valuation of supply and uses (United Nations 2003). Recalling the definition of GDP by production approach, it follows that:

$$\begin{aligned}\text{GDP} &= \text{output} + \text{taxes} - \text{subsidies} - \text{intermediate consumption}\\ &= \text{final consumption} + \text{gross capital formation} + \text{exports} - \text{imports}.\end{aligned} \tag{6.25}$$

From (6.23), it then follows that:

$$\begin{aligned}\text{GDP} &= \text{gross value added} + \text{taxes} - \text{subsidies}\\ &= \text{compensation for labor and capital services}\\ &\quad + \text{taxes} - \text{subsidies}\end{aligned} \tag{6.26}$$

From (6.25) to (6.26), the following expression holds for output:

$$\begin{aligned}\text{output} &= \text{intermediate inputs} + \text{gross value added}\\ &\quad + \text{taxes less subsidies}\end{aligned} \tag{6.27}$$

Writing L_j^* and K_j^* for the amount of labor and capital compensation generated in sector j, if each product is produced in a single and unique sector, it follows that:

$$x_j^{d*} = \sum_i \left(x_{ij}^{d*} + x_{ij}^{m*} \right) + L_j^* + K_j^* + TS_j^* \tag{6.28}$$

where TS_j^* refers to the amount of taxes minus subsidies on product j.

Monetary IO Table

Table 6.1 gives a schematic representation of a two-sector monetary IOT (MIOT). For the sake of simplicity, foreign trades (export and import) are neglected. In each of the rows referring to production sectors, the (first three) elements add up to the sectoral output (the amount of product), as in Eq. (6.22). Summing the (first four) elements of each of the columns referring to production sectors gives the sectoral output as well, as in Eq. (6.28).

Cost Equations

Labor services and capital services are called primary factors of production to refer to the fact that they are not reproduced in the system. While the items occurring in the intermediate flow matrix are being produced by endogenous production sectors, there are no endogenous sectors responsible for the provision of capital and labor services. There are many types of labor, say, manual workers, office clerks, managers, engineers, and scientists. Accordingly, L_j^* can be decomposed into its individual components:

$$L_j^* = \sum_{i=1}^{l} p_{Li} L_{ij} \tag{6.29}$$

where l refers to the number of types of labor; p_{Li} to the compensation per labor of type i, that is, wage rate; and L_{ij} to the number of labor force participants of type i in sector j. In economics, it is usual to assume that L_j^* can be further represented as a product of an aggregate of wage rates P_{Lj} and an aggregate of labor Q_{Lj} as follows (Caves et al. 1982):

$$L_j^* = \sum_{i=1}^{l} p_{Li} L_{ij} = P_{Lj} Q_{Lj} \tag{6.30}$$

Table 6.1 Schematic representation of a MIOT with two production sectors

	Sector 1	Sector 2	Final demand	Output (row sum)
Sector 1	x_{11}^*	x_{12}^*	y_1^*	x_1^*
Sector 2	x_{21}^*	x_{22}^*	y_2^*	x_2^*
Value added	v_1^*	v_2^*		
Taxes less subsidies	TS_1^*	TS_2^*		
Output (column sum)	x_1^*	x_2^*		

The aggregate Q_{Lj} is usually obtained by weighting the number of participants of each labor type with some measure of its contribution to production (Jorgenson 1988). Formally, writing g_L for an aggregator function of l types of labor, we have:

$$Q_{Lj} = g_L\left(L_{1j}, L_{2j}, \ldots, L_{lj}\right) \tag{6.31}$$

Once Q_{Lj} has been obtained, division of L_j^* by it would give the price aggregate P_{Lj}. Simply estimating Q_{Lj} on the basis of the numerical size of the labor force is equivalent to assuming homogeneity of all types of labor in sector j. Analogously, K_j^* can be represented as the product of an aggregate of prices of capital services P_{Kj} and an aggregate of capital services Q_{Kj}:

$$K_j^* = \sum_{i=1}^{k} p_{Ki} K_{ij} = P_{Kj} Q_{Kj}, \tag{6.32}$$

where k, P_{Kj}, and Q_{Kj} refer to the capital counterparts of the variables occurring in (6.30). Calculation of Q_{Kj} may be more challenging than that of labor, because measuring "capital service" is less straightforward than counting the number of workers of a given type. For further details about index numbers and aggregation in economics, see Caves et al. (1982) and Jorgenson (1988).

In the case where for any product the same price applies to all its use, it follows from (6.28) that

$$p_j^d x_j^d = \sum_i \left(p_i^d x_{ij}^d + p_i^m x_{ij}^m\right) + P_{Lj} Q_{Lj} + P_{Kj} Q_{Kj} + TS_j^* \tag{6.33}$$

Division of both the sides by x_j gives the following expression for the price of output:

$$p_j^d = \sum_i \left(p_i^d a_{ij}^d + p_i^m a_{ij}^m\right) + P_{Lj} a_{Lj} + P_{Kj} a_{Kj} + ts_j^* \tag{6.34}$$

where ts_j^* refers to the amount of taxes less subsidies on product j per unit of its quantity. Note that the right-hand side of (6.34) gives the cost of production per unit of product j. Accordingly, (6.34) can be called the unit cost function. The price of output is equal to the unit cost because profits or business surplus are included in capital compensation.

Producer and Purchaser Prices

In creating an actual MIOT, the standard price used for creating the items in Table 6.1 will become an issue. According to Lenzen et al. (2012a), the transaction price of a product has five elements: basic price, taxes, subsidies, trade margin, and transport margin. Adding taxes and subsidies to the basic price yields the producer price: the price p_j^d in (6.34) refers to the producer price. Further adding trade and transport margins yields the purchaser price.

2.3.3 The Units of Measurement: Physical Versus Monetary

IOT is designed to capture the flow of all the goods and services of the economy. Most services have no physical units, while for material products, a common physical unit is hard to conceive. Accordingly, monetary unit emerges as the most common units of measurement. With minor exceptions, the IOTs currently in use are not in physical units, but in units of value, because they are compiled using data on economic transactions measured in value units. It is therefore usually the case that the input coefficients are derived from a given IOT measured in value units.

Writing a_{ij}^* for the monetary counterpart (monetary input coefficient) of the physical input coefficient a_{ij} obtained from an MIOT like in Table 6.1 gives

$$a_{ij}^* = \frac{x_{ij}^*}{x_j^*} \tag{6.35}$$

Given that it is not physical but monetary coefficients that can be obtained from an MIOT, it is of great importance to know if the results of abovementioned analysis based on physical coefficients also hold for the monetary coefficients. Weisz and Duchin (2006) have shown that when the same input price level holds for all the users of that input.

> The two models are the same except for the change of unit operation, and the vector of unit prices provides the information needed for the change of unit (Weisz and Duchin 2006, p. 536).

In the following, their result is briefly outlined.

Case 1: Input Price Is the Same for All Users

Let x^*_{ij}, y^*_j, and x^*_j be x_{ij}, y_j, and x_j measured in monetary units (Table 6.1). Let p_i be the price of input i, and assume that it applies to all its users. It then follows that

$$x^*_{ij} = p_i x_{ij}, x^*_j = p_j x_j, y^*_j = p_j y_j \tag{6.36}$$

Accordingly, for input coefficients obtained from an MIOT (6.35), the following holds:

$$a^*_{ij} = \frac{p_i x_{ij}}{p_j x_j} = a_{ij}\frac{p_i}{p_j} \tag{6.37}$$

Now consider whether use of monetary coefficient a^*_{ij} in place of physical coefficient a_{ij} has any effects on a solution such as (6.20), that is, whether the inverse coefficients b^*_{ij} obtained from a^*_{ij} differ from those based on a_{ij}. For b^*_{12} we have from (6.11)

$$\begin{aligned} b^*_{12} &= \frac{a^*_{12}}{(1-a^*_{11})(1-a^*_{22}) - a^*_{12}a^*_{21}} \\ &= \frac{a_{12}p_1/p_2}{(1-a^*_{11}p_1/p_1)(1-a^*_{22}p_2/p_2) - a_{12}p_1/p_2 a_{21}p_2/p_1} \\ &= \frac{a_{12}}{(1-a_{11})(1-a_{22}) - a_{12}a_{21}}\frac{p_1}{p_2} \\ &= b_{12}\frac{p_1}{p_2} \end{aligned} \tag{6.38}$$

It follows that the use of monetary coefficients in place of physical coefficients leads to no changes in induced effects, the only difference being the difference in the unit of measurement (see Weisz and Duchin (2006) for more formal discussion).

Case 2: Input Price Is User Specific

If the input price is user specific, or in other words nonhomogeneous, the following holds

$$a^*_{ij} = \frac{p_{ij} x_{ij}}{p_j x_j} = a_{ij}\frac{p_{ij}}{p_j} \tag{6.39}$$

and the expression for b^*_{12} becomes

$$b_{12}^* = \frac{a_{12}p_{12}/p_2}{(1 - a_{11}p_1/p_1)(1 - a_{22}p_2/p_2) - a_{12}p_{12}/p_2 a_{21}p_{21}/p_1} = \frac{a_{12}(p_{12}/p_2)}{(1 - a_{11})(1 - a_{22}) - a_{12}a_{21}(p_{12}p_{21})/(p_1p_2)} \tag{6.40}$$

which does not reduce to (6.38) unless $p_{12} = p_1$ and $p_{21} = p_2$. The presence of user specific (heterogeneous) input prices thus has severe consequences for the use of input coefficients obtained from an MIOT. A high level of sector aggregation is a major reason for the existence of heterogeneous prices, because it results in the mixing of a large number of different products with different prices as aggregated inputs and outputs. This can be coped with by increasing the resolution of the MIOT, that is, by employing more detailed sectoral classifications (see Weisz and Duchin (2006) for further discussion).

Accommodating Price Variations Over Time by Price Indices

Product prices vary over time in response to changes in demand and supply conditions, among other things. Changes in the input coefficients taken from the MIOTs for different years, the current year, and a base year, say, reflect both changes in prices and changes in underlying physical input–output relationships. It is often important to isolate the former changes from the latter. This can be facilitated by harmonizing the price levels (i.e., converting current price levels to base or reference levels) using price indices (see Miller and Blair 2009, p. 157; Nakamura and Kondo 2009, p. 40).

Accommodating Price Variations Over Space

Product prices of products generally vary not only over time but geographically as well, making comparison of the MIOTs of different regions/countries problematical. Simple conversion using exchange rates of various currencies relative to a reference currency like the US dollar is unable to capture geographical differences in relative prices. This is an example of the user-specific price levels mentioned above. Resorting to spatial price indices, known as purchasing power parity (PPP) indices, is a possible way to address this issue (ICP, Hertwich and Peters 2010).

Conversion of MIOT to Physical Flows of Materials: WIO-MFA

If there is price homogeneity among users, an MIOT can be transformed into a physical IOT (PIOT) in a straightforward fashion. Its multiplication by the inverse of the diagonal matrix of the vector of relevant price indices gives rise to the corresponding PIOT. While this conversion gives physical flows, they refer to the weight-based aggregation of diverse materials with differing potentials for environmental effects. For an LCA, disaggregated flows of individual materials/substances would be more useful. Waste input–output material flow analysis (WIO-MFA) (Nakamura and Nakajima 2005; Nakamura et al. 2007) is a methodology that meets this requirement. At the core of this methodology is the derivation, from physical information on the sectoral use of materials, of a matrix of material composition that gives the composition of products in terms of a given set of materials. WIO-MFA can convert an MIOT into the corresponding PIOT in terms of individual materials (examples include Nakamura et al. (2009) for PVC and Nakajima et al. (2013) for elements of alloy steel such as Ni, Cr, and Mo). Furthermore, WIO-MFA enables estimation of the flow of specific materials/substances associated with particular products (see Nakamura et al. 2011) for the flows of primary and secondary steel associated with a car, and Ohno et al. (2014) for the flows of ferrous materials recovered from EoL cars).

In the following, it is assumed that the price of a product is the same for all its users. The input coefficients obtained from an MIOT are treated as equivalent to those from physical data. Accordingly, not $A^* = (a^*_{ij})$ but $A = (a_{ij})$ will be used as the matrix of input coefficients throughout the rest of this chapter, although in reality these have been mostly obtained from MIOT.

2.3.4 Imports

Imports were mentioned above in relation to national accounts, but have not yet been explicitly taken into account in the IO model. Imports can be described as competitive or complementary (noncompetitive) according to whether or not the products in question are produced domestically (Stone 1961 p. 55). In countries at high altitude, bananas and coffee are noncompetitive imports, while for many of them, textiles and electronics are competitive. The distinction is a relative one, and will change depending on the level of resolution of product definition. Many products will become noncompetitive as product classification becomes finer, while the converse will hold as classification becomes coarser.

The IO model can be broken down into a competitive import model and a noncompetitive import model, in accordance with whether products of domestic and foreign origin are regarded as competitive or noncompetitive.

Competitive IO Model

In the competitive IO model, a given product j can be supplied by both domestic and foreign sources. Accordingly, the balance of output is given by (6.22). Subtraction of imports from both sides of (6.22) gives

$$
\begin{aligned}
x_i^d &= \sum_j (x_{ij}^d + x_{ij}^m) + (y_{iC}^d + y_{iC}^m) \\
&\quad + (y_{iI}^d + y_{iI}^m) + (y_{iX}^d + y_{iX}^m) - x_i^m \\
&= \sum_j x_{ij} + y_{iC} + y_{iI} + y_{iX} - x_i^m.
\end{aligned} \tag{6.41}
$$

Henceforth, the expression $H = (h_{ij})$ refers to a matrix H with h_{ij} as its (i, j)-th element. Stacking the above equation for all n sectors and using matrix notation, this becomes

$$x^d = Ax^d + y - x^m \tag{6.42}$$

where $A = (a_{ij})$ with

$$a_{ij} = \frac{x_{ij}^d + x_{ij}^m}{x_j^d} \tag{6.43}$$

The presence of x_{ij}^m in the definition of a_{ij} implies the dependence of imports, x^m, on domestic production, x^d, which needs to be taken into account in (6.42) as in (6.20). Following the Chenery–Moses model (Moses 1955)), define the share of imports in total domestic demand, μ_i, as

$$\mu_i = \frac{x_i^m}{\sum_j x_{ij} + \sum_k y_{ik}} = \frac{x_i^m}{\sum_j a_{ij} x_j^d + \sum_k y_{ik}} \tag{6.44}$$

or in matrix notation

$$\text{diag}(\mu) = \text{diag}(x^m)\, \text{diag}(Ax^d + y)^{-1}, \tag{6.45}$$

where diag(z) refers to a diagonal matrix with the elements of vector z occurring in its diagonal elements. By use of μ, (6.42) becomes

$$x^d = Ax^d + y - \text{diag}(\mu)(Ax^d + y) \tag{6.46}$$

with solution

$$x^d = (I - (I - \text{diag}(\mu))A)^{-1}(I - \text{diag}(\mu))y \tag{6.47}$$

The demand for imports is then given by

$$\begin{aligned} x^m &= \text{diag}(\mu)\left(Ax^d + y\right) \\ &= \text{diag}(\mu)\left(A(I - (I - \text{diag}(\mu))A)^{-1}(I - \text{diag}(\mu))y + y\right) \end{aligned} \tag{6.48}$$

Noncompetitive IO Model

In the noncompetitive IO model, the counterpart of (6.41) needs to be written for domestic and imported product separately because of the absence of homogeneous counterparts:

$$\begin{aligned} x_i^d &= \sum_j x_{ij}^d + y_i^d, \quad i = 1, \ldots, n \\ x_i^m &= \sum_j x_{ij}^m + y_i^m, \quad i = 1, \ldots, n_m \end{aligned} \tag{6.49}$$

where n_m may differ from n. In matrix notation, this becomes

$$\begin{aligned} x^d &= A^d x^d + y^d \\ x^m &= A^m x^d + y^m \end{aligned} \tag{6.50}$$

where $A^m = \left(a_{ij}^m\right)$ with

$$a_{ij}^m = \frac{x_{ij}^m}{x_j^d}, \tag{6.51}$$

and the solution is given by

$$\begin{aligned} x^d &= \left(I - A^d\right)^{-1} y^d \\ x^m &= A^m \left(I - A^d\right)^{-1} y^d + y^m \end{aligned} \tag{6.52}$$

2.3.5 Multiregional Extension

The above model is a domestic country-centric model, focused solely on the domestic economy. In other words, the analysis is of one direction, with its focus on the effects on the exporting economy of the domestic economy, but not on the effects other way round. For instance, in today's globalized supply chain, an increase in the import of a group of products, say consumer products, may result in an increase in the export of other groups of products, say materials. Consideration of these mutual interdependences via foreign trade requires multiregional extension of the IO model. A multiregional input–output table (MRIO) covers a number of nations or regions with a certain geographic resolution. For a single country, the whole country serves as the system boundary, while there are MRIOs that amalgamate countries into multiple regions. MRIOs that encompass the economic activities of multiple countries, taking their borders as the system boundary, are called international IOTs.

Two-Country Case

For the sake of simplicity, consider first the case where there are only two countries (regions) in the world (a domestic country and the rest of the world), countries a and b. Furthermore, set the number of products equal to, say, n, across the two countries. Supposing that a refers to the domestic country above, Eq. (6.49) can be rewritten as

$$\begin{aligned} x_i^a &= \underbrace{\sum_j x_{ij}^{aa} + y_i^{aa}}_{\text{use in } a} + \underbrace{\sum_j x_{ij}^{ab} + y_i^{ab}}_{x_i^{ab}:\ \text{export to } b} \\ x_i^{ba} &= \underbrace{\sum_j x_{ij}^{ba} + y_i^{ba}}_{x_i^{ba}:\ \text{export to } a} \end{aligned} \tag{6.53}$$

where x_{ij}^{ab} refers to the intermediate use of i, say, steel, produced in country a to produce j, say, cars, in country b, and y_i^{ba} refers to the final use in country a of i, say, an aircraft, produced in country b.

Extending this model to a MRIO requires introduction of a component referring to the use of x_i^b in country b:

$$\begin{aligned} x_i^a &= \sum_j \left(x_{ij}^{aa} + x_{ij}^{ab}\right) + \left(y_i^{aa} + y_i^{ab}\right) \\ x_i^b &= \sum_j \left(x_{ij}^{ba} + x_{ij}^{bb}\right) + \left(y_i^{ba} + y_i^{bb}\right) \end{aligned} \tag{6.54}$$

Stacking these equations for all the products and using obvious matrix notation, it follows that

$$\begin{aligned} x^a &= A^{aa}x^a + A^{ab}x^b + y^{aa} + y^{ab} \\ x^b &= A^{ba}x^a + A^{bb}x^b + y^{ba} + y^{bb} \end{aligned} \tag{6.55}$$

or

$$\begin{pmatrix} x^a \\ x^b \end{pmatrix} = \begin{pmatrix} A^{aa} & A^{ab} \\ A^{ba} & A^{bb} \end{pmatrix} \begin{pmatrix} x^a \\ x^b \end{pmatrix} + \begin{pmatrix} y^{aa} + y^{ab} \\ y^{ba} + y^{bb} \end{pmatrix} \tag{6.56}$$

where $A^{ab} = \left(a_{ij}^{ab}\right)$ and $y^{ba} = \left(y_i^{ba}\right)$, with solution

$$\begin{aligned} \begin{pmatrix} x^a \\ x^b \end{pmatrix} &= \begin{pmatrix} I - A^{aa} & -A^{ab} \\ -A^{ba} & I - A^{bb} \end{pmatrix}^{-1} \begin{pmatrix} y^{aa} + y^{ab} \\ y^{ba} + y^{bb} \end{pmatrix} \\ &= \begin{pmatrix} B^{aa} & B^{ab} \\ B^{ba} & B^{bb} \end{pmatrix} \begin{pmatrix} y^{aa} + y^{ab} \\ y^{ba} + y^{bb} \end{pmatrix} \end{aligned} \tag{6.57}$$

This model provides an easy way to quantitatively assess the interdependence of international supply chains. Suppose, for instance, that a is intensive in manufacturing sectors, while b is intensive in resource sectors. An increase in the export of cars from a to b may result in an increase in the export of iron ore from b to a. These effects can be captured by $b_{ij}^{ba} y_j^{ab}$ with $i =$ iron ore and $j =$ car.

n Country Case

Extension of (6.56) to an arbitrary number of countries/regions, say, from $a =$ Afghanistan to $z =$ Zimbabwe, is straightforward, at least conceptually:

$$\begin{pmatrix} x^a \\ \vdots \\ x^z \end{pmatrix} = \begin{pmatrix} A^{aa} & \cdots & A^{az} \\ \vdots & \cdots & \cdots \\ A^{za} & \cdots & A^{zz} \end{pmatrix} \begin{pmatrix} x^a \\ \vdots \\ x^z \end{pmatrix} + \begin{pmatrix} \sum_{J=a,\ldots z} y^{aJ} \\ \vdots \\ \sum_{J=a,\ldots z} y^{zJ} \end{pmatrix} \tag{6.58}$$

Actual compilation of an n-country MRIO is still quite a challenging task, however. Besides the compilation of mutually consistent multiregional trade flow matrices, this applies in particular to the conversion of national/country MIOTs to common price and classification levels (Hertwich and Peters 2010; Timmer 2012; Lenzen et al. 2013) (see the above discussion about PPP in Sect. 2.3.3).

Available World MRIO Database

Several organizations are currently developing MRIO tables for the entire global economy, some of which are available to the public free of charge or for a fee. Each of these employ a different set of years, countries, and regions, as well as a different system of sectoral classification. A number of them use the compiled data to estimate sectoral environmental burdens and suchlike, which are presented in satellite accounts. The following introduces the main MRIO tables that are currently available.

The Institute of Developing Economies, Japan External Trade Organization (IDE-JETRO), has developed Asian international input–output tables (AIIOTs) encompassing 76 sectors in each of ten countries, including East Asian countries and the USA, for every 5 years between 1975 and 2005, and provides the 2005 table and those thereafter free of charge (OECD web).

The Organization for Economic Cooperation and Development (OECD) has prepared intercountry input–output (ICIO) tables comprising data for 58 countries including the "rest of the world (ROW)," which are also provided free of charge (OECD web). Tables are currently available for 1995, 2000, 2005, 2008, and 2009.

The Global Trade Analysis Project (GTAP) of the Center for Global Trade Analysis, Purdue University, offers the GTAP database for a fee (GTAP web). The current, latest version, GTAP8, has tables for 2004 and 2007, comprising 129 countries and regions, including the ROW, each of which consists of 57 endogenous sectors. These tables are supplemented by satellite data on such issues as energy consumption, CO_2 emissions, and land use. While the GTAP database is not in fact founded on the MRIO accounting framework, MRIO tables can be constructed from the data provided (Peters et al. 2011).

A New Environmental Accounting Framework Using Externality Data and Input–Output Tools for Policy Analysis (EXIOPOL), an EU research project conducted between 2007 and 2011, prepared a year 2000 MRIO table composed of 44 countries, including the ROW, each of which consists of 129 sectors (EXIOBASE web). Extensive satellite data on environmental burden and resource use were used to estimate 30 types of environmental load and 80 types of resource use for each sector.

The World Input–Output Database: Construction and Applications (WIOD) project, another EU research project, carried out between 2009 and 2012, constructed MRIO tables for each year from 1995 to 2009, which are provided free of charge (WIO Database web). The MRIO tables are defined based on a classification comprising 35 industries and 59 products and are supplemented by satellite data on energy consumption, CO_2 emissions, land use, water consumption, and other inputs as environmental burden data.

ISA (Integrated Sustainability Analysis), at the University of Sydney, has developed and made available free of charge the Eora MRIO database consisting of 15,909 sectors $\times$ 15,909 sectors in a total of 187 countries, each of which includes 25–500 sectors (Eora MRIO web). Tables were prepared for each of the years from 1990 to 2009. The satellite data are also useful, covering such issues as energy consumption, greenhouse gas emissions, ecological footprint, water footprint, and effect on endangered species listed by the International Union for Conservation of Nature (IUCN).

A number of country-based footprint analyses have been conducted using the MRIO data outlined above. Examples include studies of carbon footprints using GTAP (Hertwich and Peters 2009; Davis and Caldeira 2010), footprints of land use and water consumption (Steen-Olsen et al. 2012; Weinzettel et al. 2013), biodiversity footprints using Eora (Lenzen et al. 2012b), and worker footprints (Alsamawi et al. (2014)). Furthermore, a health impact footprint of $PM_{2.5}$ that incorporates consideration of atmospheric advection and diffusion of particles has been calculated using AIIOT (Takahashi et al. 2014).

2.4 Cost and Price Models in IO

A feature of IO that appears to be less well known, at least in the LCA and IE, is that it can also be used to consider the effects on product prices of the prices of primary factors. Let p^d be a $1 \times n$ vector with the price of domestic product i as its i-th element and p^m be the corresponding vector of import prices. Writing (6.34) in a matrix form

$$p^d = p^d A^d + p^m A^m + P_L a_L + P_K a_K + ts^*, \tag{6.59}$$

and solving for p^d gives

$$p^d = (p^m A^m + P_L a_L + P_K a_K + ts^*)(I - A^d)^{-1}. \tag{6.60}$$

The prices occurring in the first parenthesis on the right-hand side are given from outside, that is, they are exogenous to the system under consideration: there is no mechanism in the above IO model to determine these prices. Accordingly, (6.60) can be used to assess the effects on domestic prices of changes in these exogenous prices and/or the rate of taxes minus subsidies.

The above price model was derived for a noncompetitive import model. The competitive import version of (6.60) is given by

$$\begin{aligned} p^d = {} & (p^m \mathrm{diag}(\mu) A + P_L a_L + P_K a_K + ts^*) \\ & (I - (I - \mathrm{diag}(\mu) A))^{-1}. \end{aligned} \tag{6.61}$$

In essence, the price model states that the price of an endogenous product is determined by the price of exogenous inputs and technology. It may seem surprising that the model contains no variables referring to the magnitude of demand, which may appear counterintuitive to the notion that prices are determined by supply and demand conditions. This apparently counterintuitive feature of the price model follows from the assumption of the underlying technology being subject to constant returns to scale (see Sect. 2.1.2), which manifests itself in the unit cost of production being independent of the level of production (see (6.34)).

Note that this feature applies equally to the quantity model (6.20), manifesting itself in production levels being independent of price levels. In short, in the standard IO model, the levels of output and price are determined independently of one another. This feature enables one to consider the effects on production of a change in final demand without bothering about the effects of the latter on price levels: a great simplification of analysis. A brief account of the case where this simplification is not assumed is provided in Sect. 7.2.1

3 Input–Output LCA

3.1 EEIO: Environmentally Extended IO

Let there be m types of emissions such as SOx, PM, and CO_2, the amounts of which can be regarded as proportional to the use of certain inputs in production processes. Denote by $D = (d_{kj})$ an $m \times n$ matrix, with d_{kj} referring to the amount of emission (or environmental burden, to be more general) k emitted from production sector j. Division of d_{kj} by x_j gives the amount of emission k per unit of production, r_{kj}. Let R be the $m \times n$ matrix with r_{kj} as its elements that is called the unit environmental burden matrix. From (6.20), the amounts of emission, $e{=}(e_{\mathrm{k}})$, that are directly and indirectly generated in satisfying a given final demand y are given by

$$e = R(I - A)^{-1}y \tag{6.62}$$

This is the fundamental equation of IO-LCA. In the language of LCA, y refers to the functional unit. If one is concerned with the emissions originating from the production of final product j, say, a car, the components of y should be equal to zero except its j component, y_j. It is noteworthy that the matrix $(I - A)$ corresponds to the technology matrix in PLCA (Heijungs and Suh 2002).

The above calculation does not consider the use phase. As a result, IO-LCA is often designated as "cradle to gate." Consideration of the use phase is straightforward, however. The additional emission from the use phase of product j, say, a car, can be obtained by including in y the consumption volumes of all the items consumed in the use phase, such as fuel, spare parts, and repair services:

$$y = \begin{pmatrix} \text{a car} \\ \text{fuel} \\ \text{repair parts} \\ 0 \\ \vdots \end{pmatrix} \tag{6.63}$$

In contrast to the use phase, consideration of the EoL phase (including recycling of by-products/wastes) is not straightforward, and requires nontrivial extensions of the standard IO model as shown below in Sect. 3.4.2 (Nakamura and Kondo 2002). Also missing in the above calculation is the production of capital goods required for the production of goods and services. The matrix A refers to the intermediate flow of goods and services, and excludes the flow of capital goods such as machines, equipment, buildings, and infrastructures that are necessary for the realization of x. While many life cycle assessment case studies neglect the production of capital goods, it is doubtful if capital goods can be excluded per se (Frischknecht et al. 2007). Thus, it seems sensible to include capital goods production in EEIO. Before touching upon these issues, this section starts with a discussion about methodological issues associated with the integration of PLCA and IO.

3.2 Hybrid IO-LCA

As described by Suh et al. (2004), the greatest problem of using input–output tables for LCA is that inventory analysis is limited to the products and services in the sector classifications defined by the input–output table. The greatest advantage of using input–output tables for LCA is that the system boundaries for the analysis comprehensively cover a country's entire economy. The flip-side is that because it is comprehensive, the degree of sectoral resolution is low; in other words, the classification of goods and industries becomes coarser. For example, the input–output table for Japan, with among the world's highest sectoral resolutions, is composed of about 400 product sectors. The sectoral categories of an input–output table are defined on the basis of single-product groups (of goods or services) for economic activities of large monetary value or for products meriting future attention, while items of small monetary value are generally combined into aggregates. For a sector defined across multiple products, inventory analysis focused on a single product cannot be conducted.

A specific example from Japan's input–output tables is helpful here. The 2005 table has sectors that include "personal computers," "commercial residential air conditioners," and "cameras." Inventory analysis can therefore be conducted for each of these individual items. However, the "other electrical equipment and devices" sector includes silicon wafers, lighting sockets, permanent magnets, and solar photovoltaic cells. Thus, an LCA of photovoltaic cells would include the other products, too. Even for a sector defined for a single product such as personal computers, this includes desktop computers as well as laptop computers. Despite the materials and processes they embody differing, they are nonetheless aggregated into a single sector. An analysis reflecting the differences between these two products cannot therefore be performed solely on the basis of this input–output table. Although the USA and Korea have developed input–output tables comprising about 500 and 400 categories, respectively, the tables developed by the Statistical Office of the European Communities have only around 60 categories for each European country, making LCAs of individual products using these categories directly almost impossible.

Hybrid IO-LCA is an approach with which these constraints can be overcome. In this method, process data are collected separately on goods for which conventional input–output tables cannot reflect appropriate production processes because of their coarse sectoral classifications, and then added as a supplement to the input–output tables. It is called a hybrid method because it combines bottom-up LCA inventory analysis based on a stacking method with top-down input–output table inventory analysis. This hybrid method has been widely used since its initial application in energy analysis in the 1970s. There are three methodologies for hybrid LCA (Suh and Huppes 2005): tiered hybrid analysis, input–output-based hybrid analysis, and integrated hybrid analysis. In the following sections, the main features of each of these methodologies are introduced.

3.2.1 Tiered Hybrid Analysis

Tiered hybrid analysis is the simplest of the hybrid LCA methodologies; its advantage is its ease of use. For example, consider an LCA for a notebook computer. Because the sectoral classification of the input–output table has no specific category for notebook personal computers, inventory data on the production stages of such computers must be collected by the process method. The energy consumption in the use stage and the waste management fees for the end-of-life stage are then specified, and input–output table analysis is used for inventory analysis of these two stages. Thus, inventory data for several important production stages as well as the product data required for the use and end-of-life stages are gathered, and the analysis of inventory data upstream from that of the collected data is supplemented by input–output analysis.

This method can be formalized as follows. The LCA technology matrix $\widetilde{A} = (\widetilde{a_{ij}})$ comprises inventory data obtained by the process method. The element $\widetilde{a_{ij}}$ is the input or output of good i to or from process j. This corresponds to the description of the materials and energy in the production stage of the notebook computer. The emissions $\widetilde{r_{kj}}$ for environmental burden k per unit operation of process j are an element of the unit environmental burden matrix $\widetilde{R} = (\widetilde{r_{ij}})$. If $\widetilde{y} = (\widetilde{y_i})$ is the vector for the functional unit of assessment, e.g., per notebook computer, the direct and indirect emissions of environmental burden k per functional unit become $\widetilde{R}\widetilde{A}^{-1}\widetilde{y}$. Let $y = (y_i)$ be the vector determining the functional unit for the input–output table sector. The costs of energy consumption during use and end-of-life management obtained and the values of each corresponding to sector i are set to y_i. The direct and indirect environmental burden k per functional unit y becomes the emissions $R(I - A)^{-1}$. The total of the environmental burden from the process method and that from the input–output table analysis is the emissions e_{TH} from the tiered hybrid analysis, as expressed by Eq. (6.64):

$$e_{\mathrm{TH}} = \widetilde{R}\widetilde{A}^{-1}\widetilde{y} + R(I - A)^{-1}y \tag{6.64}$$

Although this method is simple, it does not take into consideration the interrelationship of $\widetilde{y}$ and y or that of $\widetilde{A}$ and A. Thus, even if the LCA system described by the process method impacts the economic system described by the input–output table, for example, that impact cannot be considered. Additionally, because the production system for the process method is, as a general rule, theoretically included in the transactions of the input–output table for the same year, one must be aware of the possibility of double counting the inventory, depending on how the functional unit is specified. This double-counting problem violates the clarity of the system boundary, which is the greatest advantage of input–output analysis. To avoid double counting, Strømman et al. (2009), Strømman (2009), and Lenzen (2009) have developed a methodology to apply structural path analysis (SPA) to input–output table analysis. Lenzen and Crawford (2009) have also proposed the path exchange method, which uses SPA, and have used it for an environmental assessment of different departments of a university (Baboulet and Lenzen 2010).

3.2.2 IO-Based Hybrid Analysis

IO-based hybrid analysis does not link inventory data obtained by the process method; rather, it subdivides the sectoral classification of the input–output table. Because this method maintains the calculation system of the input–output table, the spatial system boundary remains the whole of a single country. Subdivision of categories involves division of the personal computer sector i into a "Notebook computer" sector ia and an "Other personal computers" sector ib, for example. In this case, because the number of sectoral rows and columns in the input–output table has each increased by one, the input coefficient A is expanded in the manner of $A^{'}$ shown in Eq. (6.65):

$$A^{'} = \begin{pmatrix} a_{11} & \cdots & a_{1,ia} & a_{1,ib} & \cdots & a_{1n} \\ \vdots & & \vdots & \vdots & & \vdots \\ a_{ia,1} & \cdots & a_{ia,ia} & a_{ia,ib} & \cdots & a_{ia,n} \\ a_{ib,1} & \cdots & a_{ib,ia} & a_{ib,ib} & \cdots & a_{ib,n} \\ \vdots & & \vdots & \vdots & & \vdots \\ a_{n1} & \cdots & a_{n,ia} & a_{n,ib} & \cdots & a_{nn} \end{pmatrix} \tag{6.65}$$

Similarly, the functional unit vector y and the unit environmental burden matrix R are also subdivided for each sector. Let these be $R^{'}$ and $y^{'}$; then the environmental burden e_{IOH} for the functional unit $y^{'}$ obtained by the IO-based hybrid is given by

$$e_{\mathrm{IOH}} = R^{'}\left(I - A^{'}\right)^{-1} y^{'} \tag{6.66}$$

3.2.3 Integrated Hybrid Analysis

Integrated hybrid analysis is an analytical method that tackles the issue that becomes problematic in tiered hybrid analysis – the interaction between the LCA technical system $\tilde{A}$ described by the process method and the economic system $\mathbf{A}$ described by the input–output table. First defined are the upstream cutoff matrix C^{u}, which describes the inputs from the input–output table sector to the LCA technical system $\widetilde{A}$, and conversely the downstream cutoff matrix C^d, which describes the outputs to the input–output table from the LCA technical system $\widetilde{A}$. By using the matrices C^u and C^d to connect A and $\widetilde{A}$, respectively, the mutual interaction between the process method and input–output table analysis is incorporated. The environmental burden e_{IH} that accompanies the functional unit $\widetilde{y}$ is calculated from Eq. (6.67):

$$e_{IH} = \begin{pmatrix} \widetilde{R} & 0 \\ 0 & R \end{pmatrix} \begin{pmatrix} \widetilde{A} & -C^d \\ -C^u & I - A \end{pmatrix}^{-1} \begin{pmatrix} \widetilde{y} \\ 0 \end{pmatrix} \tag{6.67}$$

Note that if the LCA technical system has already been incorporated into the input–output table, it is necessary to use matrix A^{corr} as shown in Eq. (6.68), which has duplicated portions removed from matrix A.

$$e_{IH} = \begin{pmatrix} \widetilde{R} & 0 \\ 0 & R \end{pmatrix} \begin{pmatrix} \widetilde{A} & -C^d \\ -C^u & I - A^{corr} \end{pmatrix}^{-1} \begin{pmatrix} \widetilde{y} \\ 0 \end{pmatrix} \tag{6.68}$$

The method for generating A^{corr} is described in detail in the Appendix of Suh (2004). The path exchange method of Lenzen and Crawford 2009), which uses SPA, in addition to achieving the same results as integrated hybrid analysis, has the feature that the correction of matrix A is more flexible than that in integrated hybrid analysis. However, partial correction of the transactions between sectors of matrix A expanded under SPA does not preserve the balance of currency flow.

3.2.4 Rounding Up

Of the three hybrid IO methodologies discussed above, the IO-based and the integrated methods keep the system boundary intact, while for the tired method does not always hold. The IO-based method is straightforward when the input coefficients referring to particular sectors are to be completely replaced by extended ones based on, say, process information. The above example of the aggregated computer sector disaggregated into two types of PCs is a case in point. If replacement is not complete, however, and the extension is achieved by adding new input coefficients referring to "Note PCs," while keeping the original coefficients of "PCs" unaltered to represent "Other PCs," due care is required to adjust the latter for inclusion of the former, a formal description of which is the integrated method.

In the rest of this chapter, it is understood that the elements (6.62) are available at the desired level of resolution for use in these hybrid approaches.

3.3 Capital Goods Input

The flow of goods and services is made possible by the presence of the stock of capital goods such as machinery, buildings, and infrastructure. In IO accounts, purchases of capital goods are counted as final outputs of the economic system rather than as intermediate inputs into production, that is, they are counted as elements of gross capital formation, y_I in (6.22). Because the matrix A does not include capital goods, the impacts captured by (6.62) are limited to those associated with current production, but exclude those associated with the production of capital goods enabling current production. The same applies to PLCA as well: many PLCA case studies neglect the production of capital goods (Frischknecht et al. 2007). It was found by Frischknecht et al. (2007), however, that the contributions of capital goods predominate when it comes to the environmental impacts of products like photovoltaic and wind power, no matter which indicator is chosen.

Write I_j for the amount of purchases of capital goods, that is, fixed investment, conducted by sector j within a given year, and I_{ij} for its i-th element, mostly durables such as machinery, building, and installations, but possibly also including nonphysical items like consulting, trade, transport, and miscellaneous services related to physical investment. Noting that I_{ij} is in monetary units, it follows by definition that

$$\sum_i I_{ij} = I_j \tag{6.69}$$

On the other hand, the sum of I_{ij} over j gives the total amount of product i that was used for fixed investment. Recalling that this is equal to the final demand for product i for the purpose of gross capital formation, it follows that

$$\sum_j I_{ij} = y_{iI} \tag{6.70}$$

The matrix $I = \left(I_{ij}\right)$ is called the investment matrix and corresponds to the capital counterpart of intermediate flow matrix $X = \left(x_{ij}\right)$. Investment in capital is made for the installment of new productive capacity or for the replacement of existing productive capacity. Write $\overline{x}_j$ for the amount of productive capacity to be realized and $\overline{I}_j$ for the amount of capital investment required for it. Define the capital coefficient:

$$k_{ij} = \frac{\overline{I}_{ij}}{\overline{x}_j} \tag{6.71}$$

which gives the amount of input of product i for the expansion of a unit of productive capacity of sector j. Writing $K = \left(k_{ij}\right)$, the environmental burdens that result from not only the current production of y but also from construction of the underlying productive capacity will then be given by

$$R(I-(A+K))^{-1}y \tag{6.72}$$

Suppose that realization of this y calls for the production of steel, among other things. Using Eq. (6.62), one can assess the environmental burdens associated with the production of the intermediate goods like ore, limestone, oxygen, power, and fuels, which are required for its realization, but not those associated with the production of capital goods such as steel mills, oxygen plants, power stations, petrochemical refineries, and transport systems. Equation 6.72 enables one to consider the effects associated with the production of both intermediate and capital goods.

In reality, however, application of Eq. (6.72) is hampered by the paucity of data on K, and examples of real application are rare (e.g., see Lenzen and Treloar 2005). In particular, mere availability of the investment matrix I_{ij}, compilation of which is not a routine matter as with an IOT, does not suffice for estimating the K matrix. This is so because the K matrix calls not simply for I_{ij} but also for $\bar{I}_{ij}$ and $\bar{x}_j$, provision of which will require detailed engineering information. Alternatively, one could resort to using LCA inventory data to provide information on capital goods (Frischknecht et al. 2007).

3.4 *End of Life, By-products, Waste, and Waste Management*

IO-LCA is often criticized for its neglect of the EoL phase including recycling (Reap et al. 2008). The rest of this section is devoted to further extensions of IO-LCA that have been developed to accommodate the EoL phase, including both closed-loop and open-loop recycling of by-products and wastes. It is shown that with these extensions, IO-LCA can answer the above criticism.

3.4.1 Waste and Waste Treatment Without Recycling

We start from a simple case where wastes are disposed of without any recycling.

Landfill as the Only Way of Waste Disposal

First, consider a simple case with two producing sectors and one landfill sector, as given by Table 6.2. Landfill is the only waste disposal method, and there is no recycling. The coefficient g_j^{out} refers to the generation of waste for landfill per unit of production in sector j, y_w to the amount of waste for landfill from the final demand sector, and y_e to the amount of direct emission from the final demand sector.

Table 6.2 A two-sector model with waste generation and landfill

	Sector 1	Sector 2	Sector 3 (landfill)	Final demand
Sector 1	a_{11}	a_{12}	a_{13}	y_1
Sector 2	a_{21}	a_{22}	a_{23}	y_2
Waste	g_1^{out}	g_2^{out}	g_3^{out}	y_w
Emission	r_1	r_2	r_3	y_e

A salient feature of IOT in Table 6.2 is the occurrence of waste in place of sector 3 as a row element. In the present case, this "asymmetry" causes no problem because landfill is the only means of disposal, and hence its level of activity (the amount of waste disposed to a landfill), say, x_3, is equal to the total amount of waste generated. Accordingly, the following balance holds for waste and landfill:

$$\underbrace{x_3}_{\text{waste landfilled}} = \underbrace{g_1^{out}x_1 + g_2^{out}x_2 + g_3^{out}x_3 + y_w}_{\text{waste generated}} \tag{6.73}$$

The right-hand side of this equation gives direct or production origins of waste. In line with (6.20), the solution for output is then given by

$$\begin{pmatrix} x_1 \\ x_2 \\ x_3 \end{pmatrix} = \begin{pmatrix} 1-a_{11} & -a_{12} & -a_{13} \\ -a_{21} & 1-a_{22} & -a_{23} \\ -g_1^{out} & -g_2^{out} & 1-g_3^{out} \end{pmatrix}^{-1} \begin{pmatrix} y_1 \\ y_2 \\ y_w \end{pmatrix} \tag{6.74}$$

Writing b_{ij} for the (i,j) component of the inverse matrix on the right-hand side (6.74), the following holds for the waste balance:

$$\underbrace{x_3}_{\text{waste landfilled}} = \underbrace{b_{31}y_1 + b_{32}y_2 + b_{33}y_w}_{\text{waste generated}} \tag{6.75}$$

where $b_{31}y_1$ refers to the waste directly and indirectly generated to satisfy the final demand for product 1 and $b_{33}y_w$ to the waste directly and indirectly generated to dispose of the waste from the final demand sector. The above model was first considered by Leontief (1970). In contrast to (6.73), which gives the production origins of waste, (6.74) gives the final demand origins of waste.

Landfill and Incineration

Consider next the case where incineration occurs as an additional waste treatment process. Incineration converts the noncombustible components of waste into another waste, namely, residue (ash). Accordingly, Table 6.2 has to be extended in terms of both processes and wastes, as in Table 6.3, in which the coefficient g_{24}^{out} refers to the generation of incineration residue per unit of waste incinerated.

Table 6.3 A two-sector model with waste generation, landfill, and incineration

	Sector 1	Sector 2	Landfill	Incineration	Final demand
Sector 1	a_{11}	a_{12}	a_{13}	a_{14}	y_1
Sector 2	a_{21}	a_{22}	a_{23}	a_{24}	y_2
Waste 1	g_{11}^{out}	g_{12}^{out}	g_{13}^{out}	g_{14}^{out}	y_{w1}
Waste 2 (residue)	0	0	0	g_{24}^{out}	0
Emission	r_1	r_2	r_3	r_4	y_e

The following holds with regard to waste balances:

$$\begin{aligned} g_{11}^{\text{out}}x_1 + g_{12}^{\text{out}}x_2 + g_{13}^{\text{out}}x_3 + g_{14}^{\text{out}}x_4 + y_{w1} &= w_1 \\ g_{21}^{\text{out}}x_1 + g_{22}^{\text{out}}x_2 + g_{23}^{\text{out}}x_3 + g_{24}^{\text{out}}x_4 + y_{w2} &= w_2 \end{aligned} \tag{6.76}$$

where w_i refers to the amount of waste i to be disposed of. In the present case of no waste recycling, w_i coincides with the amount of waste i generated. Integrated with the balances for products, this becomes

$$\begin{pmatrix} a_{11} & a_{12} & a_{13} & a_{14} \\ a_{21} & a_{22} & a_{23} & a_{24} \\ g_{11}^{\text{out}} & g_{12}^{\text{out}} & g_{13}^{\text{out}} & g_{14}^{\text{out}} \\ g_{21}^{\text{out}} & g_{22}^{\text{out}} & g_{23}^{\text{out}} & g_{24}^{\text{out}} \end{pmatrix} \begin{pmatrix} x_1 \\ x_2 \\ x_3 \\ x_4 \end{pmatrix} + \begin{pmatrix} y_1 \\ y_2 \\ y_{w1} \\ y_{w2} \end{pmatrix} = \begin{pmatrix} x_1 \\ x_2 \\ w_1 \\ w_2 \end{pmatrix} \tag{6.77}$$

or, in matrix form,

$$\begin{pmatrix} A_{\text{I}} & A_{\text{II}} \\ G_{\text{I}}^{\text{out}} & G_{\text{II}}^{\text{out}} \end{pmatrix} \begin{pmatrix} x_{\text{I}} \\ x_{\text{II}} \end{pmatrix} + \begin{pmatrix} y_{\text{I}} \\ y_{\text{w}} \end{pmatrix} = \begin{pmatrix} x_{\text{I}} \\ w \end{pmatrix} \tag{6.78}$$

where $\text{I} = \{1, 2\}$ is the set of production sectors and $\text{II} = \{3, 4\}$ the set of waste treatment sectors, with $A_{\text{I}} = (a_{ij}), i, j \in \text{I}$, $A_{\text{II}} = (a_{ij}), i \in \text{I}, j \in \text{II}$, $G_{\text{I}}^{\text{out}} = (g_{ij}^{\text{out}}), j \in \text{I}$, and $G_{\text{II}}^{\text{out}} = (g_{ij}^{\text{out}}), j \in \text{II}$. This system is not solvable, because of the asymmetry; wastes occur on the right-hand side, while the levels of waste treatment occur on the left-hand side.

In the case of Table 6.2, it was possible to derive the solution (6.74) because of the one-to-one correspondence between the waste and its treatment. In the present case represented by Table 6.3, this is not the case, because waste 1 can be submitted to both landfill and incineration, while for waste 2 landfill is the only disposal option (one-to-one correspondence is available) (Nakamura and Kondo 2002). The very fact that waste 1 can be submitted to either or both the treatment processes implies that the amount of waste 1 for disposal corresponds to neither the activity level of landfill nor that of incineration, although it is true that the total activity level of the two treatment processes must equal the total amount of wastes. The fundamental difference between Tables 6.3 and 6.2 lies in the fact that a solution in the form of (6.74) is not readily available without further information on how wastes are to be allocated between the two treatment processes. The Leontief EEIO model (Leontief 1970), including its extension by Duchin (1990), is not able to deal with this situation.

3.4.2 Waste IO (WIO) Model

To cope with the above issue, Nakamura and Kondo (2002) introduced the concept of an allocation matrix $S = \left(s_{ij}\right)$, with s_{ij} giving the share of waste j submitted to treatment process i. Because waste for treatment has to be subjected to at least one treatment process, it follows that

$$\sum_{i=1}^{n_{\mathrm{W}}} s_{ij} = 1, \quad j = 1, \ldots, n_{\mathrm{T}} \tag{6.79}$$

where n_{W} refers to the number of waste types for treatment and n_{T} to the number of waste treatment processes. For the example in Table 6.3, S is a 2×2 matrix with s_{11} referring to the share of waste 1 landfilled and s_{21} to the share of waste 1 incinerated. Since landfill is the only treatment option applicable to waste 2 (incineration residue), it follows $s_{12} = 1$ and $s_{22} = 0$. Multiplication of the last two rows of (6.77) by S gives a symmetric representation:

$$\begin{pmatrix} a_{11} & a_{12} & a_{13} & a_{14} \\ a_{21} & a_{22} & a_{23} & a_{24} \\ s_{11}g_{11}^{\mathrm{out}} + s_{12}g_{21}^{\mathrm{out}} & s_{11}g_{12}^{\mathrm{out}} + s_{12}g_{22}^{\mathrm{out}} & s_{11}g_{13}^{\mathrm{out}} + s_{12}g_{23}^{\mathrm{out}} & s_{11}g_{14}^{\mathrm{out}} + s_{12}g_{24}^{\mathrm{out}} \\ s_{21}g_{11}^{\mathrm{out}} + s_{22}g_{21}^{\mathrm{out}} & s_{21}g_{12}^{\mathrm{out}} + s_{22}g_{22}^{\mathrm{out}} & s_{21}g_{13}^{\mathrm{out}} + s_{22}g_{23}^{\mathrm{out}} & s_{21}g_{14}^{\mathrm{out}} + s_{22}g_{24}^{\mathrm{out}} \end{pmatrix} \begin{pmatrix} x_1 \\ x_2 \\ x_3 \\ x_4 \end{pmatrix} + \begin{pmatrix} y_1 \\ y_2 \\ s_{11}y_{w1} + s_{12}y_{w2} \\ s_{21}y_{w1} + s_{22}y_{w2} \end{pmatrix} = \begin{pmatrix} x_1 \\ x_2 \\ s_{11}w_1 + s_{12}w_2 \\ s_{21}w_1 + s_{22}w_2 \end{pmatrix} = \begin{pmatrix} x_1 \\ x_2 \\ x_3 \\ x_4 \end{pmatrix} \tag{6.80}$$

or in matrix form

$$\begin{pmatrix} A_{\mathrm{I}} & A_{\mathrm{II}} \\ SG_{\mathrm{I}}^{\mathrm{out}} & SG_{\mathrm{II}}^{\mathrm{out}} \end{pmatrix} \begin{pmatrix} x_{\mathrm{I}} \\ x_{\mathrm{II}} \end{pmatrix} + \begin{pmatrix} y_{\mathrm{I}} \\ Sy_{\mathrm{w}} \end{pmatrix} = \begin{pmatrix} x_{\mathrm{I}} \\ Sw \end{pmatrix} = \begin{pmatrix} x_{\mathrm{I}} \\ x_{\mathrm{II}} \end{pmatrix} \tag{6.81}$$

the solution of which is analogous to (6.20):

$$\begin{pmatrix} x_{\mathrm{I}} \\ x_{\mathrm{II}} \end{pmatrix} = \begin{pmatrix} I - A_{\mathrm{I}} & -A_{\mathrm{II}} \\ -SG_{\mathrm{I}}^{\mathrm{out}} & I - SG_{\mathrm{II}}^{\mathrm{out}} \end{pmatrix}^{-1} \begin{pmatrix} y_{\mathrm{I}} \\ Sy_{\mathrm{w}} \end{pmatrix} \tag{6.82}$$

Writing R_{I} for the direct emission coefficients from producing sectors and R_{II} for the direct emission coefficients form waste treatment sectors, the equivalent of (6.62) becomes

$$e = \begin{pmatrix} R_{\mathrm{I}} \\ R_{\mathrm{II}} \end{pmatrix} \begin{pmatrix} I - A_{\mathrm{I}} & -A_{\mathrm{II}} \\ -SG_{\mathrm{I}}^{\mathrm{out}} & I - SG_{\mathrm{II}}^{\mathrm{out}} \end{pmatrix}^{-1} \begin{pmatrix} y_{\mathrm{I}} \\ Sy_{\mathrm{w}} \end{pmatrix} \tag{6.83}$$

This gives the basic IO equation for LCA from cradle to grave of a product in the absence of recycling.

3.5 *Recycling*

The above model is now extended to incorporate recycling (and reuse) of by-products including waste. The distinction between a by-product and a waste is a fuzzy one, and is affected, among others, by economic conditions: a waste may have a positive price when the economy is booming, but a negative one when the economy is just slugging along. In the following discussion, by-products are broadly defined to include wastes. Following Nakamura and Kondo (2009), it is useful to distinguish two types of by-product, termed by-product I and by-product II, depending on whether there is a production sector where the by-product is obtained as the primary product (Nakamura and Kondo 2009). First, we consider the case of by-product I, for which there is a primary producer, and then proceed to the case of by-product II, for which there is no primary producer.

3.5.1 The Recycling of By-product I

Recycling of by-product I can be accounted for as a negative input from the sector where the product is primarily produced into the sector where it is obtained as a by-product. A typical example of by-product I associated with waste treatment is electricity generated using the heat produced in the incineration process (the heat being used to generate steam that is used to drive a turbine to produce electricity).

In Table 6.4, the generation of electricity as a by-product of waste incineration occurs as a negative input coefficient$-a_{24}$, shown within a box. The same EEIO calculation such as (6.83) can be applied, with A_{II} given by:

$$A_{\mathrm{II}} = \begin{pmatrix} a_{13} & a_{14} \\ a_{23} & -a_{24} \end{pmatrix} \tag{6.84}$$

This method of introducing by-products into the IO model via negative input coefficients is called the negative input method of Stone, taking the name of its inventor (Stone 1961, p. 19). In this method, an increase in the final demand for the

Table 6.4 A simple WIO model with power generation from waste heat: an example of recycling of by-product for which there is a primary producer

	Sector 1	Electricity	Landfill	Incineration	Final demand
Sector 1	a_{11}	a_{12}	a_{13}	a_{14}	y_1
Electricity	a_{21}	a_{22}	a_{23}	$\boxed{-a_{24}}$	y_2
Waste 1	g_{11}^{out}	g_{12}^{out}	g_{13}^{out}	0	y_{w1}
Waste 2	0	0	0	g_{24}^{out}	
Emission	r_1	r_2	r_3	r_4	y_e

primary product, waste incineration in the current example, would increase the supply of its by-product, electricity, and would reduce its supply from its primary producer, the power sector.

While this procedure represents a straightforward way to introduce by-products into the IO model, it can cause problems. The point is that the matrix A now contains negative elements, and hence Theorem 1 is no longer applicable. Accordingly, the nonnegativity of the solution (6.20) is no longer guaranteed, because the Leontief inverse matrix $(I - A)^{-1}$ now contains negative elements. A negative output of a product occurs when its supply from the sectors other than the primary production sector exceeds the demand for it. This can happen because the supply of a by-product is driven not by demand for it, but by demand for the primary product from which the by-product is obtained: there is no mechanism in the model to restore the balance of supply of and demand for the by-product. In reality, waste management takes care of an excess supply of by-products. The weakness of the model does not consist in the occurrence of negative coefficients, that is, the Stone method, but in the neglect of this important adjustment mechanism that is embodied in waste management. In other words, the Stone method should be used with great caution in cases where there is any possibility of this problem occurring (see Nakamura and Kondo (2009) for further details). In such cases, the modeling applicable to by-product II should be applied, to which we now turn.

3.5.2 The Recycling of By-product II

Ash (incineration residue) can be cited as a typical example of by-product II which is associated with waste treatment. Ash qualifies as by-product II because there is no sector that produces it as primary product. A typical way of ash recycling is to use it as a material for cement production (as a substitute for clay and silica stone). Let sector 2 be the cement-producing sector, where ash is used as an input (Table 6.5). Write g_{22}^{in} for the input of ash per unit of cement. Subtracting the amount of waste recycled from the amount generated gives the net amount of waste generated that is submitted to waste treatment. Assuming that the cement process is the only process where ash from the incineration process is used, and that the incineration process is the only process where ash is generated as by-product, the balancing equation for ash is given by

Table 6.5 A simple WIO model of recycling of ash in cement production: an example of recycling of by-product for which there is no primary producer

	Sector 1	Cement	Landfill	Incineration	Final demand
Sector 1	a_{11}	a_{12}	a_{13}	a_{14}	y_1
Cement	a_{21}	a_{22}	a_{23}	a_{24}	y_2
Waste 1	g_{11}^{out}	g_{12}^{out}	g_{13}^{out}	0	y_{w1}
Waste 2 (ash)	0	$\boxed{-g_{22}^{\text{in}}}$	0	g_{24}^{out}	0
Emission	r_1	r_2	r_3	r_4	y_e

$$\underbrace{-g_{22}^{\text{in}}x_2}_{\text{ash recycled}} + \underbrace{g_{24}^{\text{out}}x_4}_{\text{ash generated}} = \underbrace{w_2}_{\text{ash for waste treatment}} \tag{6.85}$$

Recall that, if not recycled, ash is disposed to landfill. The above equation implies that if there is an excess supply of ash over the amount that can be absorbed by the cement process, the excess is treated as waste and landfilled. This is in sharp contrast to the method for by-product I mentioned above, where it is assumed that the supply of a by-product is always met by the equal amount of demand for it. Because of the explicit consideration of the adjustment mechanism of waste treatment sectors in dealing with an excess supply of a by-product, the occurrence of a negative production, as mentioned above, can be avoided. In the above example, an excess supply of ash would not lead to a negative production of clay or silica stone, but to the disposal of ash to a landfill.

The same balance equation as (6.77) applies to the present case, the only difference that g_{22}^{out} is now replaced by $-g_{22}^{\text{in}}$. Replacing the waste generation coefficients G^{out} by the net waste input coefficients$G = (g_{ij})$, where

$$g_{ij} = g_{ij}^{\text{out}} - g_{ij}^{\text{in}}, \tag{6.86}$$

results in a general expression for the balance equation

$$\begin{pmatrix} x_{\text{I}} \\ x_{\text{II}} \end{pmatrix} = \begin{pmatrix} A_{\text{I}} & A_{\text{II}} \\ SG_{\text{I}} & SG_{\text{II}} \end{pmatrix} \begin{pmatrix} x_{\text{I}} \\ x_{\text{II}} \end{pmatrix} + \begin{pmatrix} y_{\text{I}} \\ Sy_{\text{w}} \end{pmatrix}, \tag{6.87}$$

with the solution

$$\begin{pmatrix} x_{\text{I}} \\ x_{\text{II}} \end{pmatrix} = \begin{pmatrix} I - A_{\text{I}} & -A_{\text{II}} \\ -SG_{\text{I}} & I - SG_{\text{II}} \end{pmatrix}^{-1} \begin{pmatrix} y_{\text{I}} \\ Sy_{\text{w}} \end{pmatrix} \tag{6.88}$$

Equation 6.88 represents an extended version of the Leontief EIO model, which is applicable to waste management issues including recycling where there is no one-to-one correspondence between waste types and treatment processes. Because of this distinguishing feature, the approach is called the waste input–output (WIO) method (Nakamura 1999; Nakamura and Kondo 2002). See Nakamura and Kondo (2009) for further details of WIO.

3.5.3 Supplementary Issues Associated with Recycling

Recycling of waste results in the substitution of secondary materials for primary materials, and hence in the alteration of the relevant elements of a_{ij}, g_{ij}, and r_j. On the other hand, recycling of EoL products will call for them to be submitted to a series of pretreatment processes such as collection, separation, dismantling, shredding, etc., before they can be used as secondary materials. These processes should occur as column elements referring to waste treatment, with the current inputs such as utilities and chemicals occurring as elements of A_{II}, and generated scrap and waste occurring as elements of $G_{\mathrm{II}}^{\mathrm{out}}$ (Kondo and Nakamura 2004).

3.5.4 Closed- and Open-Loop Recycling: Numerical Examples

Recycling is classified as either closed loop or open loop, depending on whether an EoL product is used for production of the same product or other products. Recycling an aluminum can back into another aluminum can is an example of closed-loop recycling, while recycling it into cast materials for a car engine is an example of open-loop recycling. In the case of construction and demolition waste, their use in civil engineering (earthworks and road construction sector) refers to open-loop recycling, while their use for concrete production in the structural engineering sector refers to closed-loop recycling (Weil et al. (2006). Recycling of ash for cement production, discussed above (Table 6.5), is another example of open-loop recycling.

In reality, open-loop recycling is more common than closed-loop recycling, even for metals (Nakamura et al. 2012). In PLCA, open-loop recycling involves major methodological problems in inventory analysis, that is, in the calculation of environmental burdens (Klöpffer 1996). So far, there has been no standardized procedure in LCA for open-loop recycling (Shen et al. 2010), though system expansion is regarded as the most desirable course from a scientific point of view (Klöpffer 1996). System expansion may be difficult to perform in PLCA (Klöpffer 1996; Shen et al. 2010). In contrast, a system expansion is rather straightforward in EEIO in general and in WIO in particular, because of the inclusion in their system boundary of almost all the activities in a given economy.

This subsection illustrates the implementation of closed-loop and open-loop recycling in IO-LCA based on WIO, using recycling of aluminum cans as an example. Simplified numerical examples are used for illustration. These examples will also be useful to familiarize readers with computational aspects of IO-LCA, albeit in a highly simplified manner.

Table 6.6 WIO with closed-loop recycling: nonsymmetric table, numerical example

	Al can	Al wrought	Al casting	Others	Landfill	Final demand
	A_{I}				A_{II}	y_{I}
Al can	0.00	0.00	0.00	0.00	0.00	1.00
Al wrought	0.90	0.00	0.00	0.00	0.00	0.00
Al casting	0.00	0.00	0.00	0.00	0.00	0.00
Others	0.20	0.20	0.20	0.00	0.20	0.00
	G_{I}				G_{II}	y_{w}
Al can scrap	0.00	−0.80	−0.80	0.00	0.00	1.00
Other waste	0.10	0.00	0.00	0.20	0.00	0.00
	SG_{I}				SG_{II}	Sy_{w}
Landfill	0.10	−0.80	−0.80	0.20	0.00	1.00

Units: Al products, landfill activity, wastes in physical units, and others in monetary units

Closed-Loop Recycling

We first consider closed-loop recycling of Al can, where EoL Al can is recycled back into a new Al can after being converted into Al wrought, the major Al feed for Al can, a numerical example of which is given by Table 6.6. The negative entry of EoL can into Al wrought $g_{12}^{\mathrm{in}} = -0.80$ represents its recycling into Al wrought in the same manner as in Table 6.5. While EoL Al can also be used to produce Al casting, $g_{13}^{\mathrm{in}} = -0.80$, this channel of recycling does not occur due to the absence of demand for Al casting. The column termed "Final demand" corresponds to the functional unit in the present case, where one unit of Al can is produced and disposed of.

This example concerns one form of waste treatment (landfill) and two types of waste: EoL Al can and Other waste. Because landfill is assumed to be the only waste treatment process available, the allocation matrix S is of order 1×2, and given by

$$S = \begin{pmatrix} 1 & 1 \end{pmatrix} \tag{6.89}$$

Multiplying this S, the 2×5 matrix referring to waste flows (G_{I}, G_{II}, and y_{W}) in Table 6.6 is converted into a 1×5 row referring to the input of landfill at the bottom of the table, which is simply the sum of its column elements.

From (6.88), the solution for x is given by

$$\begin{aligned} x &= \begin{pmatrix} I - A_{\mathrm{I}} & -A_{\mathrm{II}} \\ -SG_{\mathrm{I}} & I - SG_{\mathrm{II}} \end{pmatrix}^{-1} \begin{pmatrix} y_{\mathrm{I}} \\ Sy_{\mathrm{w}} \end{pmatrix} \\ &= \begin{pmatrix} 1.00 & 0.00 & 0.00 & 0.00 & 0.00 \\ 0.90 & 1.00 & 0.00 & 0.00 & 0.00 \\ 0.00 & 0.00 & 1.00 & 0.00 & 0.00 \\ 0.71 & 0.11 & 0.11 & 2.78 & 0.56 \\ -0.48 & -0.78 & -0.78 & 0.56 & 1.11 \end{pmatrix} \begin{pmatrix} 1 \\ 0 \\ 0 \\ 1 \end{pmatrix} = \begin{pmatrix} 1.00 \\ 0.90 \\ 0.00 \\ 1.27 \\ 0.63 \end{pmatrix} \end{aligned} \tag{6.90}$$

with the net waste generation given by

$$w = Gx + y_w$$
$$= \begin{pmatrix} 0.00 & -0.80 & -0.80 & 0.00 & 0.00 \\ 0.10 & 0.00 & 0.00 & 0.20 & 0.00 \end{pmatrix} \begin{pmatrix} 1.00 \\ 0.90 \\ 0.00 \\ 1.27 \\ 0.63 \end{pmatrix} \tag{6.91}$$
$$+ \begin{pmatrix} 1.0 \\ 0.0 \end{pmatrix} = \begin{pmatrix} 0.28 \\ 0.35 \end{pmatrix}$$

It follows that of one unit of EoL Al can, 0.28 units ends up in landfill, with the rest being recycled into new cans. The output of Al casting is zero, because of the absence of demand for it.

Open-Loop Recycling

We now turn to the numerical example of open-loop recycling represented in Table 6.7. EoL Al can is no longer used in Al wrought production, $g_{12}^{\text{in}=0}$, while it can still be used to produce Al casting, $g_{13}^{\text{in}} = -0.8$. Al casting cannot be used to produce a new can and hence has to be put to use for other purposes such as car engines. Accordingly, for EoL Al can to be recycled, the demand for Al casting has to come from somewhere, that is, it has to occur in the final demand. In other words, the system under consideration needs to be expanded to accommodate the production of Al casting.

The amount of EoL Al can recycled depends on the final demand for Al casting, y_3. Based on a preliminary WIO calculation, it was obtained that $y_3 = 0.90$ is needed to realize the equal amount of Al can recycling as in the above case of closed-loop recycling. It then follows from (6.88) that

$$x = \begin{pmatrix} I - A_{\text{I}} & -A_{\text{II}} \\ -SG_{\text{I}} & I - SG_{\text{II}} \end{pmatrix}^{-1} \begin{pmatrix} y_{\text{I}} \\ Sy_{\text{w}} \end{pmatrix}$$
$$= \begin{pmatrix} 1.00 & 0.00 & 0.00 & 0.00 & 0.00 \\ 0.90 & 1.00 & 0.00 & 0.00 & 0.00 \\ 0.00 & 0.00 & 1.00 & 0.00 & 0.00 \\ 1.28 & 1.17 & 0.04 & 1.04 & 0.21 \\ 0.45 & 0.33 & -0.79 & 0.21 & 1.04 \end{pmatrix} \begin{pmatrix} 1 \\ 0 \\ 0.90 \\ 0 \\ 1 \end{pmatrix} = \begin{pmatrix} 1.00 \\ 0.90 \\ 0.90 \\ 1.53 \\ 0.78 \end{pmatrix} \tag{6.92}$$

Table 6.7 WIO with open-loop recycling: nonsymmetric table, numerical example

	Al can	Al wrought	Al casting	Others	Landfill	Final demand
	A_I				A_{II}	y_I
Al can	0.00	0.00	0.00	0.00	0.00	1.00
Al wrought	0.90	0.00	0.00	0.00	0.00	0.00
Al casting	0.00	0.00	0.00	0.00	0.00	$\boxed{y_3}$
Others	0.20	1.10	0.20	0.60	0.20	0.00
	G_I				G_{II}	y_w
Al can scrap	0.00	0.00	−0.80	0.00	0.00	1.00

$$w = Gx + y_w = \begin{pmatrix} 0.0 & -1.0 & 0.0 & 0.0 \\ 0.1 & 0.0 & 0.2 & 0.0 \end{pmatrix} \begin{pmatrix} 1.00 \\ 0.90 \\ 0.90 \\ 1.53 \\ 0.98 \end{pmatrix} + \begin{pmatrix} 1.0 \\ 0.0 \end{pmatrix} = \begin{pmatrix} 0.28 \\ 0.50 \end{pmatrix} \tag{6.93}$$

Because of the occurrence of additional demand for Al casting in the case of open-loop recycling considered above, it is no surprise that in terms of both production and waste (Other waste), the open-loop case proves to have larger impacts than the closed-loop case. However, these results are not comparable, because of the use of different functional units.

We now therefore consider applying the same functional unit to both the closed- and open-loop cases. Applying the functional unit referring to the open-loop case, (6.92), to the Leontief inverse matrix referring to the closed-loop case, (6.90), results in

$$x = \begin{pmatrix} 1.00 & 0.00 & 0.00 & 0.00 & 0.00 \\ 0.90 & 1.00 & 0.00 & 0.00 & 0.00 \\ 0.00 & 0.00 & 1.00 & 0.00 & 0.00 \\ 0.71 & 0.11 & 0.11 & 2.78 & 0.56 \\ -0.48 & -0.78 & -0.78 & 0.56 & 1.11 \end{pmatrix} \begin{pmatrix} 1.0 \\ 0.0 \\ 0.90 \\ 0.0 \\ 1.0 \end{pmatrix} = \begin{pmatrix} 1.00 \\ 0.90 \\ 0.90 \\ 1.37 \\ -0.07 \end{pmatrix} \tag{6.94}$$

The occurrence of a negative output of landfill implies the presence of excess demand for EoL Al can over its supply; the amount of EoL Al can available is not large enough to meet the final demand.

A preliminary WIO calculation showed that the excess demand for EoL Al can become zero with $y_3 = 0.35$, which corresponds to 100 % recycling of Al can, resulting in a new functional unit that is feasible for both closed- and open-loop recycling:

$$\begin{pmatrix} y_{\mathrm{I}} \\ y_{\mathrm{W}} \end{pmatrix} = \begin{pmatrix} 1.00 \\ 0.00 \\ 0.35 \\ 0.00 \\ 1.00 \\ 0.00 \end{pmatrix} \tag{6.95}$$

Application of this functional unit (final demand) to the Leontief inverse matrix in (6.90) and in (6.92) results in

$$\begin{pmatrix} x \\ \hline w \end{pmatrix}_{\text{Closed}} = \begin{pmatrix} 1.00 \\ 0.93 \\ 0.35 \\ 1.31 \\ 0.36 \\ 0.00 \\ 0.36 \end{pmatrix}, \quad \begin{pmatrix} x \\ \hline w \end{pmatrix}_{\text{Open}} = \begin{pmatrix} 1.00 \\ 0.90 \\ 0.35 \\ 1.50 \\ 1.21 \\ 0.72 \\ 0.49 \end{pmatrix} \tag{6.96}$$

where "Closed" refers to closed-loop recycling and "Open" to open-loop recycling. As far as this simple numerical example is concerned, closed-loop recycling is found to have a smaller environmental burden than open-loop recycling because it is characterized by smaller amounts of production and wastes.

3.6 Input–Output Table Based on Supply and Use Tables

3.6.1 Industries and Commodities

The supply and use table is a calculation system that describes transactions between input–output table sectors defined and classified as industries and commodities. Using the industry (row) × commodities (column) $V = \left(v_{ij}\right)$ matrix and the commodities (row) × industry (column) $U = \left(u_{ij}\right)$ matrix, the transactions within a single country's economy can be expressed. v_{ij} is the annual production of good i from industry j, while u_{ij} is the annual input of good j into industry i. This division into industry and commodities enables the description of joint production, in which two or more commodities are produced from a single production process.

By using matrices U and V, the annual commodities output q and the industry output g can be calculated with Eqs. (6.97) and (6.98), respectively. Here, vector ι is a summation vector with all its elements unity, and "'" attached to a matrix refers to its transpose:

$$q = V'\iota \tag{6.97}$$

$$g = V\iota \tag{6.98}$$

In addition, the unit environmental burden matrix R *is* determined by dividing the annual environmental burden D by g:

$$R = D\,\mathrm{diag}(g)^{-1} \tag{6.99}$$

The element D_{kj} represents the environmental burden k per unit output of industry j.

3.6.2 The Supply and Use Framework for IO-LCA

Following Suh et al. (2010), three main modeling methods based on the supply and use framework can be considered for IO-LCA:

1. The industry-technology model
2. The commodity-technology model
3. The by-product-technology model

First, in the industry-technology model, each industry is assumed to have a single, inherent technology that cannot be broken down according to the multiple products produced by that industry. Hence, the inputs to and outputs from the industry are allocated directly to the multiple products in proportion to the output of each product. In accordance with this assumption, formalizing the input coefficient matrix A_I and the unit environmental burden matrix R_I yields Eqs. (6.100) and (6.101):

$$A_I = U\,\mathrm{diag}(g)^{-1}V\,\mathrm{diag}(q)^{-1} \tag{6.100}$$

$$R_I = R\,\mathrm{diag}(g)^{-1}V\,\mathrm{diag}(q)^{-1} \tag{6.101}$$

Therefore, the direct and indirect environmental burden e_I with y as the functional unit vector for IO-LCA can be obtained by using Eq. (6.102):

$$e_I = R_I(I - A_I)^{-1}y \tag{6.102}$$

This model is synonymous with the partitioning method Suh et al. (2010) for addressing the LCA allocation issue.

If one can assume that the number of industries is equal to the number of commodities, that is, both U and V are square, an alternative model, the commodity-technology model, can be derived.

The commodity-technology model assumes that the production of each particular commodity is associated with an inherent technology and that the inputs and outputs are determined by what is produced rather than by which industry produces it. Formalizing the input coefficient matrix A_c and the unit environmental burden matrix R_c then yields Eqs. (6.103) and (6.104):

$$A_c = UV'^{-1} \tag{6.103}$$

$$R_c = RV'^{-1} \tag{6.104}$$

The environmental burden vector e_c from the IO-LCA is given by Eq. (6.105):

$$e_c = R_c(I - A_c)^{-1}y \tag{6.105}$$

The third model, the by-product-technology model, assumes that the production of by-products is entirely dependent on the production of the industry's primary product; the model treats the generation of by-products as negative inputs for that industry. The equality of the number of industries and commodities is kept assumed. For this model, formalizing the input coefficient matrix A_B and unit environmental burden matrix R_B yields Eqs. (6.106) and (6.107), respectively. V_d is a matrix with V partitioned to contain only the diagonal components of V, and V_{od} is the matrix that contains only the non-diagonal components of V:

$$A_B = \left(U - V'_{od}\right)V_d^{-1} \tag{6.106}$$

$$R_B = RV_d^{-1} \tag{6.107}$$

The environmental burden vector e_B from the IO-LCA is obtained using Eq. (6.108):

$$e_B = R_B(I - A_B)^{-1}y \tag{6.108}$$

This model is synonymous with the system expansion method for addressing the LCA allocation issue and is identical to the second, commodity-technology model for applications limited to IO-LCA. Both models can be transposed to Eq. (6.109):

$$e_C = e_B = D\left(V' - U\right)^{-1}y \tag{6.109}$$

The reader is referred to Suh et al. (2010) for details, including an interpretation of WIO within a supply and use framework.

3.6.3 Extension of WIO Based on Supply and Use Framework

Lenzen and Reynolds (2014) proposed a novel extension of the WIO framework, a waste supply–use table (WSUT) model, by incorporating a supply–use formalism.

The WSUT enables explicit and simultaneous representation of multiple waste types and treatment methods. Using the same notations as in the WIO model above, the balance equations of the WSUT model are obtained by extending (6.87) to a supply and use framework as follows:

$$\begin{pmatrix} x_{\mathrm{I}} \\ x_{\mathrm{II}} \\ w \end{pmatrix} = \begin{pmatrix} A_{\mathrm{I}} & A_{\mathrm{II}} & 0 \\ 0 & 0 & S \\ G_{\mathrm{I}} & G_{\mathrm{II}} & 0 \end{pmatrix} \begin{pmatrix} x_{\mathrm{I}} \\ x_{\mathrm{II}} \\ w \end{pmatrix} + \begin{pmatrix} f_{\mathrm{I}} \\ 0 \\ f_{\mathrm{w}} \end{pmatrix}, \tag{6.110}$$

with the solution

$$\begin{pmatrix} x_{\mathrm{I}} \\ x_{\mathrm{II}} \\ w \end{pmatrix} = \begin{pmatrix} I - A_{\mathrm{I}} & -A_{\mathrm{II}} & 0 \\ 0 & I & S \\ -G_{\mathrm{I}} & -G_{\mathrm{II}} & I \end{pmatrix}^{-1} \begin{pmatrix} f_{\mathrm{I}} \\ 0 \\ f_{\mathrm{w}} \end{pmatrix}. \tag{6.111}$$

Among other things, the WUST model enables one to obtain the solution for w in one step together with x_{I} and x_{II}, whereas in the original WIO, it had to be calculated in two steps, as described in 3.5. Still, in terms of the implications derived from the Leontief inverse matrix, the original WIO and the WSUT are equivalent (see Lenzen and Reynolds 2014 for details).

3.7 *Tools of IO-LCA*

The strength of IO-LCA consists in its well-established mathematical foundations of IO economics, which is not the case with PLCA. On the other hand, physical conditions impose limitations on the extent to which these economic tools can be applied to LCA.

3.7.1 Contributions Analysis: Quantifying Impact of Sources

The contribution of the goods and services required by the functional unit vector y to the environmental burden e can be quantified using the following contributions analysis. Taylor expansion of $(I - A)^{-1}$ from Eq. (6.62) yields

$$R(I - A)^{-1} = R\left(I + A + A^2 + \cdots + A^n\right) \tag{6.112}$$

that can be disaggregated as in Eq. (6.113) to Ry and eAy:

$$
\begin{aligned}
e &= R(I - A)^{-1} \\
&= R\left(I + A + A^2 + \ldots\right)y \\
&= Ry + R\left(I + A + A^2 + \ldots\right)Ay \\
&= Ry + R(I - A)^{-1}Ay \\
&= Ry + eAy
\end{aligned}
\tag{6.113}
$$

Ry is the burden directly emitted from production of the goods required by functional unit y. By disaggregating this as $R^* = (r_j^*) = R\ \mathrm{diag}(y)$, the burden r_j^* directly emitted from the production of good j required by element j of the functional unit can be known. eAy is the burden induced by the goods and services directly input to produce the goods required by functional unit y. Here, by disaggregating $E^* = (e_{ij}^*) = \mathrm{diag(e)}A\,\mathrm{diag(y)}$, the burden e_{ij}^* induced by the good i input to produce good j required by element j of the functional unit can be found. Reducing the input of product i with a large burden e_{ij}^* will make a large contribution to reducing the burden emitted to satisfy the functional unit.

3.7.2 Structural Path Analysis (SPA): Quantifying Emissions for Each Path

Structural path analysis (Defourny and Thorbecke (1984)) is a method for structural analysis of input–output tables that is extremely useful in energy analysis applications (Treloar (1997)) and for interpreting the structure of IO-LCA. For SPA, power series expansion of the environmental burden e in IO-LCA as shown in Eq. (6.113) allows disaggregation of the configuration of e into directRy, primary indirect RAy, and secondary indirect RA^2y contributions, as follows:

$$
\begin{aligned}
e &= R(I - A)^{-1}y \\
&= R\left(I + A + A^2 + \cdots + A^\infty\right)y \\
&= Ry + RAf + RA^2y + \cdots + \cdots RA^\infty y
\end{aligned}
\tag{6.114}
$$

Next, expression of the matrix elements of Eq. (6.113) results in Eq. (6.115). The respective components (called the paths) a_{ij}, $a_{ij}a_{ki}$, $a_{ji}a_{lk}a_{ki}$ of the IO-LCA configuration can be disaggregated and their effects studied:

$$
\begin{aligned}
e &= \sum_i^n y_i \left(\sum_j^n r_j \left(\delta_{ji} + a_{ji} + \left(A^2\right)_{ji} + + \left(A^3\right)_{ji} + \cdots \right) \right) \\
&= \sum_i^n y_i \left(\sum_j^n r_j \left(\delta_{ji} + a_{ji} + \sum_k a_{jk}a_{ki} + \sum_l \sum_k a_{jl}a_{lk}a_{ki} + \cdots \right) \right)
\end{aligned}
\tag{6.115}
$$

Here, i, j, k, and l are sectors of the input–output table; if $i = j$, $\delta_{ji} = 1$; otherwise, $\delta_{ji} = 0$. n is the number of sectors. For disaggregation in SPA, n paths exist for primary indirect, n^2 paths for secondary indirect, and n^3 paths for tertiary indirect burdens. Thus, the path of focus needs to be selected in actual analysis.

3.7.3 Structural Decomposition Analysis (SDA): Quantifying Factors That Change Emissions

Structural decomposition analysis (SDA) is a useful tool in IO-LCA for quantifying the factors driving environmental burdens that have changed between two points in time. Let the IO-LCA result for environmental burden in year $t=1$ from Eq. (6.116) be $e^{(1)}$ and that in year $t=2$ be $e^{(2)}$. The change in environmental burden Δe between the 2 years is obtained using Eq. (6.118), while the changed factors can be disaggregated into the functional unit change Δy, the supply chain change ΔL, and unit environmental burden change ΔR:

$$e^{(1)} = R^{(1)}\left(I - A^{(1)}\right)^{-1} y^{(1)} = R^{(1)}L^{(1)}y^{(1)} \tag{6.116}$$

$$e^{(2)} = R^{(2)}\left(I - A^{(2)}\right)^{-1} y^{(2)} = R^{(2)}L^{(2)}y^{(2)} \tag{6.117}$$

$$\Delta e = e^{(2)} - e^{(1)} = R^{(2)}L^{(2)}y^{(2)} - R^{(1)}L^{(1)}y^{(1)} \tag{6.118}$$

Here, the changed factors must be variables assumed to change independently of one another. Using the full decomposition method of Dietzenbacher and Los (1998), (6.118) can be rewritten as

$$\begin{aligned}\Delta e &= \Delta R L^{(2)}y^{(2)} + R^{(1)}\Delta L_y{}^{(2)} + R^{(1)}L^{(1)}\Delta y\\ &= \Delta R L^{(1)}y^{(1)} + R^{(2)}\Delta L y^{(1)} + R^{(2)}L^{(2)}\Delta y\\ &= R^{(1)}\Delta L y^{(1)} + \Delta R L^{(2)}y^{(1)} + R^{(2)}L^{(2)}\Delta y\\ &= R^{(2)}\Delta L y^{(2)} + \Delta R L^{(1)}y^{(1)} + R^{(1)}L^{(1)}\Delta y\\ &= R^{(1)}L^{(1)}\Delta y + R^{(1)}\Delta L y^{(2)} + \Delta R L^{(2)}y^{(2)}\\ &= R^{(2)}L^{(2)}\Delta y + R^{(2)}\Delta L y^{(1)} + \Delta R L^{(1)}y^{(1)}\end{aligned} \tag{6.119}$$

In Eq. (6.119), if the change amounts ΔR, ΔL, and Δy are averaged as shown in Eqs. (6.120), (6.121), and (6.122), the contributions of $\Delta\bar{R}$, ΔL^{-}, and $\Delta\bar{y}$ to Δe can be allocated. More specifically, $\Delta e = \Delta\bar{R} + \Delta\bar{L} + \Delta\bar{y}$ holds:

$$\begin{aligned}\Delta\bar{R} = &\frac{1}{6}(\Delta R L^{(2)}y^{(2)} + \Delta R L^{(1)}y^{(1)}\\ &+ \Delta R L^{(2)}y^{(1)} + \Delta R L^{(1)}y^{(1)} + \Delta R L^{(2)}y^{(2)} + \Delta R L^{(1)}y^{(1)})\end{aligned} \tag{6.120}$$

$$\begin{aligned}\Delta\bar{L} = &\frac{1}{6}(R^{(1)}\Delta L y^{(2)} + R^{(2)}\Delta L y^{(1)}\\ &+ R^{(1)}\Delta L y^{(1)} + R^{(2)}\Delta L y^{(2)} + R^{(1)}\Delta L y^{(2)} + R^{(2)}\Delta L y^{(1)})\end{aligned} \tag{6.121}$$

$$\Delta y^{-} = \frac{1}{6}(R^{(1)}L^{(1)}\Delta y + r^{(2)}L^{(2)}\Delta y + R^{(2)}L^{(2)}\Delta y + R^{(1)}L^{(1)}\Delta y + R^{(1)}L^{(1)}\Delta y + R^{(2)}L^{(2)}\Delta y) \tag{6.122}$$

In the full decomposition method, if n is the number of changed factors, the contribution of any such factor to the IO-LCA is determined by the overall average of the $n!$ expressions in which the factor is included when written out in disaggregated form. In the above case, the changed factors are R, L, and y, that is, $n = 3$, which means that in disaggregated form, there are $3! = 3 \times 2 \times 1 = 6$ expressions in which each factor is included. While the full decomposition method has the advantage of considering all combinations in disaggregated form, it brings with it a substantial calculation burden, as the number of combinations rises dramatically as n increases. With $n = 5$, for instance, the total number of combinations comes to $5! = 5 \times 4 \times 3 \times 2 \times 1 = 120$, and with $n = 10$, it increases to $10! = 103,628,800$.

The calculation burden of the full decomposition method can be reduced by using an alternative SDA method known as the polar decomposition method. In this method, rather than considering all the n! combinations, only two arbitrary combinations are selected, in which the order of the changed factors is reversed: $\Delta R \rightarrow \Delta L \rightarrow \Delta y$ and $\Delta y \rightarrow \Delta L \rightarrow \Delta R$, for example. In the case of Eq. (6.118), the first and sixth, second and fifth, and third and fourth combinations are disaggregated in reverse order, and in the polar decomposition method, any of these pairs could therefore be taken; see Dietzenbacher and Los (1998) for more details.

When Δe is defined as the ratio of $e^{(2)}$ to $e^{(1)}$ as in Eq. (6.123), following Dietzenbacher's multiplicative decomposition method, the six alternative structural disaggregation paths for Δe shown in Eq. (6.124) can be considered:

$$\Delta e = \frac{e^{(2)}}{e^{(1)}} = \frac{R^{(2)}L^{(2)}y^{(2)}}{R^{(1)}L^{(1)}y^{(1)}} \tag{6.123}$$

$$\begin{aligned} \Delta e &= \frac{R^{(2)}L^{(2)}y^{(2)}}{R^{(1)}L^{(2)}y^{(2)}} \times \frac{R^{(1)}L^{(2)}y^{(2)}}{R^{(1)}L^{(1)}y^{(2)}} \times \frac{R^{(1)}L^{(1)}y^{(2)}}{R^{(1)}L^{(1)}y^{(1)}} \\ &= \frac{R^{(2)}L^{(2)}y^{(2)}}{R^{(1)}L^{(2)}y^{(2)}} \times \frac{R^{(1)}L^{(2)}y^{(2)}}{R^{(1)}L^{(2)}y^{(1)}} \times \frac{R^{(1)}L^{(2)}y^{(1)}}{R^{(1)}L^{(1)}y^{(1)}} \\ &= \frac{R^{(2)}L^{(2)}y^{(2)}}{R^{(2)}L^{(1)}y^{(2)}} \times \frac{R^{(2)}L^{(1)}y^{(2)}}{R^{(1)}L^{(1)}y^{(2)}} \times \frac{R^{(1)}L^{(1)}y^{(2)}}{R^{(1)}L^{(1)}y^{(1)}} \\ &= \frac{R^{(2)}L^{(2)}y^{(2)}}{R^{(2)}L^{(1)}y^{(2)}} \times \frac{R^{(2)}L^{(1)}y^{(2)}}{R^{(2)}L^{(1)}y^{(1)}} \times \frac{R^{(2)}L^{(1)}y^{(2)}}{R^{(1)}L^{(1)}y^{(1)}} \\ &= \frac{R^{(2)}L^{(2)}y^{(2)}}{R^{(2)}L^{(2)}y^{(1)}} \times \frac{R^{(2)}L^{(2)}y^{(1)}}{R^{(1)}L^{(2)}y^{(2)}} \times \frac{R^{(1)}L^{(2)}y^{(1)}}{R^{(1)}L^{(1)}y^{(1)}} \\ &= \frac{R^{(2)}L^{(2)}y^{(2)}}{R^{(2)}L^{(2)}y^{(1)}} \times \frac{R^{(2)}L^{(2)}y^{(1)}}{R^{(2)}L^{(1)}y^{(1)}} \times \frac{R^{(2)}L^{(1)}y^{(1)}}{R^{(1)}L^{(1)}y^{(1)}} \end{aligned} \tag{6.124}$$

From Eq. (6.124), the contribution of ΔR^{-} to Δe can be measured by obtaining the geometric mean of terms that include $\frac{R^{(2)}}{R^{(1)}}$ as in Eq. (6.125). Similarly, Eqs. (6.126) and (6.127) can be used to obtain ΔL^{-} and Δy^{-}, respectively. Here, $\Delta e = \Delta\overline{R} \times \Delta\overline{L} \times \Delta\overline{y}$ holds:

$$\Delta R = \left(\frac{R^{(2)}L^{(2)}y^{(2)}}{R^{(1)}L^{(2)}y^{(2)}} \times \frac{R^{(2)}L^{(2)}y^{(2)}}{R^{(1)}L^{(2)}y^{(2)}} \times \frac{R^{(2)}L^{(1)}y^{(2)}}{R^{(1)}L^{(1)}y^{(2)}} \times \frac{R^{(2)}L^{(1)}y^{(1)}}{R^{(1)}L^{(1)}y^{(1)}} \times \frac{R^{(2)}L^{(2)}y^{(1)}}{R^{(1)}L^{(2)}y^{(1)}} \times \frac{R^{(2)}L^{(1)}y^{(1)}}{R^{(1)}L^{(1)}y^{(1)}}\right)^{\frac{1}{6}} \tag{6.125}$$

$$\Delta L = \left(\frac{R^{(1)}L^{(2)}y^{(2)}}{R^{(1)}L^{(1)}y^{(2)}} \times \frac{R^{(1)}L^{(2)}y^{(1)}}{R^{(1)}L^{(1)}y^{(1)}} \times \frac{R^{(2)}L^{(2)}y^{(2)}}{R^{(2)}L^{(1)}y^{(2)}} \times \frac{R^{(2)}L^{(2)}y^{(2)}}{R^{(2)}L^{(1)}y^{(2)}} \times \frac{R^{(1)}L^{(2)}y^{(1)}}{R^{(1)}L^{(1)}y^{(1)}} \times \frac{R^{(2)}L^{(2)}y^{(1)}}{R^{(2)}L^{(1)}y^{(1)}}\right)^{\frac{1}{6}} \tag{6.126}$$

$$\Delta y = \left(\frac{R^{(1)}L^{(1)}y^{(2)}}{R^{(1)}L^{(1)}y^{(1)}} \times \frac{R^{(1)}L^{(2)}y^{(2)}}{R^{(1)}L^{(2)}y^{(1)}} \times \frac{R^{(1)}L^{(1)}y^{(2)}}{R^{(1)}L^{(1)}y^{(1)}} \times \frac{R^{(2)}L^{(1)}y^{(2)}}{R^{(2)}L^{(1)}y^{(1)}} \times \frac{R^{(2)}L^{(2)}y^{(2)}}{R^{(2)}L^{(2)}y^{(1)}} \times \frac{R^{(2)}L^{(2)}y^{(2)}}{R^{(2)}L^{(2)}y^{(1)}}\right)^{\frac{1}{6}} \tag{6.127}$$

In the above-described additive decomposition and multiplicative decomposition, the structural decomposition is conducted by assuming that the unit environmental burden vector R and the total requirement matrix L vary independently. However, $L = (I - A)^{-1}$, and the input structure for goods described in matrix A and environmental burden emissions are intimately related. For example, CO_2 emissions, a component of R, depend on the type of fossil fuel and amount consumed, a component of A. Users of these decomposition methods should therefore always bear in mind the underlying strong assumption of independence of R and L.

4 Extensions of Input–Output-Life Cycle Assessment

4.1 Multiregional IO-LCA

By using a multiregional IOT (MRIO), discussed in Sect. 2.3.5, the system boundaries for IO-LCA can be expanded from one country to multiple countries, thus including international trade. Conversely, it is possible to divide a single country into multiple regions. MRIO-LCA using MRIO is conducted by a method similar to that of IO-LCA. When international input–output tables (MRIO in terms of nations/countries as regional units) are used in IO-LCA, this allows the system boundaries of the IO-LCA to be expanded from a single country and makes it possible to appropriately reflect the environmental burden associated with imported goods.

As a simple example, consider the case in which the system boundary comprises two countries. Equation 6.57 provides the basic quantity model. Write R^{a} and R^{b} for the matrix of environmental burden k from sector j per unit of output for country a and country b. Following (6.62), the two-country MRIO model to compute induced esmissions, $e = (e_{\mathrm{k}})$, is then given by

$$e = \begin{pmatrix} R^a & R^b \end{pmatrix} \begin{pmatrix} I - A^{aa} & -A^{ab} \\ -A^{ba} & I - A^{bb} \end{pmatrix}^{-1} \begin{pmatrix} y^a \\ y^b \end{pmatrix} \tag{6.128}$$

where

$$\begin{aligned} y^a &= y^{aa} + y^{ab} \\ y^b &= y^{bb} + y^{ba} \end{aligned} \tag{6.129}$$

See Wiedmann (2009) for a recent review of MRIO with environmental extension.

In IO-LCA the total environmental burden k from sector i of country a is allocated to sector j of each country in proportion to the size x_{ij}^{ab} of the output of sector i for country a to b. Thus, depending on whether the basic price, producer price, or purchaser price is used in the analysis, the amount of environmental burden to be allocated will differ. Especially for international IOTs, not only do taxes, subsidies, or both vary greatly among countries, but differences in trade margins and transport margins will also occur, depending on the transport distances and modes involved. Thus, even when importing the same good from a certain exporting country, the higher the import cost is for an importing country, the greater will be the environmental burden of the exporting country allocated to the importing country. This issue is identical to that of allocating environmental burden by cost in LCA. Incorporating international trade in an MRIO has a large impact, especially on the interpretation of results.

4.2 *Environmental Life Cycle Costing*[1]

It was shown in Sect. 2.4 that corresponding to a Leontief quantity model, there exists a Leontief cost/price model. The same applies to WIO as well: corresponding to a WIO quantity model (6.82), one can derive a WIO cost/price model (Nakamura and Kondo 2006).

First consider the case where there is no recycling of waste materials. Writing p_{I} for an $n \times 1$ vector of the prices of goods and services and p_{II} for a $n_T \times 1$ vector of the prices of waste treatment services, the unit cost equations will be given by

$$\begin{aligned} p_{\mathrm{I}} &= p_{\mathrm{I}} A_{\mathrm{I}} + p_{\mathrm{II}} S G_{\mathrm{I}}^{\mathrm{out}} + v_{\mathrm{I}} \\ p_{\mathrm{II}} &= p_{\mathrm{I}} A_{\mathrm{II}} + p_{\mathrm{II}} S G_{\mathrm{II}}^{\mathrm{out}} + v_{\mathrm{II}} \end{aligned} \tag{6.130}$$

[1] "Environmental life-cycle costing" (and not environmental LCC) means the environmental-oriented LCC, which is compatible with LCA, in contrast to the (old) purely economic LCC (which is not compatible with LCA). "A primary motivation for LCC studies is to fully account for the financial costs of life-cycle environmental aspects and impacts that ultimately result from a decision" (Swarr et. al. 2011).

where v_{I} and v_{II} refer, respectively, to the row vectors of value added ratios of the goods and service sectors and the waste treatment sectors, which are exogenously given. The second term on the right-hand side of each equation refers to the cost associated with the treatment of waste generated in production. Implicit in this formulation of the cost is the notion of the polluter pays principle (PPP) or extended producer responsibility (EPR), under which the cost of waste disposal is internalized into the cost of the product. Solving for the product prices yields

$$\begin{pmatrix} p_{\mathrm{I}} & p_{\mathrm{II}} \end{pmatrix} = \begin{pmatrix} v_{\mathrm{I}} & v_{\mathrm{II}} \end{pmatrix} \begin{pmatrix} I - A_{\mathrm{I}} & -A_{\mathrm{II}} \\ -S\,G_{\mathrm{I}}^{\mathrm{out}} & I - S\,G_{\mathrm{II}}^{\mathrm{out}} \end{pmatrix}^{-1} \tag{6.131}$$

This generalizes the Leontief EIO price model (Leontief 1970) to the case where there is no one-to-one correspondence between waste (pollutant) and its treatment (abatement).

The case with recycling, where $G_{\mathrm{I}^{\mathrm{out}}}$ and $G_{\mathrm{II}^{\mathrm{out}}}$ are replaced by G_{I} and G_{II}, calls for considering the following additional items:

(a) The cost of the input of waste materials
(b) The revenue from the sale of waste materials

Item (a) refers to the expenditure incurred by users of a waste material for its acquisition and item (b) to the revenue that is passed back to its suppliers. It should be noted that mention of these items here does not mean that they are not considered in the standard IOT. They are certainly covered in the standard IOT as well, but in an implicit way, with no decomposition into components, as above (see (Nakamura and Kondo 2009, section 5.2.3.1) for further details).

When considering recycling, it is important to distinguish whether this involves a by-product I or by-product II (see Sect. 3.5). In the case of by-product I, if its price were equal to its primary counterpart, the terms (a) and (b) would require no consideration, because they would be subsumed in the usual cost term represented by $p_{\mathrm{I}}A_{\mathrm{I}}$ and $p_{\mathrm{II}}A_{\mathrm{II}}$. In this case, for the users of the relevant input, its origin (by-product or primary product) would be indistinguishable. For its suppliers, the term (b) is accounted for by the negative entry of the by-product as its input: the cost of production is reduced by the amount equal to the revenue from sale of the by-product. If, however, the price of by-product I is not equal to its primary counterpart, a different procedure needs to be used, which explicitly takes separate account of the terms (a) and (b).

Extension of the cost/price model (6.131) along these lines with additional consideration of the costs in the use phase is detailed in Nakamura and Kondo (2009, Chap. 6). The resulting extended cost/price model has been applied to environmental life cycle costing of air conditioners (Nakamura and Kondo 2006) and washing machines (Rebitzer and Nakamura 2008). A recent discussion of the use of IO in environmental life cycle costing is given by Heijungs et al. (2013).

4.3 From Square to Rectangular Matrices: The Choice of Technology, LP-IO Model

The assumption of A being square implies the absence of alternative technologies or processes to choose from for satisfying a given y. A one-to-one correspondence between products and processes is assumed, and the possibility of substitution among alternative processes is ruled out. The possibility of substitution can be taken into account by replacing the square matrix A with a rectangular matrix in which the number of columns referring to available processes is larger than the number of rows referring to products.

For a hypothetical two-sector model in which two processes are available for sector 1 and three processes for sector 2, the rectangular technology matrix A^+ is given by

$$A^+ = \begin{pmatrix} a_{11}^{(1)} & a_{11}^{(2)} & a_{12}^{(1)} & a_{12}^{(2)} & a_{12}^{(3)} \\ a_{21}^{(1)} & a_{21}^{(2)} & a_{22}^{(1)} & a_{22}^{(2)} & a_{22}^{(3)} \end{pmatrix} \tag{6.132}$$

where the numbers in () refer to alternative processes, and the extended balance equation is given by

$$(I^+ - A^+)x^+ = y \tag{6.133}$$

where x^+ refers to a 5×1 vector of the activity (production) levels of each of the process, and

$$I^+ = \begin{pmatrix} 1 & 1 & 0 & 0 & 0 \\ 0 & 0 & 1 & 1 & 1 \end{pmatrix} \tag{6.134}$$

Equation 6.133 states that whatever choice of processes is to be made, it has to satisfy the given demand y. The problem for solving x^+, that is, the choice of processes, can be formulated as the following linear programming problem (LP) (Koopmans 1951; Dorfman et al. 1958):

$$\min_{x^+} cx^+ \text{ subject to } (133) \tag{6.135}$$

where c refers to a 1×5 vector of constants representing the relative importance of individual processes according to some criteria, say, environmental burdens. Besides the usual nonnegativity constraints, $x^+ \geq 0$, constraints with regard to the availability of productive capacity, factor inputs, and resources, among other things, can also be taken into account in a similar way.

LP-based extensions of IO-LCA has been carried out, among others, by Kondo and Nakamura (2004) for the treatment of waste electronics and by Lin (2009,

2011) for waste water treatment. Duchin (2005) is an example of LP-IO applied to a global MRIO with explicit focus on resource constraints.

5 Case Studies

5.1 Frontiers of Hybrid LCA

The following introduces some of the latest work currently advancing the frontiers in the methodology and application of hybrid LCA.

5.1.1 Biodiesel Supply Chain

In Acquaye et al. (2011), a hybrid LCA methodology was used to evaluate the life cycle CO_2 equivalent emissions of rape methyl ester (RME) biodiesel. The methodology used the integrated hybrid framework discussed in Sect. 3.2.3 that combines IO with traditional PLCA. Besides upstream indirect impacts in the LCA system, emissions due to direct and indirect land use change and N_2 O emissions from fertilizer applications were also calculated. Structural path analysis (SPA), discussed in Sect. 3.7.2, was used to decompose the upstream indirect emission paths of the biodiesel supply chain in order to identify, quantify, and rank high-carbon emissions paths or "hotspots" in the biodiesel supply chain. It was the first time that an integrated hybrid LCA and SPA had been used to analyze the biodiesel supply chain. The aim of this detailed analysis was to help tailor and prioritize mitigation efforts through the use of biofuels.

The integrated hybrid LCA method applied in the study combined an 11×11 process matrix with a 178×178 sector input–output matrix from the UK in 2004. The 2010 ecoinvent database v2.2 was used to compile the process analysis life cycle inventory, described as unit process exchanges. Uncertainty in upstream emissions was estimated by including the maximum/minimum IO upstream cutoffs into the LCA system. To account for the maximum IO upstream cutoffs, all sectoral products potentially serving as indirect input requirements to biodiesel production were included. Similarly, to account for minimum IO upstream cutoffs, only those sectoral products that most probably serve as indirect input requirements for the biodiesel production supply chain were included. Besides its mathematical consistency, integrated hybrid LCA provides a comprehensive framework because all the inputs associated with the biodiesel supply chain can be expressed by the combination of process and IO matrices.

5.1.2 Wind Power

Wiedmann et al. (2011) explored methodological options for hybrid LCA to account for the indirect GHG emissions of energy technologies using wind power generation in the UK as a case study. Using a multi-region input–output modeling framework, they developed and compared two different hybrid approaches: input–output-based hybrid LCA and integrated hybrid LCA. The latter utilized the full-sized ecoinvent process database. The basic layout of the input–output framework was a two-region model based on supply and use tables for the UK and the rest of the world (ROW).

This was the first time a fully integrated hybrid system was described for a biregional supply and use framework. As real company or process data for the analysis were lacking, relevant data on wind turbine manufacture and operation were taken from the ecoinvent database.

For the IO-based hybrid LCA, the actual input requirements of the desired wind power subsector were approximated by using data from a process analysis by Joshi (1999) (EIO-LCA models III to VI). To this end, the electricity industry and product sectors were split into 11 subsectors, including the desired subsector of electricity generation by wind power. This kind of disaggregation procedure had also not previously been used widely for hybrid LCA. For the integrated hybrid LCA, they followed the approach of combining the full-sized process matrix from the ecoinvent LCA database, which distinguishes almost 4000 goods and processes, with the SUT framework for the UK and ROW, with 224 sectors. Finally, the path exchange method (PXC) (Lenzen and Crawford 2009) was used in this work to explore the possibility of adjusting the life cycle inventory obtained by IO-based hybrid LCA.

Uncertainty in hybrid LCA modeling arises from a number of factors, including uncertainties in source data, imputation and balancing, allocation, assuming proportionality and homogeneity, concordance, sectoral aggregation, regional aggregation, temporal discrepancies, representativeness of model data, monetary exchange rates (for multiregional models), and price conversion. It was argued that the conversion from physical to monetary units for direct inputs to IO-based Hybrid LCA and upstream inputs in the C^{u} matrix of integrated hybrid LCA (see (6.67)) is likely to constitute a major reason for uncertainty, and a simple sensitivity analysis was therefore performed by varying prices in both hybrid models by 20 %, respectively. Integrated hybrid LCA showed a narrow sensitivity range for CO_2 (10.8 %) than IO-based hybrid LCA (19.6 %). This result is due to the fact that in integrated hybrid LCA, it is only upstream inputs that are affected by price variation.

It was concluded that the IO-based hybrid LCA approach is easier to implement. It requires less effort to compile the model framework because only input–output matrices are required, not process and upstream (downstream) matrices. This means far less data processing up front and less complicated updating procedures, making IO-based hybrid LCA a more efficient and less expensive alternative to the integrated approach. Furthermore, national input–output tables automatically represent country-specific products and industries and are generally more up to date than data on specific processes.

Integrated hybrid LCA, on the other hand, has the advantage that no monetization of physical flows is necessary and that there are numerous specific processes available with which to match primary data if – as in this case – detailed LCA databases such as ecoinvent have been integrated. This is an area where IO-based hybrid LCA can be improved by employing the path exchange method (PXC) based on structural path analysis. Identifying and replacing specific path information are a helpful technique for making explicit adjustments and increasing accuracy.

5.2 Examples of Hybrid LCA Application

Hybrid LCA has been used extensively in many fields, as illustrated by the following list of case studies classified by area of application:

1. *Transportation*
   (a) Electric vehicles (EVs) (Moriguchi et al. 1993; Nansai et al. 2002; Matsuhashi et al. 2000)
   (b) Plug-in hybrid vehicles (Samaras and Meisterling 2008)
   (c) Charging infrastructure for EV (Nansai et al. 2001)
   (d) Diesel and electric urban density trucks (Lee et al. 2013)
   (e) Diesel and electric public transportation buses (Cooney et al. 2013)
   (f) Freight transportation (Facanha and Horvath 2007)
2. *Fuels*
   (a) Biodiesel (Acquaye et al. 2011)
   (b) Cellulosic ethanol (Baral et al. 2012)
   (c) Corn ethanol (Yang et al. 2012)
   (d) Fuels and propulsion technologies (Lave et al. 2000)
   (e) Fuel options (gasoline, diesel, ethanol, hydrogen, and electricity) (Maclean and Lave 2003)
   (f) Natural gas, methanol, and hydrogen for transportation fuels (Strømman et al. 2006)
   (g) Low-carbon energy sources for transportation (corn-based ethanol, soybean biodiesel, cellulosic ethanol from switch grass, microbial biodiesel, coal with carbon sequestration, photovoltaic cells, and solar concentrators) (Harto et al. 2010)
   (h) Biofuels, gasoline, and electricity for transportation (Scown et al. 2011)
   (i) Shale gas (Jiang and Hendrickson 2014)
   (j) Village-level biomass gasification projects (Wang et al. 2012)

3. *Wind and photovoltaic power*
   (a) Onshore and offshore wind power (Arvesen and Hertwich 2011)
   (b) Offshore wind power (Arvesen et al. 2013)
   (c) Wind power in China (Li et al. 2012)
   (d) Wind power in the UK (Wiedmann et al. Wiedmann et al. 2011)
   (e) Wind turbines in Brazil and Germany (Lenzen and Wachsmann 2004)
   (f) Multicrystalline silicon photovoltaic systems (Zhai and Williams 2010)
   (g) Parabolic trough concentrating solar power plant (Lenzen and Wachsmann 2004)
4. *New technologies*
   (a) Carbon capture and storage (CCS) (Singh et al. 2011, 2012)
   (b) Disassembly technology (Nakamura and Yamasue 2010)
   (c) Colloidal silica and cement grouted soil barrier remediation technologies (Gallagher et al. 2013)
   (d) Water loss control using pressure management (Stokes et al. 2013)
   (e) Carbon nanotubes (Gavankar et al. 2014)
5. *Water supply and waste treatment*
   (a) Water supply (Stokes and Horvath 2006, 2009; Mo et al. 2010)
   (b) Wastewater treatment (Lin 2009; Alvarez-Gaitan et al. 2013; Shao and Chen 2013)
   (c) Household food waste (Inaba et al. 2010)
   (d) Crop residues as an energy source (Lu and Zhang 2010)
6. *Food*
   (a) Meat (Peters et al. 2010a, b)
   (b) Wheat (Piringer and Steinberg 2006)
7. *Electrical equipment and products*
   (a) Desktop personal computers (Williams 2004; Yao et al. 2010)
   (b) Laptop computers (Deng et al. 2011)
   (c) Semiconductor chips (Boyd et al. 2009; Krishnan et al. 2008)
   (d) Print circuit boards (Lee and Ma 2013)
   (e) Primary and rechargeable batteries (Lankey and McMichael 2000)
   (f) Home appliances (Kondo and Nakamura 2004)
8. *Construction and building*
   (a) Airports (Lenzen et al. 2003)
   (b) Wood and concrete buildings (Lenzen and Treloar 2002)
   (c) Commercial office and residential buildings (Crawford 2008)
   (d) Concrete buildings in the USA (Vieira and Horvath 2008)

(e) High-rise buildings in China (Chang et al. 2012)
(f) Residential and commercial buildings in the USA (Onat et al. 2014)
(g) Commercial office buildings in Thailand (Kofoworola and Gheewala 2008)
(h) Green building codes and certification systems (Suh et al. 2014)
(i) Building construction phase (Sharrard et al. 2008)
(j) Asphalt mixtures with high reclaimed asphalt pavement (Aurangzeb et al. 2014)
(k) Reinforced concrete and hot-mix asphalt pavements (Kucukvar and Tatari 2012)
(l) Reinforced concrete and timber railway sleepers (Crawford 2009)
(m) Road construction and use (Trelor et al. 2004)
(n) Hybrid LCA database for building products (Sangwon and Lippiatt 2012)

9. *Services*

(a) Print and online advertising (McKenzie and Durango-Cohen 2010)
(b) Traditional and E-commerce DVD rental networks (Sivaraman et al. 2007)
(c) Service industries (Berners-Lee et al. 2011; Shrake et al. 2013)

10. *Regions*

(a) European Union (Huppes et al. 2006; Scholer et al. 2012; Schoer et al. 2013)
(b) A US city (Hillman and Ramaswami 2010; Ramaswami et al. 2008)
(c) A Finnish city (Heinonen and Junnila 2011)
(d) A Norwegian city (Larsen and Hertwich 2009)
(e) Industrial parks (Mattila et al. 2010; Dong et al. 2013)

11. *Lifestyle and residence*

(a) Rural and urban lifestyles (Heinone and Junnila 2011; Heinone et al. 2011b)
(b) Diets (Meier and Christen 2013)
(c) Residential wood-based heating (Solli et al. 2009)
(d) Residential systems (Heinone et al. 2012; Urban and Bakshi 2013)
(e) Residential swimming pools (Forrest and Williams 2010)

12. *Institutions*

(a) A water corporation in Sydney (Rowley et al. 2009)
(b) Energy-consuming products (Junnila 2008)
(c) An inland marine freight transportation company (Ewing et al. 2011)
(d) A university (Baboulet and Lenzen 2010)

6 IO-LCA Database

To support IO-LCA and hybrid LCA, LCA analysts can use databases developed by many institutions, which are provided for a fee or free of charge. Although databases intended for application in IO-LCA often hold the economy of a single country as the system boundary, multiregional input–output tables (MRIO) have also been published in recent years, to which environmental load data have been appended, facilitating the application of the data to IO-LCA based on global system boundaries, including international trade. Those new databases with MRIO are expected to be useful for various LCA purposes in the future just like IO-LCA databases with a system boundary of a single country which have practically built upon myriad examples of LCA. However, MRIO construction requires strong assumptions and simplifications to complete the input–output tables of each country and trade data constructed under different definitions, methods, and standards of quality, which are often self-contradictory. Along with the numerical increase of countries and sectors, precise estimates of environmental burdens for each sector naturally become difficult. Robustness for processes and environmental inventories is the bread and butter for LCA. Thereby the fundamental challenges for future IO-LCA are enhancement of MRIO reliability and establishing an environmental database for MRIO with a scope sufficiently broad to cover multiple impact categories in life cycle impact assessment. Some major databases for LCA using a country as a system boundary and some of the MRIO database for LCA purposes now being developed are presented below.

6.1 Databases Using a Country as a System Boundary

6.1.1 3EID

Embodied **E**nergy and **E**mission **I**ntensity **D**ata for Japan Using Input–Output Tables (3EID) provided by the National Institute for Environmental Studies of Japan (NIES), Japan, is a database for IO-LCA that uses Japan's input–output tables. The database is obtainable from the NIES website free of charge. The sectoral resolution of the latest data, derived from the year 2005 tables, includes 403 sectors. The year 2005 tables report energy consumption and greenhouse gases as environmental load factors. The embodied intensity of each sector is provided on the basis of both producer and consumer price. The embodied intensity is broken down by type of fuel and raw materials, by sector inducing environmental load, and by type of input into sectors, which facilitates an understanding of the structural drivers of environmental loads.

6.1.2 CEDA

CEDA developed and distributed by IERS LLC is a professional-quality input–output LCA database based on expert analysis of input–output tables of the USA, China, the UK, and other European countries. It was first released in 2000, and it has been continued to be updated to the current, fourth version, where the US year 2002 tables have been analyzed using environmental load data of various types. It is currently being updated to its version 5, where geographical boundary has been extended into other regions. From version 5, annual update will be provided. Academic license is provided for free of charge, while license fees are charged for nonacademic uses to fund update and maintenance. CEDA uses multiple allocation techniques based on standard supply and use framework and incorporates various emission estimation techniques and validation procedures. Feedback and error reporting by CEDA users over the last 14 years also helped increase the reliability of CEDA. The database has been utilized in various high-profile projects commissioned by the US Environmental Protection Agency (EPA), National Institute of Standards and Technology (NIST), and General Services Administration (GSA).

6.1.3 EIO-LCA

EIO-LCA is a database for IO-LCA provided by the Green Design Team at Carnegie Mellon University. It is based on the US input–output tables. Fee payment is required for commercial use, but for other purposes, copies are available free of charge on the EIO-LCA website. Data involving the input–output tables of Canada, Germany, and China have recently been added.

6.1.4 OPEN IO

OPEN IO is a database based on the US input–output tables developed by The Sustainability Consortium at the University of Arkansas. The database, analyzing 417 sectors based on 2002 tables, is provided entirely free of charge.

6.1.5 Supply Chain (Scope 3) Greenhouse Gas Emission Factors

This is a database provided by the Centre for Sustainability Accounting Ltd. based on the UK input–output tables, which supports the calculation of scope 3 emissions. The estimation uses 75 sectors from the year 2009 data.

6.1.6 Balancing Act

The Balancing Act report is a database developed by Integrated Sustainability Analysis at the University of Sydney by expanding sectors from the Australian input–output tables. It is useful for triple bottom-line assessment of the environment, economy, and society. The analysis of year 2005 data is based on a classification into 135 sectors.

6.1.7 SimaPro

SimaPro is comprehensive LCA software sold by Pre Consultants. Its inventory data includes IO-LCA-based data, in addition to process-based databases such as ecoinvent. SimaPro8 facilitates IO-LCA based on input–output tables from the USA (using CEDA), Denmark, and Switzerland.

6.2 *Databases Using Multiple Countries as System Boundary*

6.2.1 EXIOPOL

A New Environmental Accounting Framework Using Externality Data and Input–Output Tools for Policy Analysis (EXIOPOL) was an EU research project during 2007–2011 which developed the MRIO (EXIOBASE) for the year 2000 consisting of 44 countries including the "rest of the world (ROW)," each comprising 129 sectors (products × industries) (see EXIOBASE web). Sample data are provided free of charge, while the entire database is available on purchase of a license. Extensive satellite data on environmental loads and resource use were used to estimate 30 types of environmental load and 80 types of resource use for each sector. The subsequent project CREEA (2011–2014) developed time-series MRIO data and extended the types of environmental load and resource use further (see Tukker et al. 2015).

6.2.2 WIOD

The World Input–Output Database: Construction and Applications (WIOD) project implemented as an EU research project during 2009–2012 constructed MRIO data for each year between 1995 and 2009, which are provided free of charge (see WIOD web). The MRIO data are organized as a supply and use table defined according to a classification into 35 industries and 59 products. The square input–output tables of 35 × 35 industrial sectors are also provided for the purpose of input–output analysis. The environmental satellite data include energy consumption, CO_2 emissions, resource consumption, land use, water consumption, and other inputs.

6.2.3 Eora

The University of Sydney's Integrated Sustainability Analysis (ISA) unit has developed and made available free of charge the Eora MRIO database, comprising 15,909 sectors × 15,909 sectors in a total of 187 countries, each of which includes 25–500 sectors (see Eora web). The outstanding feature of Eora MRIO is that it maintains, to the greatest extent possible, the original sectoral classification resolution of each country's input–output tables. The project developed a software package called AISHA (An Integrated System for Building Harmonized Accounts) that semiautomated the construction of an MRIO and achieved low-cost MRIO estimation and fast reporting. Compiling the data with this many sectors and countries in the absence of harmonized global data to support them has certainly necessitated the use of assumptions and simplifications (Lenzen et al. 2012a). Nevertheless, the level of resolution and the wide coverage of Eora have proven to be extremely valuable for many studies. To date, tables for the years 1990–2009 have been estimated. The extensive satellite data include energy consumption, greenhouse gas emissions, air pollutant emissions, ecological footprint, water footprint, and effect on endangered species listed by the International Union for Conservation of Nature (IUCN).

6.2.4 GLIO

The global link input–output model (GLIO) (Nansai et al. 2009) is an MRIO model that globally expands the system boundary from the detailed input–output tables of Japan. It reduces the amount of labor and time required for developing the tables by simplifying the MRIO structure and incorporates 231 countries and regions in its system boundary while maintaining the resolution of Japan's sectoral classification, which at 406 sectors is as high as the classification of Japanese sectors in Eora.

Simply put, this approach builds the intermediate demand sector of the GLIO model for the year 2005 using 1042 × 1042 sectors and divides these sectors into three categories. The first category is the domestic product sector (406 sectors) of the focal country (here, Japan), which is defined in detail using the sectoral classification of the original input–output table and describes the output of each sector. The second category is the sector of direct imports (406 sectors) to the final demand sector of the focal country (Japan). No intermediate demand from the direct import sector to the final demand sector is included. The final category is the overseas sector, which defines each of 230 countries and regions as having one sector and which indicates their output values to Japan's domestic product sector (first category) and sector of direct imports to the final demand sector (second category). In the environmental input–output analysis, the GLIO model is characterized above all by its conversion of the output from the overseas sector to the size of the domestic environmental footprint.

For the carbon footprint, for instance, a MRIO is developed by converting the output from the overseas sector into the amount of direct and indirect GHG emissions in the export country (domestic carbon footprint) caused by its export, considering the differences in carbon footprint in the exported products. This mixed-unit MRIO, expressed in monetarily for the flow from Japan and in physical units of domestic environmental footprint for the flow from foreign countries, permits IO-LCA using conventional input–output analysis, with Japanese sectors defined in detail and 231 countries as the system boundary. The characteristics of simplifying MRIO using GLIO are detailed in Nansai et al. (2013).

To date, GLIO has enabled calculation of the environmental loads (energy consumption, GHG, air pollutants) generated in and outside Japan per unit of production (1 million yen) of Japanese domestic products (406 sectors) (Nansai et al. 2012). These data are provided free of charge on the 3EID website (National Institute for Environmental Studies) and are widely used as IO-LCA data for Japan's scope 3 calculations.

7 Resolved and Unresolved Issues

7.1 Weak Points of IO-LCA from the Point of View of PLCA

In their review of IO-LCA, Reap et al. (2008), while acknowledging its advantages over PLCA in regard to its comprehensiveness and speed of boundary selection, pointed out the following problematical issues:

1. *Imports*: The IO portion assumes that the amounts of imported commodities to the product system under study are negligible or that they come from countries with similar production technologies and economic structures.
2. *Unbalanced data*: For most countries, there is a lack of applicable, well-balanced sectoral environmental data that can be correlated with economic data.
3. *Old data*: The IO-based data is usually several years older than the process-based data.
4. *Resolution*: The IO-based data is usually aggregated for industries and commodities, thus diminishing the resolution capability of the IO analysis when compared with more detailed LCAs. The IO-based data assumes companies are perfectly homogeneous, meaning that they produce only one commodity and as such there is an allocation error.
5. *EoL*: Many IO-LCA-based methods do not consider cradle-to-grave industrial processes, thus themselves introducing a truncation error.
6. *Recycling*: Many IO-LCA-based methods do not consider the recycling or remanufacturing industry sectors.

This chapter has shown that attempts have been made to cope with these difficulties with different degrees of achievements, which can be summarized as follows:

1. *Imports*: As mentioned in 2, there has been a remarkable increase in the development of MRIO databases and their use in hybrid LCA. Among other things, the development and use of MRIO have resulted in an explosion of studies on environmental footprints, as surveyed in Hubacek et al. (2014) and Lenzen (2013).
2. *Unbalanced data*: As we have seen in Sect. 6.1, there has been considerable progress in developing consistent IO-based databases incorporating environmental data. Thanks to its well-defined yet flexible accounting and modeling framework, IO moreover has the potential to serve as a basis for developing a platform for storing these data (Nakamura 2011).

3/4. *Old data/low resolution*: As key components of the systems of national accounts, IOTs are developed and published on a regular basis, but with some delay (up to 3 years or more), owing to the considerable resources required for their development. This oft-voiced criticism of IO data ensues partly from the fact that the date of their development is explicitly stated, which is not the case for PLCA data. Furthermore, new IO data represent a full revision of previous data, which is also not the case for PLCA data.

5/6. *EoL/recycling*: These issues have been resolved by WIO, at least conceptually; see Sects. 3.4 and 3.5 above. Development of the data base required for the implementation of WIO is still at a preliminary stage, however. In particular, there are as yet no international IOTs incorporating waste flows.

In summary, it seems safe to conclude that the major weakness of IO-LCA compared with PLCA consists in not conceptual, but in data-related issues. Integrated use of IO data and updated data on processes of interest, as discussed in Sect. 5, will be a productive strategy to pursue.

7.2 Limitations of IO Relevant to LCA

The following issues are often cited as major conceptual limitations of IO that should be taken into consideration when using IO models in an LCA context (NWT 2006, p. 8):

1. *Fixed coefficients*: The technical coefficients, the elements of the A and R matrices in the fundamental EEIO Eq. (6.62), are assumed to be fixed. Price effects, substitution, changing technology, and economies of scale are not considered.
2. *No supply constraints*: There are no constraints on the supply of primary factors of production, the supply of resources, or productive capacity.
3. *Capital goods*: Goods can be produced without additions to capital stock.

It is noteworthy that these limitations apply almost equally to PLCA as well. As mentioned above in Sect. 4.3 with regard to the rectangular extension of the A matrix to accommodate the presence of alternative processes, the presence of supply constrains can be taken into consideration by resorting to LP-IO where they occur as additional linear constraints. It is to be noted that consideration of these constraints in LP-IO can be achieved without rectangular extension of the A matrix; see Nakamura and Kondo (2009), Sect. 3.5.1 for further details. The issue of capital goods has already been discussed above, in Sect. 3.3. The first point is discussed below.

7.2.1 Fixed Coefficients

Many factors are involved behind the assumption of fixed coefficients, which can be divided into the following groups:

1. Linear technology: The independence of A from x.
2. No alternative processes: The matrix A is square.
3. No price effects: The independence of A from prices of inputs.

Linear Technology

The independence of A from x, that is, the independence of technology from the scale of production, is justified by assuming the technology to follow constant returns to scale (Sect. 2.1.2). If this assumption is not used, A depends on x, and the balance Eq. (6.17) becomes

$$x = A(x)x + y, \tag{6.136}$$

which requires information about the functional form of $A(x)$ for a solution to be obtained. Because x is determined by y, the dependence of A on x is tantamount to its dependence on y. Scale effects represented by the size of y are not the only factor contributing to this dependence. The composition of y may also affect A. For instance, a change in the composition of waste being submitted to a waste incineration process can result in changes in certain input coefficients relating to utilities and chemicals (Nakamura and Kondo, 2009, Sect. 6.2). Nonlinearity in technical input–output relationships can be incorporated into IO once the underlying process knowledge is available. For example, Nakamura and Kondo (2002) extended the standard IO model by introducing a system engineering model of waste treatment to consider the effects of changes in the scale of operation of incinerators and in the composition of waste fed into the incinerators.

No Alternative Processes

This has already been discussed above in Sect. 4.3.

No Price Effects

Many LCA studies are concerned with comparing alternative production processes to meet the requirements of a given functional unit y. As discussed in Sect. 2.4, changes in certain elements of A can lead to changes in product prices. Accordingly, different A can result in different sets of product prices. It then follows that if y depends on product prices, that is, $y = y(p)$, a change in A can affect y via its effects on p, that is, y can no longer remain at its initial level and composition before the change took place. Introduction of (certain) market mechanisms is a feature that distinguished consequential LCA (CLCA) from other models of LCA, albeit not endogenously at present, as the market mechanisms are derived exogenously from economic models (Hertwich 2005; Zamagni et al. 2012).

Computable general equilibrium (CGE) models, such as the GTAP model (Klijn et al. 2005), are economic models that are frequently used in LCA to consider market mechanisms (Earles and Halog 2011). CGE models are closely related to IO in terms of both concepts and data, although there are also several important differences between them (Rose 1995). CGE models, being economic models, are based on a set of economic assumptions in order to make them operational. Some of these assumptions are concerned with specification of technology and hence are of considerable relevance to LCA. To our understanding, however, these assumptions and their implications for LCA do not seem well known in the LCA community, and for this reason the topic will now be elaborated on.

7.2.2 CES Functions and Separability: The Way CGE Incorporates Substitution Based on Price Mechanisms

As the name indicates, a general equilibrium (GE) model can be quite general, at least conceptually. Among others things, it is free of stringent restrictions such as fixed coefficients on technical input–output relationships. For the case of a single-output j and n types of input, the technical input–output relationships can be represented in its general form by the "production function" f_j:

$$x_j = f_j(x_{1j}, \ldots, x_{1n}, K_{1j}, \ldots, K_{kj}, L_{1j}, \ldots, L_{lj}) \tag{6.137}$$

For a GE model to be computational, however, f_j must be functionally specified and its parameters quantified using real-world data.

In LCA and IO, f_j is specified by a set of fixed coefficients obtained from process information, operational data, and/or transaction data, that is, in a bottom-up fashion. In CGE, a different approach is used. The functional form of f_j is chosen from among the arsenal of economic tools in a top-down fashion, based on a balance between analytical convenience and the generality with regard to the potential for substitution among inputs. Process or technological information are not generally used in its specification (Schumacher and Sands 2007).

CES Functions

The constant elasticity of substitution (CES) production function with two inputs given below is the one most widely used in CGE models:

$$x_j = \alpha_j \left[\beta_j x_{1j}^{\rho_j} + (1 - \beta_j) x_{2j}^{\rho_j}\right]^{1/\rho_j} \tag{6.138}$$

where $\beta \in [0, 1]$ is the share parameter, α is the efficiency parameter, and $\rho_j \leq 1$ is the parameter referring to the degree of substitutability of the inputs. This specification is general in regard to substitution possibilities, because no a priori restrictions are imposed on ρ_j. A fixed-coefficient, zero-substitution model follows as $\rho_j \to -\infty$, while the opposite of infinite or complete substitutability follows if $\rho_j \to 1$. The CES function has a fundamental weakness, however. Extending it to include more than two inputs is difficult except for the uninteresting case where the substitution potential is the same among all the inputs occurring in the function (see Nakamura and Kondo 2009, Sect. 4.3 for further details). It is thus no mere coincidence that the CES function given by Eq. (6.138) contains only two inputs. Because the number of inputs in CGE models is generally greater than two, this feature is problematic.

Separability of Inputs into Groups

This shortcoming of the CES function is coped with by assuming a tree structure of production, or nested groupings (nested CES functions) of inputs into several groups, with each group consisting of a "small number," mostly two, of inputs such that a CES function can be applied to its components (Paltsev et al. 2005; Qi et al. 2014). For the case of $n = 6$ and three groupings, this tree/nested structure can be given by

$$\begin{aligned} x_j &= f_j(x_{1j}, x_{2j}, x_{3j}, x_{4j}, x_{5j}, x_{6j}) \\ &= F_j\Big(g_{1j}(x_{1j}, x_{2j}), g_{2j}(x_{3j}, x_{4j}), g_{3j}(x_{5j}, x_{6j})\Big) \\ &= F_j(G_{1j}, G_{2j}, G_{3j}) \end{aligned} \tag{6.139}$$

with

$$G_{ij} = g_{ij}\left(x_{ij}, x_{i+1,j}\right) = \alpha_{ij}\left[\beta_{ij}x_{ij}^{\rho_{ij}} + \left(1 - \beta_{ij}\right)x_{i+1j}^{\rho_{ij}}\right]^{1/\rho_{ij}} \tag{6.140}$$

where g_{ij} refers to the sub-production or aggregator function, G_{ij} to the resulting aggregate, and F_j to the production function in terms of the three aggregates. Substitutability among the three aggregate inputs can be taken into account by introducing additional grouping of the first two aggregates $\left(G_{1j}\text{ and }G_{2j}\right)$ into one using an aggregator function g_{Aj}:

$$\begin{aligned} x_j &= F_j\left(g_{Aj}(G_{1j}, G_{2j}), G_{3j}\right) \\ &= \bar{F}_j(G_{Aj}, G_{3j}) \end{aligned} \tag{6.141}$$

with g_{Aj} a CES function in G_{1j} and G_{2j}, and $\bar{F}_j$ a CES function in G_{Aj} and G_{3j}.

Figure 6.1 gives a visual representation of the nested structure or production tree (for real examples, see Paltsev et al. 2005 and Qi et al. 2014). In sharp contrast to LCA, industrial process knowledge or information on technological changes are not generally used to parameterize the CES function (Schumacher and Sands 2007). Instead, this is usually performed using statistical (econometric) estimates taken from the literature or provided a priori in a top-down fashion.

Implications of Separability

The seemingly convenient practice of using nested CES functions based on the grouping of inputs is not without problems, however. The grouping of inputs into several groups is possible because the inputs are assumed to be "separable" into

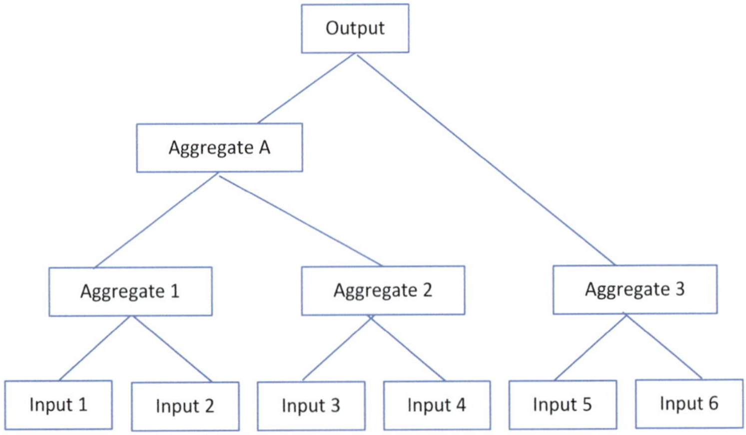

Fig. 6.1 The nesting structure of production sectors: an example

mutually exclusive groups of inputs. This assumption of separability implies that the composition of individual components within a group can be altered independently of the compositions of the items occurring in other input groups (Leontief 1947; Sono 1961). Accordingly, (6.139) implies that the combination of the individual inputs in each of the three groups can be altered independently of one another. For instance, a simplified version of the nested CES model of Qi et al. (2014) corresponds to the case where inputs 1 and 2 refer to capital and labor, which are aggregated by a CES function to value added, inputs 3 and 4 refer to energy sources such as oil and gas, which are aggregated to an energy aggregate, and inputs 5 and 6 refer to other intermediate inputs, for which fixed coefficients are used. This implies that the composition of the energy aggregate can be altered independently of the capital–labor combination. It is noteworthy that "capital" is itself an aggregate of individual capital goods and needs to be constructed using an aggregator function (see (6.32)). Accordingly, representation of the production structure in terms of "capital" implies the separability of its components, capital goods, from the remaining inputs. In economic analysis focusing on highly aggregated macroeconomic subjects in monetary terms, this point will not be of great importance. In LCA, however, use of this concept automatically implies the independence of energy mix from the composition of capital goods, which is at odds with the fact that alternation of energy sources usually requires alternation of energy-related machines and equipment.

It is also noteworthy that, in reality, substitution of technology may require the installation or alteration of processes, which involve considerable amounts of investment expenditure in relevant productive facilities. Accommodation of this point in a CGE model based on the assumption of separability of capital goods from other inputs, that is, the existence of an aggregated capital, appears conceptually difficult. Extension of LP-IO with explicit consideration of capital goods 3.3 would be conceptually more straightforward and simpler to implement.

7.2.3 A Closing Remark

Any analytical tool has its strength and weakness, and IO-LCA is no exception. When using such tools, it is important to be aware of their limitations and weakness and refrain from applying them in a merely mechanical fashion, with no consideration given to the constraints that will automatically be imposed on the results they yield.

Acknowledgments We would like to thank Tomy Wiedmann, Manfred Lenzen, Edgar Hertwich, Sangwon Suh, and Klaus Hubacek for providing us with the latest information about their research and Yosuke Shigetomi for the preparation of the extensive list of references for case studies. We are also grateful to Nigel Harle of Gronsveld, the Netherlands, for his careful revision of our English.

References

Acquaye AA, Wiedmann T, Feng K, Crawford RH, Barrett J, Kuylenstierna J, Duffy AP, Koh SL, McQueen-Mason S (2011) Identification of carbon hot-spots and quantification of GHG intensities in the biodiesel supply chain using hybrid LCA and structural path analysis. Environ Sci Technol 45(6):2471–2478

Alsamawi A, Murray J, Lenzen M (2014) The employment footprints of nations. J Ind Ecol 18 (1):59–70

Alvarez-Gaitan PJ, Peters MG, Rowley VH, Moore S, Short DM (2013) A hybrid life cycle assessment of water treatment chemicals: an Australian experience. Int J Life Cycle Assess 18:1291–1301

Arvesen A, Hertwich GE (2011) Corrigendum: environmental implications of large-scale adoption of wind power: a scenario-based life cycle assessment. Environ Res Lett 6:1–3

Arvesen A, Birkeland C, Hertwich GE (2013) The importance of ships and spare parts in lcas of offshore wind power. Environ Sci Technol 47:2948–2956

Aurangzeb Q, Al-Qadi LI, Ozer H, Yang R (2014) Hybrid life cycle assessment for asphalt mixtures with high rap content. Resour Conserv Recycl 83:77–86

Baboulet O, Lenzen M (2010) Evaluating the environmental performance of a university. J Clean Prod 18:1134–1141

Baral A, Bakshi RB, Smith LR (2012) Assessing resource intensity and renewability of cellulosic ethanol technologies using eco-lca. Environ Sci Technol 46:2436–2444

Berners-Lee M, Howard CD, Moss J, Kaivanto K, Scott AW (2011) Greenhouse gas footprinting for small business -the use of input-output data. Sci Total Environ 409:883–891

Boyd S, Horvath A, Dornfeld D (2009) Life-cycle energy demand and global warming potential of computational logic. Environ Sci Technol 43:7303–7309

Bullard C, Herendeen R (1975) The energy cost of goods and services. Energy Policy 3:268–278

Bureau of Economic Analyses. Concepts and Methods of the U.S. Input-Output Accounts, 2009. URL http://www.bea.gov/papers/pdf/IOmanual_092906.pdf

Caves DW, Christensen LR, Diewert WE (1982) The economic theory of index numbers and the measurement of input, output, and productivity. Econometrica (J Econ Soc) 50:1393–1414

Chang Y, Ries JR, Lei JR (2012) The embodied energy and emissions of a high-rise education building: a quantification using process-based hybrid life cycle inventory model. Energy Build 55:790–798

Cooney G, Hawkins RT, Marriott J (2013) Life cycle assessment of diesel and electric public transportation buses. J Ind Ecol 17(5):689–699

Crawford HR (2008) Validation of a hybrid life-cycle inventory analysis method. J Environ Manage 88:496–506

Crawford HR (2009) Greenhouse gas emissions embodied in reinforced concrete and timber railway sleepers. Environ Sci Technol 47(3):3885–3890

Davis SJ, Caldeira K (2010) Consumption-based accounting of CO_2 emissions. Proc Natl Acad Sci U S A 107:5687–5692

Defourny J, Thorbecke E (1984) Structural path-analysis and multiplier decomposition within a social accounting matrix framework. Econ J 94(373):111–136

Deng L, Babbitt WC, Williams DE (2011) Economic-balance hybrid LCA extended with uncertainty analysis: case study of a laptop computer. J Clean Prod 19:1198–1206

Dietzenbacher E, Los B (1998) Structural decomposition techniques: sense and sensitivity. Econ Syst Res 10(4):307–324

Dong H, Geng Y, Xi F, Fujita T (2013) Carbon footprint evaluation at industrial park level: a hybrid life cycle assessment approach. Energy Policy 57:298–307

Dorfman R, Samuelson P, Solow R (1958) Linear programming & economic analysis. McGraw Hill, New York

Duchin F (1990) The conversion of biological materials and wastes to useful products. Struct Chang Econ Dyn 1:243–261

Duchin F (2005) A world trade model based on comparative advantage with *m* regions, *n* goods, and *k* factors. Econ Syst Res 17(2):141–162

Earles JM, Halog A (2011) Consequential life cycle assessment: a review. Int J Life Cycle Assess 16:445–453

Eora MRIO Database. http://worldmrio.com

Ewing A, Thabrew L, Perrone D, Abkowitz M, Hornberger G (2011) Insights on the use of hybrid life cycle assessment for environmental footprinting -a case study of an inland marine freight transportation company. J Ind Ecol 15(6):937–950

EXIOBASE. http://www.exiobase.eu

Facanha C, Horvath A (2007) Evaluation of life-cycle air emission factors of freight transportation. Environ Sci Technol 41:7138–7144

Forrest N, Williams E (2010) Life cycle environmental implications of residential swimming pools. Environ Sci Technol 44:5601–5607

Frischknecht R, Althaus H-J, Bauer C, Doka G, Heck T, Jungbluth N, Kellenberger D, Nemecek T (2007) The environmental relevance of capital goods in life cycle assessments of products and services. Int J Life Cycle Assess 7(Special Issue 1):7–17

Gallagher MP, Spatari S, Cucura J (2013) Hybrid life cycle assessment comparison of colloidal silica and cement grouted soil barrier remediation technologies. J Hazard Mater 250–251, 421–430

Gavankar S, Suh S, Keller AA (2014) The role of scale and technology maturity in life cycle assessment of emerging technologies: a case study on carbon nanotubes. J Ind Ecol 19 (1):51–60

GTAP. https://www.gtap.agecon.purdue.edu/default.asp

Harto C, Meyers R, Williams E (2010) Life cycle water use of low-carbon transport fuels. Energy Policy 28:4933–4944

Heijungs R, Suh S (2002) The computational structure of life cycle assessment. Kluwer, Dordrecht

Heijungs R, Settanni E, Guinée J (2013) Toward a computational structure for life cycle sustainability analysis: unifying LCA and LCC. Int J Life Cycle Assess 18(9):1722–1733

Heinone J, Junnila S (2011) A carbon consumption comparison of rural and urban lifestyles. Sustainability 3:1234–1249

Heinone J, Saynajoki J-A, Kurornen M, Junnila S (2012) Are the greenhouse gas implications of new residential developments understood wrongly? Energies 5:2874–2893

Heinonen J, Junnila S (2011) Case study on the carbon consumption of two metropolitan cities. Int J Life Cycle Assess 16:569–579

Hertwich EG (2005) Consumption and the rebound effect: an industrial ecology perspective. J Ind Ecol 9(1–2):85–98

Hertwich E, Peters G (2009) Carbon footprint of nations: a global, trade-linked analysis. Environ Sci Technol 43:6414–6420

Hertwich EG, Peters GP (2010) Multiregional input-output database. Technical report, OPEN: EU technical document. http://www.oneplaneteconomynetwork.org/resources/work-package-1.html

Hillman T, Ramaswami A (2010) Greenhouse gas emission footprints and energy use benchmarks for eight U.S. cities. Environ Sci Technol 44:1902–1910

Horowitz KJ, Planting MA (2009) Concepts and methods of the U.S. input-output accounts. Bureau of Economic Analysis, U.S. Dept of Commerce (BEA). http://www.bea.gov/papers/pdf/IOmanual_092906.pdf

Hubacek K, Feng K, Minx JC, Pfister S, Zhou N (2014) Teleconnecting consumption to environmental impacts at multiple spatial scales. J Ind Ecol 18(1):7–9

Huppes G, De Koning A, Suh S, Heijungs R, Van Oers L, Nielsen P, Guinée J (2006) Environmental impacts of consumption in the European Union – high-resolution input-output tables with detailed environmental extensions. J Ind Ecol 10(3):129–146

Inaba R, Nansai K, Fujii M, Hashimoto S (2010) Hybrid life-cycle assessment (LCA) of CO_2 emission with management alternatives for household food wastes in Japan. Waste Manag Res 28:496–507

Jiang M, Hendrickson TC (2014) Life cycle water consumption and wastewater generation impacts of a Marcellus shale gas well. Environ Sci Technol 48:1911–1920

Jorgenson DW (1988) Productivity and postwar U.S. economic growth productivity and postwar U.S. economic growth. J Econ Perspect 2(4):23–41

Joshi S (1999) Product environmental life-cycle assessment using input-output techniques. J Ind Ecol 2–3:95–120

Junnila S (2008) Life cycle management of energy-consuming products in companies using io-lca. Int J Life Cycle Assess 13:432–439

Klijn JA, Vullings L, Van den Berg M, Van Meijl H, Van Lammeren R, Van Rheenen T, Veldkamp A, Verburg P, Westhoek H, Eickhout B, Tabeau AA (2005) The eururalis study: technical document. Technical report. https://www.wageningenur.nl/en/Publication-details.htm?publicationId=publication-way-333433363432

Klöpffer W (1996) Allocation rule for open-loop recycling in life cycle assessment. Int J Life Cycle Assess 1(1):27–31

Kofoworola FO, Gheewala S (2008) Environmental life cycle assessment building in Thailand. Int J Life Cycle Assess 13:498–511

Kondo Y, Nakamura S (2004) Evaluating alternative life-cycle strategies for electrical appliances by the waste input-output model. Int J Life Cycle Assess 9(4):236–246

Koopmans TC (ed) (1951) Activity analysis of production and allocation. Wiley, New York

Krishnan N, Boyd S, Somani A, Raoux S, Clark D, Dornfeld D (2008) A hybrid life cycle inventory of nano-scale semiconductor manufacturing. Environ Sci Technol 42:3069–3075

Kucukvar M, Tatari O (2012) Ecologically based hybrid life cycle analysis of continuously reinforced concrete and hot-mix asphalt pavements. Transp Res D 17:86–90

Lankey LR, McMichael CF (2000) Life-cycle methods for comparing primary and rechargeable batteries. Environ Sci Technol 34:2299–2304

Larsen NH, Hertwich EG (2009) The case for consumption-based accounting of greenhouse gas emissions to promote local climate action. Environ Sci Policy 12:791–798

Lave L, Maclean H, Hendrickson C, Lankey R (2000) Life-cycle analysis of alternative automobile fuel/propulsion technologies. Environ Sci Technol 34:3598–3605

Lee HC, Ma WH (2013) Improving the integrated hybrid lca in the upstream scope 3 emissions inventory analysis. Int J Life Cycle Assess 18:17–23

Lee YD, Thomas MV, Brown AM (2013) Electric urban delivery trucks: energy use, greenhouse gas emissions, and cost-effectiveness. Environ Sci Technol 47:8022–8030

Lenzen M (2009) Dealing with double-counting in tiered hybrid life-cycle inventories: a few comments. J Clean Prod 17:1382–13848

Lenzen M (2013) An outlook into a possible future of footprint research. J Ind Ecol 18(1):4–6

Lenzen M, Crawford R (2009) The path exchange method for hybrid lca. Environ Sci Technol 43:8251–8256

Lenzen M, Reynolds JC (2014) A supply-use approach to waste input-output analysis. J Ind Ecol 18(2):212–226

Lenzen M, Treloar G (2002) Embodied energy in buildings: wood versus concrete – reply to Börjesson and Gustavsson. Energy Policy 30:249–255

Lenzen M, Treloar G (2005) Endogenising capital: a comparison of two methods. J Appl Input-Output Anal 10:1–11

Lenzen M, Wachsmann U (2004) Wind turbines in Brazil and Germany: an example of geographical variability in life-cycle assessment. Appl Energy 77:119–130

Lenzen M, Murray AS, Korte B, Dey JC (2003) Environmental impact assessment including indirect effects -a case study using input-output analysis. Impact Assess Rev 23:263–282

Lenzen M, Kanemoto K, Moran D, Geschke A (2012a) Mapping the structure of the world economy. Environ Sci Technol 46(15):8374–8381

Lenzen M, Moran D, Kanemoto K, Foran B, Lobefaro L, Geschke A (2012b) International trade drives biodiversity threats in developing nations. Nature 486:109–112

Lenzen M, Moran D, Kanemoto K, Geschke A (2013) Building eora: a global multi-region input-output database at high country and sector resolution. Econ Syst Res 25:20–49

Leontief W (1947) A note on the interrelation of subsets of independent variables of a continuous function with continuous first derivatives. Bull Am Math Soc 53:343–350

Leontief W (1970) Environmental repercussions and the economics structure: an input-output approach. Rev Econ Stat 52:262–271

Li X, Feng K, Siu LY, Hubacek K (2012) Energy-water nexus of wind power in China: the balancing act between CO_2 emissions and water consumption. Energy Policy 45:440–448

Lin C (2009) Hybrid input-output analysis of wastewater treatment and environmental impacts: a case study for the tokyo metropolis. Ecol Econ 68:2096–2105

Lin C (2011) Identifying lowest-emission choices and environmental pareto frontiers for wastewater treatment wastewater treatment input-output model based linear programming. J Ind Ecol 15(3):367–380

Lu W, Zhang T (2010) Life-cycle implications of using crop residues for various energy demands in china. Environ Sci Technol 44:4026–4032

Maclean L, Lave BL (2003) Life cycle assessment of automobile/fuel options. Environ Sci Technol 37:5445–5452

Matsuhashi R, Kudoh Y, Yoshida Y, Ishitani H, Yoshioka M, Yoshioka K (2000) Life cycle of CO_2-emissions from electric vehicles and gasoline vehicles utilizing a process-relational model. Int J Life Cycle Assess 5(5):306–312

Mattila JT, Pakarinen S, Sokka L (2010) Quantifying the total environmental impacts of an industrial symbiosis – a comparison of process-, hybrid and input-output life cycle assessment. Environ Sci Technol 44:4309–4314

McKenzie CE, Durango-Cohen LP (2010) An input-output approach for the efficient design of sustainable goods and services. Int J Life Cycle Assess 15:946–961

Meier T, Christen O (2013) Environmental impacts of dietary recommendations and dietary styles: Germany as an example. Environ Sci Technol 47:877–888

Miller RE, Blair PD (2009) Input-output analysis: foundations and extensions. Cambridge University Press, Cambridge

Minx J, Wiedmann T, Barrett J, Suh S (2008) Methods review to support the PAS process for the calculation of the greenhouse gas emissions embodied in goods and services. Technical report, Dept for Environment, Food and Rural Affairs

Mo W, Nasiri F, Eckelman JM, Zhang Q, Zimmerman BJ (2010) Measuring the embodied energy in drinking water supply systems: a case study in the great lakes region. Environ Sci Technol 44:9516–9521

Moriguchi Y, Kondo Y, Shimizu H (1993) Analysing the life cycle impacts of cars: the case of CO_2. Ind Environ 16:42–45

Moses L (1955) The stability of interregional trading patterns and input-output analysis. Am Econ Rev 45:803–826

Nakajima K, Ohno H, Kondo Y, Matsubae K, Takeda O, Miki T, Nakamura S, Nagasaka T (2013) Simultaneous MFA of nickel, chromium and molybdenum used in alloy steel by means of input-output analysis. Environ Sci Technol 47:4563–4660

Nakamura S (1999) Input-output analysis of waste cycles. In: Environmentally conscious design and inverse manufacturing, 1999. In: Proceedings EcoDesign'99: first international symposium, pp 475–480

Nakamura S (2011) Hybrid input-output analysis as a tool for communication among scientists of different disciplines. J Ind Ecol 15(5):661–663

Nakamura S, Kondo Y (2002) Input-output analysis of waste management. J Ind Ecol 6(1):39–63

Nakamura S, Kondo Y (2006) Hybrid LCC of appliances with different energy efficiency. Int J Life Cycle Assess 11:305–314

Nakamura S, Kondo Y (2009) Waste input-output analysis: concepts and application to industrial ecology. Eco-efficiency in industry and science. Springer, New York

Nakamura S, Nakajima K (2005) Waste input-output material flow analysis of metals in the Japanese economy. Mater Trans 46:2550–2553

Nakamura S, Yamasue S (2010) Hybrid lca of a design for disassembly technology: active disassembling fasteners of hydrogen storage alloys for home appliances. Environ Sci Technol 44:4402–4408

Nakamura S, Nakajima K, Kondo Y, Nagasaka T (2007) The waste input-output approach to materials flow analysis. J Ind Ecol 11(4):50–63

Nakamura S, Nakajima K, Yoshizawa Y, Matsubae-Yokoyama K, Nagasaka T (2009) Analyzing polyvinyl chloride in Japan with the waste input-output material flow analysis model. J Ind Ecol 13:706–717

Nakamura S, Kondo K, Matsubae K, Nakajima K, Nagasaka T (2011) Upiom: a new tool of MFA and its application to the flow of iron and steel associated with car production. Environ Sci Technol 45:1114–1120

Nakamura S, Kondo Y, Matsubae K, Nakajima K, Tasaki T, Nagasaka T (2012) Quality-and dilution losses in the recycling of ferrous materials from end-of-life passenger cars: input-output analysis under explicit consideration of scrap quality. Environ Sci Technol 46 (17):9266–9273

Nansai K, Tohno S, Kono M, Kasahara M, Moriguchi Y (2001) Life-cycle analysis of charging infrastructure for electric vehicles. Appl Energy 70:251–265

Nansai K, Tohno S, Kono M, Kasahara M (2002) Effects of electric vehicles (ev) on environmental loads with consideration of regional differences of electric power generation and charging characteristic of ev users in Japan. Appl Energy 71:111–125

Nansai K, Kagawa S, Kondo Y, Suh S, Nakajima K (2009) Improving the completeness of product carbon footprints using a global link input-output model: the case of Japan. Econ Syst Res 21:267–290

Nansai K, Kondo Y, Kagawa S, Suh S, Nakajima K, Inaba R, Tohno S (2012) Estimates of embodied global energy and air-emission intensities of Japanese products for building a Japanese input–output life cycle assessment database with a global system boundary. Environ Sci Technol 46(16):9146–9154

Nansai K, Kagawa S, Kondo Y, Suh S (2013) Simplification of multi-regional input-output structure with a global system boundary: Global link input-output model (GLIO). In: Murray J, Lenzen M (eds) The sustainability practitioner's guide to multiregional input-output analysis. Common Ground, Champaign

National Institute for Environmental Studies. Embodied Energy and Emission Intensity Data for Japan Using Input-Output Tables (3EID). http://www.cger.nies.go.jp/publications/report/d031/3eid_e.html

Nikaido H (1970) Introduction to sets and mappings in modern economics. North-Holland, Amsterdam

Northwest Territories Bureau of Statistics (2006) NWT input-output model – an overview. http://www.statsnwt.ca/economy/multipliers/NWT

OECD web: input-output tables. http://www.oecd.org/trade/input-outputtables.htm

Ohno H, Matsubae K, Nakajima K, Nakamura S, Nagasaka T (2014) Unintentional flow of alloying elements in steel during recycling of end-of-life vehicles. J Ind Ecol 18:242–253

Onat CN, Kucukvar M, Tatari O (2014) Scope-based carbon footprint analysis of U.S. residential and commercial buildings: an input-output hybrid life cycle assessment approach. Build Environ 72:53–62

Paltsev S, Reilly JM, Jacoby HD, Eckaus RS, McFarland J, Sarofim M, Asadoorian M, Babiker M (2005) The mit emissions prediction and policy analysis (EPPA) model: version 4. Technical report, MIT Joint Program on the Science and Policy of Global Change

Peters GP, Andrew R, Lennox J (2011) Constructing an environmentally-extended multi-regional input-output table using the gtap database. Econ Syst Res 23(2):131–152

Peters MG, Rowley VH, Wiedmann S, Tucker R, Short DM, Schulz M (2010a) Red meat production in Australia: life cycle assessment and comparison with overseas studies. Environ Sci Technol 44:1327–1332

Peters MG, Wiedmann GS, Rowley VH, Tucker WR (2010b) Accounting for water use in Australian red meat production. Int J Life Cycle Assess 15:311–320

Piringer G, Steinberg H (2006) Reevaluation of energy use in wheat production in the United States. J Ind Ecol 10(1–2):149–167

Qi T, Winchester N, Karplus V, Zhang X (2014) Will economic restructuring in China reduce trade-embodied CO_2 emissions? Energy Econ 42:204–212

Ramaswami A, Hillman T, Janson B, Reiner M, Thomas G (2008) A demand-centered, hybrid life-cycle methodology for city-scale greenhouse gas inventories. Environ Sci Technol 42 (17):6455–6461

Reap J, Roman F, Duncan S, Bras B (2008) A survey of unresolved problems in life cycle assessment. Int J Life Cycle Assess 13(5):374–388

Rebitzer G, Nakamura S (2008) Environmental life cycle costing. In: Hunkeler D, Lichtenvort K, Rebitzer G (eds) Environmental life cycle costing. SETAC Press, Boca Raton

Rose A (1995) Input-output economics and computable general equilibrium models. Struct Chang Econ Dyn 6(3):295–304

Rowley VH, Lundie S, Peters MG (2009) A hybrid life cycle assessment model for comparison with conventional methodologies in Australia. Int J Life Cycle Assess 14:508–516

Samaras C, Meisterling K (2008) Life cycle assessment of greenhouse gas emissions from plug-in hybrid vehicles: implications for policy. Environ Sci Technol 42:3170–3176

S. Sangwon and B. C. Lippiatt. Framework for hybrid life cycle inventory databases: a case study on the Building for Environmental and Economic Sustainability (BEES) database. The International Journal of Life Cycle Assessment 17.5 (2012): 604-612

Schoer K, Wood R, Arto I, Weinzettel J (2013) Estimating raw material equivalents on a macro-level: comparison of multi-regional input-output analysis and hybrid LCI-IO. Environ Sci Technol 47:14282–14289

Scholer K, Weinzettel J, Kovanda J, Giegrich J, Lauwigi C (2012) Raw material consumption of the European Union – concept, calculation method, and results. Environ Sci Technol 46:8903–8909

Schumacher K, Sands RD (2007) Where are the industrial technologies in energy-economy models? An innovative cge approach for steel production in Germany. Energy Econ 29 (4):799–825

Scown DC, Horvath A, McKone ET (2011) Water footprint of U.S. transportation fuels. Environ Sci Technol 45:2541–2553

Shao L, Chen GQ (2013) Water footprint assessment for wastewater treatment: method, indicator, and application. Environ Sci Technol 47:7787–7794

Sharrard LA, Matthews SH, Ries JR (2008) Estimating construction project environmental effects using an input-output-based hybrid life-cycle assessment model. J Infrastruct Syst 14 (4):327–336

Shen L, Worrell E, Patel MK (2010) Open-loop recycling: a LCA case study of pet bottle-to-re recycling. Resour Conserv Recycl 55(1):34–52

Shrake OS, Bilec MM, Landis EA (2013) The application of a multi-faceted approach for evaluating and improving the life cycle environmental performance of service industries. J Clean Prod 42:263–276

Singh B, Strømman AH, Hertwich GE (2011) Life cycle assessment of natural gas combined cycle power plant with post-combustion carbon capture, transport and storage. Int J Greenh Gas Control 5:457–466

Singh B, Strømman AH, Hertwich GE (2012) Environmental damage assessment of carbon capture and storage -application of end-point indicators. J Ind Ecol 16(3):407–419

Sivaraman D, Pacca S, Mueller K, Lin J (2007) Comparative energy, environmental, and economic analysis of traditional and e-commerce DVD rental networks. J Ind Ecol 11(3):77–91

Solli C, Reenaas M, Strømman AH, Hertwich GE (2009) Life cycle assessment of wood-based heating in Norway. Int J Life Cycle Assess 14:517–528
Sono M (1961) The effect of price changes on the demand and supply of separable goods. International economic review (English translation of the original Japanese publication in Kokumin keizai zassi 74–3, 1943, 261–311) 2:239–271
Steen-Olsen K, Weinzettel J, Cranston G, Ercin A, Hertwich EG (2012) Carbon, land, and water footprint accounts for the European Union: consumption, production, and displacements through international trade. Environ Sci Technol 46:10883–10891
Stokes RJ, Horvath A (2006) Life cycle energy assessment of alternative water supply systems. Int J Life Cycle Assess 11(5):335–343
Stokes RJ, Horvath A (2009) Energy and air emission effects of water supply. Environ Sci Technol 43:2680–2687
Stokes RJ, Horvath A, Sturm R (2013) Water loss control using pressure management: life-cycle energy and air emission effects. Environ Sci Technol 47:10771–10780
Stone R (1961) Input-output and national accounts. The Organization for European Economic Development, Paris
Strømman AH (2009) Dealing with double-counting in tiered hybrid life-cycle inventories: a few comments – response. J Clean Prod 17:1607–1609
Strømman AH, Solli C, Hertwich GE (2006) Hybrid life-cycle assessment of natural gas based fuel chains for transportation. Environ Sci Technol 40:2797–2804
Strømman AH, Peters GP, Hertwich EG (2009) Approaches to correct for double counting in tiered hybrid life cycle inventories. J Clean Prod 17:248–254
Suh S (2004) Functions, commodities and environmental impacts in an ecological-economic model. Ecol Econ 48:451–467
Suh S (ed) (2009) Handbook of input-output economics in industrial ecology. Springer, Dordrecht
Suh S, Huppes G (2005) Methods for life cycle inventory of a product. J Clean Prod 13:687–697
Suh S, Lenzen M, Treloar GJ, Hondo H, Horvath A, Huppes G, Jolliet O, Klann U, Krewitt W, Moriguchi Y, Munksgaard J, Norris G (2004) System boundary selection in life-cycle inventories using hybrid approaches. Environ Sci Technol 38:657–664
Suh S, Weidema B, Schmidt JH, Heijungs R (2010) Generalized make and use framework for allocation in life cycle assessment. J Ind Ecol 14:335–353
Suh S, Tomar S, Leighton M, Kneifel J (2014) Environmental performance of green building code and certification systems. Environ Sci Technol 48:2551–2560
Swarr TE, Hunkeler D, Klöpffer W, Pesonen HL, Ciroth A, Brent AC, Pagan R (2011) Environmental life-cycle costing: a code of practice. Int J Life Cycle Assess 16:389–391
Takahashi K, Nansai K, Tohno S, Nishizawa M, Kurokawa J, Ohara T (2014) Production-based and consumption-based emissions and health impacts of PM2.5 carbonaceous aerosols in Asia. Atmos Environ 97:406–415
Timmer MP (2012) World input-output database. WIOD working paper 10. http://www.wiod.org/publications/papers/wiod10.pdf
Treloar GJ (1997) Extracting embodied energy paths from input-output tables: towards an input-output-based hybrid energy analysis method. Econ Syst Res 9:375–391
Trelor GJ, Love PED, Crawford RH (2004) Hybrid life-cycle inventory for road construction and use. J Constr Eng Manag 130(1):43–49
Tukker A, Bulavskaya T, Giljum S, De Koning A, Lutter S, Simas M, Stadler K, Wood R (2015) The global resource footprint of nations. Carbon, water, land and materials embodied in trade and final consumption. http://www.truthstudio.com/content/CREEA_Global_Resource_Footprint_of_Nations.pdf
United Nations (2003) National accounts: a practical introduction. Studies in methods Series F 85. Department of Economic and Social Affairs, Statistics Division
Urban AR, Bakshi RB (2013) Techno-ecological synergy as a path toward sustainability of a North American residential systems. Environ Sci Technol 47:1985–1993

Vieira SP, Horvath A (2008) Assessing the end-of-life impacts of buildings. Environ Sci Technol 42:4663–4669
Wang C, Zhang L, Yang A, Pang M (2012) A hybrid life-cycle assessment of nonrenewable energy and greenhouse-gas emissions of a village-level biomass gasification project in China. Energies 5:2708–2723
Weil M, Jeske U, Schebek L (2006) Closed-loop recycling of construction and demolition waste in Germany in view of stricter environmental threshold values. Waste Manag Res 24(3):197–206
Weinzettel J, Hertwich EG, Peters GP, Steen-Olsen K, Galli A (2013) Affluence drives the global displacement of land use. Glob Environ Chang 23(2):433–438
Weisz H, Duchin F (2006) Physical and monetary input-output analysis: what makes the difference? Ecol Econ 57:534–541
Wiedmann T (2009) A review of recent multi-region input-output models used for consumption-based emission and resource accounting. Ecol Econ 69(2):211–222
Wiedmann T, Suh S, Feng K, Lenzen M, Acquaye A, Scott K, Barrett JR (2011) Application of hybrid life cycle approaches to emerging energy technologies-the case of wind power in the UK. Environ Sci Technol 45(13):5900–5907
Williams E (2004) Energy intensity of computer manufacturing: hybrid assessment combining process and economic input-output methods. Environ Sci Technol 38:6166–6174
Williams ED, Weber CL, Hawkins TR (2009) Hybrid framework for managing uncertainty in life cycle inventories. J Ind Ecol 13(6):928–944
World Bank. The International Comparison Program (ICP). http://icp.worldbank.org/
Wright D (1974) Goods and services: an input-output analysis. Energy Policy 2:307–315
Yang Y, Bae J, Kim J, Suh S (2012) Replacing gasoline with corn ethanol results in significant environmental problem-shifting. Environ Sci Technol 46:3671–3678
Yao MA, Higgs GT, Cullen JM, Stewart S, Brady AT (2010) Comparative assessment of lifecycle assessment methods used for personal computers. Environ Sci Technol 44:7335–7346
Zamagni A, Guinée J, Heijungs R, Masoni P, Raggi A (2012) Lights and shadows in consequential LCA. Int J Life Cycle Assess 17(7):904–918
Zhai P, Williams DE (2010) Dynamic hybrid assessment of energy and carbon of multicrystalline silicon photovoltaic systems. Environ Sci Technol 44:7950–7955

Chapter 7
Material Flow Analysis

David Laner and Helmut Rechberger

Abstract Material flow analysis (MFA) is a tool to quantify the flows and stocks of materials in arbitrarily complex systems. MFA has been widely applied to material systems in providing useful information regarding the patterns of resource use and the losses of materials entering the environment. MFA and life cycle assessment (LCA) are traditionally different tools for environmental decision support. The two methods are basically different with respect to the definition of system boundaries and the actual subject of investigation. However, there are also overlaps between the tools. These overlaps highlight that MFA and LCA can complement each other and thereby increase the quality of studies in both domains. Thus, the combination of these tools offers the potential for more consistent and reliable decision support in environmental and resource management.

In this chapter, the authors aim at describing the state of the art in MFA and at highlighting the intertwined characters of MFA and LCA when it comes to the investigation of environmentally relevant material systems. Therefore, the main procedures, and the most important methodological approaches of MFA, are described in Sect. 2. Main applications of MFA to different problems and for different purposes based on selected cases from literature are dealt with in Sect. 3. In Sect. 4, the authors discuss the benefits of combining MFA and LCA including a brief outlook on the combined use of MFA and LCA in integrated assessments of environmentally relevant systems.

D. Laner (✉)
Institute for Water Quality, Resource and Waste Management, Vienna University of Technology, Karlsplatz 13, 1040 Vienna, Austria

Christian Doppler Laboratory for Anthropogenic Resources, Vienna University of Technology, Karlsplatz 13, 1040 Vienna, Austria
e-mail: david.laner@tuwien.ac.at

H. Rechberger
Institute for Water Quality, Resource and Waste Management, Vienna University of Technology, Karlsplatz 13, 1040 Vienna, Austria

M. Finkbeiner (ed.), *Special Types of Life Cycle Assessment*, LCA Compendium – The Complete World of Life Cycle Assessment,
DOI 10.1007/978-94-017-7610-3_7

Keywords Application of material flow analysis • Combining MFA and LCA • Dynamic material flow analysis • Eco-factors • Ecological scarcity method • LCA • LCIA • Life cycle assessment • Life cycle impact assessment • MFA • Mass conservation • Material flow analysis • SFA • Static material flow analysis • Statistical entropy • Substance flow analysis • Uncertainty analysis

Acronyms

AP	Acidification potential
APC	Air pollution control
CED	Cumulative energy demand
CFCs	Chlorofluorocarbons
CO_2	Carbon dioxide
COD	Cumulative energy demand
EOL	End-of-life
GWP	Global warming potential
HCl	Hydrogen chloride
HF	Hydrogen fluoride
LCA	Life cycle assessment
LCIA	Life cycle impact assessment
MFA	Material flow analysis
MSW	Municipal solid waste
ODP	Ozone depletion potential
OSR	Old scrap ratio
PUR	Polyurethane
RDF	Residue derived fuel
SE	Statistical entropy
SEA	Statistical entropy analysis
SFA	Substance flow analysis
VOCs	Volatile organic compounds

1 Introduction: Material Flow Analysis

Material flow analysis (MFA) is a tool to quantify the flows and stocks of materials in arbitrarily complex systems (Baccini and Brunner 1991). MFA has been widely applied to material systems in providing useful information regarding the patterns of resource use and the losses of materials entering the environment (e.g., Chen and Graedel 2012). It is concerned with gathering, harmonizing, and analyzing data about physical flows and stocks from different sources with varying qualities to gain an understanding about the stocks and flows of materials in the investigated system. The material flow data has to comply with the mass balance constraints defined in the model, because of the fundamental principle of mass conservation. The use of this principle to investigate metabolic systems has a long tradition, dating back to studies by the medical doctor Santorio Santorio, who established an input-output balance of his physical metabolism already 400 years ago. In this way, he discovered that around half of the input was not to be found in the outputs he measured (body stock, urine, feces). Even though he was not aware of the mass of the respiration leaving his body, he knew that there was a missing term in his balance, which he called "insensible perspiration" (Baccini and Brunner 2012). Around two centuries later, Antoine Laurent Lavoisier finally stated the law of mass conservation, with mass balances being the basic foundation of chemical experiments from thereon. Though the principle of mass conservation is fundamental to all kinds of MFA, different MFA types can be distinguished according to the major goals of each type of approach (cf. Bringezu and Moriguchi 2002). However, in practice no sharp borders exist between these types and many characteristics are common to all types of MFA. An important distinction nevertheless can be made between black box-type material balances and refined material flow analyses distinguishing various elements within the system. The former is often referred to as "material flow accounting," where the system under investigation is treated as a black box, for which material inputs and outputs are balanced. The latter is typically referred to as "substance flow analysis" or "material flow analysis," building on material flow models to describe the pattern of material use in the system under study in some detail. In this section, the latter type of MFA studies is in the focus. This type of MFA has been described in detail in textbooks by Baccini and Brunner (1991, 2012), Baccini and Bader (1996), and Brunner and Rechberger (2004).

Material flow analysis and life cycle assessment (LCA) are traditionally different tools for environmental decision support. MFA and LCA are basically different with respect to the definition of system boundaries (geographical and temporal scope) and the actual subject of investigation (tracking materials in a given system vs. the provision of a specific product or service) (cf. Udo de Haes et al. 1997). In MFA, system boundaries are fixed in space and time. Hence, unlike an life cycle inventory (LCI), which includes all relevant flows associated with a specific product or service, no matter where or when they occur, an MFA is limited to a certain geographical entity (e.g., company, country, world) for a certain time period (days, years, etc.). Because the quantity of interest in MFA is the mass of a material, MFA keeps track of a material within the defined system and can be used to identify sources, uses, and sinks. Because an LCA is related to fulfilling a specific function, the accounting of flows is not limited to one or a few materials, but aims at including all relevant flows of materials and energy associated with the provision of the function. However, there are also overlaps between the tools, as an MFA study may aim to analyze a service (e.g., treatment of a specific amount of waste) or an LCA study may be built on material flow results derived from mass balances for a defined system (e.g., regional reference flows, emission factors or transfer coefficients, process efficiencies). These overlaps highlight that MFA and LCA can also be used complementary to provide consistent and reliable decision support in the field of environmental and resource management. If experts in the fields of MFA and LCA become increasingly aware of the benefits the combination of these tools may offer with respect to decision support in environmental and resource management, this would increase the quality of many studies in both fields.

In this chapter, the authors describe the state of the art in MFA and highlight the intertwined characters of MFA and LCA when it comes to the investigation of environmentally relevant material systems. Therefore, they define the basic terms, the main procedures, and the most important methodological approaches of MFA, in the second section. Subsequently, they describe main applications of MFA to different problems and for different purposes based on selected cases from literature in Sect. 3. In Sect. 4, they discuss the benefits of combining MFA and LCA including a brief outlook on the combined use of MFA and LCA in integrated assessments of environmentally relevant systems.

2 Methodology of Material Flow Analysis

2.1 Basic Terms and Procedure

2.1.1 Basic Terms of MFA

As any scientific field, the methodology of MFA comprises terms which are precisely defined and have to be understood in order to produce sound MFA results and to be able to communicate them within the MFA community. In this section, the most important terms are briefly described; for more detailed information, see Baccini and Brunner (2012) or Brunner and Rechberger (2004).

Materials, Substances, and Goods

In MFA the term "substance" is defined, like in chemistry, as a single type of stuff consisting of uniform units. If the units are atoms, the substance is called "element," such as carbon or iron; if they are molecules, it is called "chemical compound," such as carbon dioxide or iron chloride. A substance is designated "conservative" when it is not destroyed or transformed in a process or by any event during its life cycle. The term "good" describes merchandise and wares. It is mostly used as plural, "goods," and describes an economic entity of a stuff with economic value. There are only a few goods that have no economic value, e.g., rainwater, or a negative value, e.g., waste. Goods are made up of one or several substances. The economic turnover of goods is usually reported in all kinds of statistics, and their production figures are mostly known, as they are of economic relevance. Such information is a prerequisite for establishing material budgets. "Material" serves as an umbrella term for both substances and goods.

MFA is an analytical tool to describe the flows of substances and goods. Even if a substance is in the focus of an MFA study, such as phosphorus (P) in Sect. 3.4, it does not make sense to designate such a study as substance flow analysis (SFA) because the P-containing flows of goods are instrumental. In most cases the purpose of system design is to optimize substance flows, but this is usually done by changing flows of goods, because the substance is contained in a good. Therefore the investigation of the level of goods is a central part of any MFA study.

Flows, Processes, Stocks, and Transfer Coefficients

A *process* is defined as a transport, transformation, or storage of materials. The transport process can be a natural process, such as the movement of dissolved P in a river, or it can be man-made, such as the shipping of mineral fertilizer. The same applies to transformations and storages. The oxidation of carbon to carbon dioxide may happen in natural forest fires as well as in man-made heating systems. Storage of materials happens by natural sedimentation and landfilling. A s*tock* is defined as a material reservoir and is allocated to a process comprising the mass of materials that is stored within the process. It can stay constant (steady state), or it can increase (accumulation of materials) or decrease (depletion of materials). Processes are linked by "flows" of materials. Flows entering or leaving the system are called "imports" or "exports"; flows entering or leaving a process are called "inputs" or "outputs." "Transfer coefficients" describe the partitioning of a substance by a process. They give the percentage of the total throughput of a substance that is transferred into a specific output "good"; hence, they are substance-specific and technology-specific values and stand for the characteristics of a process. Sometimes transfer coefficients can be regarded as constant within certain ranges, which makes them useful for sensitivity and scenario analysis.

System and System Boundaries

The "system" is the actual object of any MFA investigation. A system is defined by a group of MFA elements (processes, flows, etc.), the interaction between them, and the boundaries between these and other elements in space and time. An open system interacts with its surroundings; it has either material or energy imports and exports or both, while a closed system is conceived as a system in complete isolation, preventing material and energy flows across the system boundary. The actual definition of the system is a decisive and demanding task and requires some experience. Everything within the system is part of the investigation. Everything outside the system should not be considered. Therefore, the golden rule is to keep the system as small and simple as possible, still conveying a reliable and valid result. Poor results of MFA studies can often be traced back to an inadequate system definition.

For material balances of larger systems such as nations or continents, the term "cycle" has been frequently used. This term stems from the grand biochemical cycles of carbon, oxygen, hydrogen, sulfur, and phosphorus that drive the biosphere. While these natural cycles indeed show cyclic behavior, anthropogenic systems hardly do (cf. Sect. 3.4 on p. 315). Therefore, applying the term "cycle" to anthropogenic systems could be misleading, at least considered euphemistic.

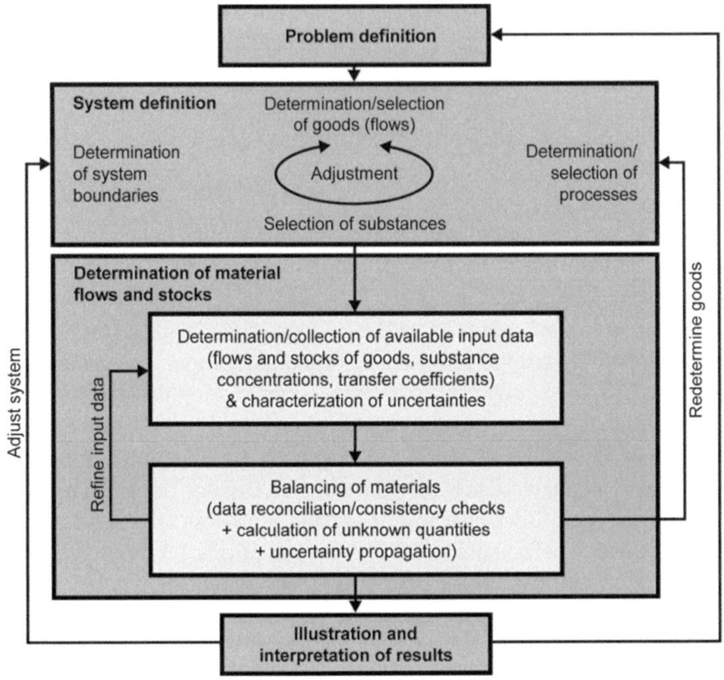

Fig. 7.1 Course of action to perform material flow analysis (Based on Brunner and Rechberger 2004)

Another term that is frequently used instead of "system" or "cycle" is "budget," indicating the input/output and balance character of MFA.

2.1.2 The Principle of Mass Conservation and MFA Procedures

MFA consists of several steps that have slightly changed since the employment of MFA software facilitating the application of error propagation and data reconciliation (cf. Fig. 7.1).

Selection of System Boundaries

The spatial system boundary is usually determined by the scope of the study. It coincides often with a politically defined region (administrative regions such as nations, states, or cities), the premises of a company, or a hydrologically defined region such as the catchment area of a river. The temporal system boundary is most often a year because many data are reported in this time resolution, and the yearly basis helps to outweigh momentary unsteadiness of flows. If smaller objects such as a technology or factory are studied, shorter system boundaries from hours to days might be adequate.

Selection of Flows and Processes

The number of processes necessary to describe the system depends on the objective of the study and on the complexity of the system. Generally, processes can be divided into subprocesses, so that handy systems of maximally 10–15 processes (not including subprocesses) can be realized. The selection of processes is a result of the course of understanding the system. The flows of goods are determined by the inputs and outputs of the processes.

Selection of Substances

The selection of substances depends on the purpose of the study and on the kind of system to be studied. Often legislation such as the Clean Air Act or standards and safety codes provide listings of relevant substances that have to be part of an investigation. An easy way to determine relevant substances in a good is to establish the ratio of substance concentrations in the selected good to geogenic reference materials. Substances with a ratio >10 are candidates for the study. Note that these rules will only assist in the selection process, while the final selection has to be checked for consistency and usefulness during the entire course of the MFA.

Balancing of Processes and the System

According to the mass balance principle (Eq. 7.1), the mass of all inputs (imports) into a process (system) equals the mass of all outputs (exports) of this process (system) plus the change of stock (ΔS) that considers accumulation ($\Delta S > 0$) or depletion ($\Delta S < 0$) of material in the process (system):

$$\sum\nolimits_{i=1}^{j} \dot{m}_i = \sum\nolimits_{i=1}^{k} \dot{m}_i + \Delta S \tag{7.1}$$

where j = number of inputs, k = number of outputs, and $\dot{m}$ = substance or total material flow.

The mass balance has to be fulfilled for each process of the system for the total material flows (goods) as well as for each single selected substance. This may result in a set of some equations. Before software had been applied to solve these balance equations, MFA was an iterative process where mass flows and concentrations were adjusted manually to make the system as consistent as possible. Nowadays, this step-by-step analysis and adjustment (improvement) of data and maybe system structure (e.g., adding new flows) are supported by MFA software, where the user gets information which flows (data) produce inconsistencies and might be erroneous. However, software can only support, but does not deliver corrected data or other information. This is still to be achieved by the user and requires some general MFA and modeling experience as well as deep understanding of the system of study (cf. Laner and Cencic 2013).

Presentation of Results

It is important to present the results of an MFA in an appropriate way. The relevant results of a study have to be condensed into a clear message that can be presented in an easily comprehensible manner which is understandable, reproducible, and trustworthy. The main results of MFA are flow diagrams in Sankey style which means that the width of the arrows indicating the flows is proportional to their value (e.g., produced with the MFA software STAN; cf. Sect. 2.2 and Figs. 7.8, 7.9, and 7.11). Some simple rules help to make such diagrams readable. First, the number of processes should be kept below 10 (max. 15). If the system is more complex, subprocesses should be introduced. The processes should be arranged in a way that crossing of flows is minimized and that all imports come from the left side and all exports leave on the right side of the system (not always possible). Quantities of flows should be rounded to two significant digits (cf. Rechberger et al. 2014). This guarantees that the number of digits is reduced as much as possible, giving shorter numbers, without losing quantitative information. The uncertainties should be expressed in % cutting off positions after the decimal point. A weak point of many MFA studies is the documentation of data and data sources. This can be adequately done by structuring a study's data in a so-called data characterization matrix. MFA studies, which have been performed with STAN, can be communicated to the global MFA community by the upload to a tailor-made platform (www.stan2web.net). The platform allows to search all STAN files after keywords and to examine MFA models.

2.2 *Data Reconciliation and Uncertainty Analysis*

Due to paucity of data and limited system understanding, MFA is naturally confronted with uncertainty. Data for material flow analysis originate from different sources and vary in terms of availability and quality, particularly if material stocks and flows of large-scale systems, such as regions or whole economies, are investigated. Data typically stems from various disciplines and comes from sources such as official statistics, scientific reports, market studies, expert estimates, and others (e.g., Chen and Graedel 2012; Hedbrant and Sörme 2001; Laner et al. 2014). In addition, the quantity of interest may not be directly extractable from existing data, creating the need for estimates based on empirical evidence (i.e., measurements), up- or downscaling, transformation of observations for similar systems, expert solicitation, and assumptions based on plausible reasoning (e.g., Björklund 2002; Danius and Burström 2001; Montangero and Belevi 2007). In addition to model parameters as major sources of uncertainty in MFA, uncertainties concerning model structure may be particularly important in case of very limited information about the system, as it is common in developing countries (cf. Do-Thu et al. 2011; Do et al. 2014). Consequently, various approaches at different levels of sophistication have been used to analyze uncertainty in MFA studies, aiming at evaluating and improving MFA input data, enabling the reconciliation of conflicting material flow data, identifying the uncertainty of material flow model results, and facilitating the assessment of the results' sensitivity. An overview of existing approaches for considering uncertainty and their main applications in MFA is given in Table 7.1, where three types of approaches are distinguished (cf. Laner et al. 2014). Qualitative and semiquantitative approaches use categories to express the uncertainty of MFA results. Data classification approaches focus on formal concepts to characterize data quality and parameter uncertainty in combination with simple mathematical methods, while statistical approaches use rigorous mathematical methods to evaluate the sensitivity and/or uncertainty of model outputs in addition to parameter uncertainty characterization.

In order to use MFA as a robust tool for decision support in resource management and environmental pollution control, rigorous uncertainty analysis should be an integral part of MFA with the level of sophistication depending on the data situation and the purpose of the study (cf. Laner et al. 2014; Wu et al. 2014). Mathematically simpler concepts with a focus on data uncertainty characterization are well suited for descriptive MFA, where the major goal is to quantify the material turnover in the system. In contrast statistical approaches are important for evaluating uncertainty and modeling sensitivity in exploratory MFA, where the major goal is to identify critical parameters and understand system behavior. Building on the existing approaches shown in Table 7.1, Laner et al. (2014) presented a framework for the systematic consideration of uncertainty in MFA consisting of four steps for descriptive MFA and five steps for exploratory MFA, respectively. In general, uncertainty analysis is embedded in the stepwise, iterative procedure for performing MFA (cf. Fig. 7.1), starting with the system definition of the MFA study. Based on the definition of the mathematical model in the first step, data quality is evaluated and model parameter uncertainty is characterized in the second step. Parameter characterization should be consistent with the knowledge about the various quantities, e.g., the nonnegativity of a quantity is reflected by the choice of characterization function. In the third step, the material flows are balanced, potentially reconciling conflicting input data with the mass balance constraints of the model. Based on the extent of data reconciliation (cf. Laner et al. 2015) or the compliance with predefined plausibility criteria (cf. Do et al. 2014), the quality of input data is evaluated. Consequently, in case of large deviations between input values and reconciled values or noncompliance of the resulting flows with the plausibility criteria, the parameter uncertainty may have to be improved. This is typically an iterative procedure with several loops of database improvement before arriving at a final model. In the fourth step, the final model (satisfying agreement

Table 7.1 Types of approaches for considering uncertainty in MFA

Parameter uncertainty characterization	Data reconciliation and uncertainty propagation	Application
Qualitative and semiquantitative approaches		
No formal way of characterizing parameter uncertainty	Uncertainty is not considered in the model	Uncertainty scores are derived to express the author's confidence in the model results [1, 2]
		Used in descriptive MFA
Data classification approaches		
Based on an evaluation of data quality, the uncertainty of model input parameters is determined	Mathematical models are simple and uncertainty treatment includes data cross-checking and interval arithmetic [3] as well as Gaussian error propagation [4]	Uncertainty ranges are derived for the model results, but interpretation is often not straightforward due to missing mathematical stringency
		Mainly used in descriptive MFA
Statistical approaches		
Data are characterized using specific probability distributions (e.g., normal distributions [5] or various distributions [6]) or fuzzy sets [7]	Inconsistent data can be reconciled in the STAN software [5] or within fuzzy set theory [7, 8]. Uncertainty propagation is typically performed using Monte Carlo simulation [6, 9, and 10] or analytically [5]	Model results come with corresponding uncertainty ranges. The extent of data reconciliation can be used as an indication of model quality [8]. Sensitivity analysis is performed to identify critical model parameters and develop scenarios [9, 10]
		Used in descriptive and exploratory MFA

Based on Laner et al. (2014)
Literature: [1], Graedel et al. 2004; [2], Andersson et al. 2012; [3], Lassen and Hansen 2000; [4], Hedbrant and Sörme 2001; [5], Cencic and Rechberger 2008; [6], Gottschalk et al. 2010; [7], Dubois et al. 2014; [8], Laner et al. 2015; [9], Schaffner et al. 2009; [10], Glöser et al. 2013

between data and model results has been achieved) is used to determine the material flows and their associated uncertainty either analytically or using Monte Carlo simulation. While this is typically the last step for descriptive MFA, for the exploratory MFA, sensitivity analysis is used in a fifth step to evaluate the effects of parameter variation on the model outputs and identify critical parameters as a basis to develop scenarios.

A popular approach for addressing uncertainty in MFA has been to assume uncertain quantities to be normally distributed given as mean values and standard deviations (e.g., Andersen et al. 2010; Bonnin et al. 2013; Dos Santos et al. 2012; Graedel et al. 2004; Ott and Rechberger 2012). This approach has been implemented in the MFA software STAN (Cencic and Rechberger 2008), the most comprehensive and widely used software tool to perform MFA in consideration of uncertainty. The STAN software offers a graphical user interface to build MFA systems with different layers (goods, substance, energy) and for multiple time periods (free download from http://www.stan2web.net/). The resulting material flows can be displayed in Sankey style. If more balance equations can be established than there are unknown quantities, conflicting data have to be reconciled with the mass balance constraints defined by the material flow model. The software can perform data reconciliation using a Gauss-Jordan elimination process to iteratively solve the reconciliation problem (cf. Fellner et al. 2011). This is a weighted least-squares optimization with errors being minimized under the constraints given by the mass balance equations of the model. The higher the standard deviation of an uncertain quantity, the stronger it may be reconciled in case of conflicts with other given quantities. The reconciled values are subsequently used to calculate the unknown quantities (e.g., material flows) with their associated uncertainties using the method of Gaussian error propagation.

Given the lack of information about MFA data as the major challenge for meaningful uncertainty analysis in MFA (cf. Danius and Burström 2001; Laner et al. 2014), practitioners should take particular care about the consistent characterization of uncertain MFA data. While the lack of formal procedures for uncertainty analysis impairs statements about the reliability of MFA results, systematic uncertainty analysis produces results as precise as the data warrant (Laner et al. 2014). In the light of these findings, comprehensive approaches to characterize material flow data are being developed (e.g., Feketitsch et al. 2013) as a basis to evaluate and improve the consistency of the database via data reconciliation procedures (Laner et al. 2015). In addition to the classical approach of data reconciliation in MFA based on least-squares optimization implemented in the STAN software, recent studies have focused on data reconciliation under fuzzy constraints in MFA (e.g., Dubois et al. 2014) and data reconciliation of non-normally distributed quantities (Cencic and Frühwirt 2015). In fuzzy set theory, nonprecise information is expressed via membership functions, which represent the degree of truth and not the likelihood of an event or condition. A simple fuzzy set is an interval, which defines the possible range containing the true value. Because fuzzy sets do not make assumptions about the probability of a specific condition, but can be used to define possible ranges for a quantity, they are particularly well suited to express expert judgment based on scarce information or to describe vague knowledge. Figure 7.2 shows two different characterization functions for an uncertain quantity described by the normal distribution approach used in the STAN software (mean = 50, standard deviation = 2.25) and by the fuzzy approach using a trapezoidal membership function (lower limit, 45.5; lower core, 47.75; upper core, 52.25; upper limit, 54.5), respectively. Laner et al. (2015) conclude that the normal (probabilistic) approach is often the default choice for uncertainty analysis in MFA without a critical assessment of its appropriateness, but in case of poor databases, fuzzy sets may be better suited to express the effect of limited knowledge on the MFA results. In general, data reconciliation offers the potential to reduce the uncertainty of the model results and to conduct plausibility checks. Therefore, the generation of as many independent estimates for material flows as possible is the best way to produce reliable results of MFA studies.

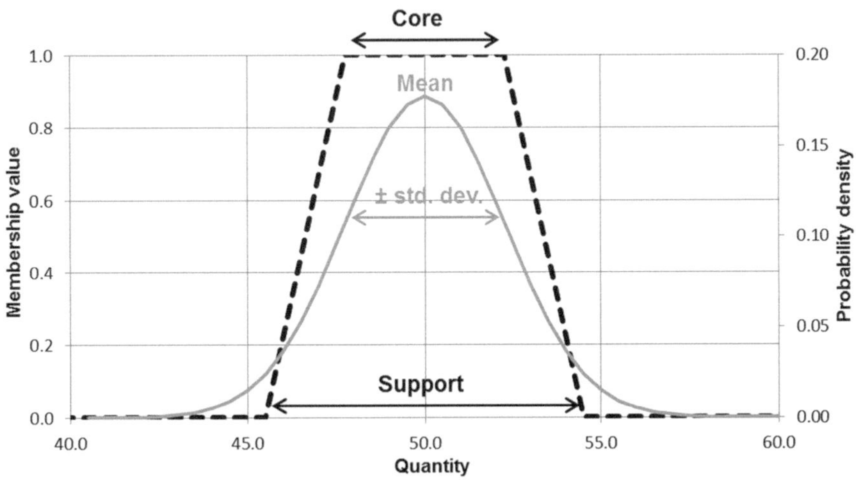

Fig. 7.2 Illustration of characterization functions of a specific quantity using the approach of the STAN software assuming normal distributions (*solid gray line*) and the fuzzy approach using trapezoidal membership functions (*dotted line*) (Based on Laner et al. 2015)

2.3 *Static and Dynamic MFA*

Static material flow analyses are established for a certain balancing period in time, for instance, for 1 year. They provide a snapshot of a system in time and are done at different levels of sophistication to investigate the patterns of material use and material losses in the system. Static MFA is typically used to generate a quantitative understanding of material systems and develop alternative management scenarios (cf. Brunner and Rechberger 2004). Dynamic material flow analyses describe the behavior of a system over several time increments. Thus, dynamic MFA provides information about material usage over time and consequent changes in stocks and flows within the system. While material flows in a static MFA are time independent (i.e., they are not related to material flows at another point in time), material flows in a dynamic model at the time T can potentially depend on all previous states of the system (cf. Baccini and Bader 1996). In the past, most MFA studies used static models to investigate material flow systems, but since the late 1990s, dynamic models have become increasingly popular with the primary focus on the investigation of material stocks in society (e.g., Zeltner et al. 1999). Metals, in particular, have been subject to dynamic MFA due to the large accumulated metal stocks in society and their potential value for society as secondary raw materials (cf. Chen and Graedel 2012; Müller et al. 2014). Typically stock estimates refer to the in-use stock without making a clear distinction between materials actually in use (i.e., product still fulfills its original function), materials in hibernation (i.e., products, which remain in storage but are not used any more), and materials contained in obsolete products, which are not discarded or demolished. However, as the availability of materials is dependent on the type of stock, current research has focused on the characterization and evaluation of the anthropogenic material stock (cf. Johansson et al. 2013; Lederer et al. 2014). Even if hibernating stock is included in stock estimates, two different approaches can be used to estimate material stocks in society (cf. Fig. 7.3). The top-down approach is used in dynamic MFA and is based on accounting of the net flows into or out of the stock over the past. The bottom-up approach consists of summing up the amount of material contained in all relevant products (or wastes) in the various sectors. While the latter approach is typically used in static MFA, due to the lack of time series data in the model, it can also be used in dynamic MFA to directly investigate the material stock over time (cf. Müller et al. 2014). The two different approaches are schematically illustrated in Fig. 7.3 for determining the stock in buildings of a specific material at time T. For the top-down estimate, a time series of input-output balances is used to calculate the total stock, while the bottom-up estimate is based on deriving the total stock from the material intensities in all relevant products (buildings in case of Fig. 7.3).

The formula for estimating the material stock $S(T)$ at time T using the bottom-up approach is shown in Eq. 7.2. The stock is the sum of the contents of the material of interest c_i for all relevant products or product groups P_i at the time T. This approach appears to be particularly useful, if the material is mainly used in a few different applications, for which sufficient data are available (e.g., platinum group metals in vehicle catalytic converters). For materials with diverse fields of application, stock estimates using the bottom-up approach are often incomplete and therefore tend to underestimate the actual anthropogenic stock, as discussed for aluminum by Buchner et al. (2014):

$$S(T) = \sum_{i=1}^{I} P_i(T) \cdot c_i(T) \tag{7.2}$$

$$S(T) = S(0) + \sum_{t=1}^{T} \dot{m}_I(t) - \dot{m}_O(t) \tag{7.3}$$

The basic formula for the top-down approach is described in Eq. 7.3 for the stock of a material $S(T)$ at the time T as the sum of the net additions to stock at previous years. The net addition to stock is calculated by subtracting the output mass in a specific year $t(\dot{m}_O(t))$ from the input mass in that year $(\dot{m}_I(t))$. To calculate the total material stock, also the initial stock at the beginning of the time series $S(0)$ is included (see Eq. 7.3). This approach has been popular to derive stock estimates of current in-use metal stocks based on historic production and consumption data and also to make projections on future secondary resource availability (e.g., Müller 2006; Glöser et al. 2013; Pauliuk et al. 2013). However, as historic data on the outputs from use sectors are rarely available, the output of obsolete products is often calculated based on lifetime functions. Such functions are defined for specific products and end-use sectors, with outputs being calculated by the accumulation of the fraction of all former inputs becoming obsolete in the respective year (cf. Müller et al. 2014). Lifetime distribution functions also frequently used in the

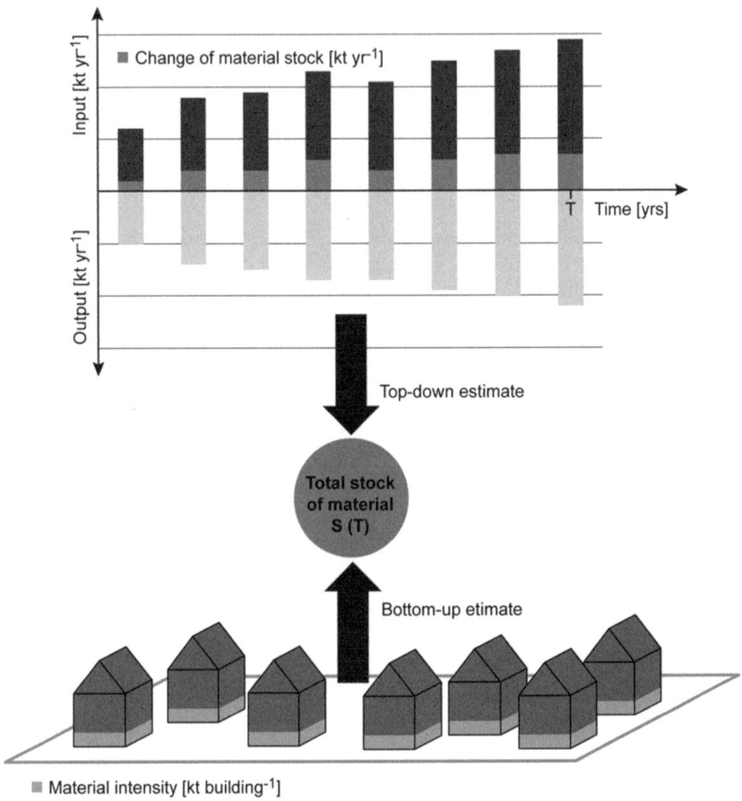

Fig. 7.3 Course of action to perform material flow analysis (Based on Brunner and Rechberger 2004)

field of system reliability are the Dirac delta distribution and the Weibull distribution. Furthermore, normal, log-normal, beta, and gamma distributions are used in dynamic MFA to derive output flows based on the residence time of products in the stock. Though the lifetime function method is most widely applied to calculate outflows from different stocks, for some stocks an alternative method of using leaching coefficients may be better suited. The coefficients determine the fraction of stock leaving the stock in a specific year, which means that all the material in the stock has an equal chance of exiting the stock (cf. van der Voet 2002). This method could be used to describe the leaching of metals from a landfill or metal emissions due to the weathering of surfaces.

Apart from typical approaches to estimate anthropogenic stocks, dynamic and static models also differ with respect to the main purpose of their application. Static MFA is concerned with generating a better understanding of a material system based on simple accounting principles (i.e., mass balance equations) or stationary models. The mathematically simpler form of the models puts more focus on the underlying data and can therefore serve as a basis improving material flow databases. Static analyses are well suited to identify general patterns of materials use and losses for a specific time increment (typically 1 year) to characterize sources, pathways, and sinks of materials. Based on the insights gained from the analysis, alternative management scenarios can be defined using MFA and consequently evaluated (cf. Sect. 3.6). Dynamic MFA is primarily used to investigate the stock buildup of materials in society (i.e., secondary resources) and in the environment (i.e., dissipative losses) based on the investigation of material flows over time. Dynamic models are more complex and have a higher data demand than stationary models, which typically poses a challenge with respect to checking the plausibility of results. In this respect the combination of static and dynamic material flow analysis offers the chance of checking results of the dynamic model for specific points in time, for which detailed snapshots of the material flows and stocks have been established. The advantage of dynamic material flow analysis is that it offers more explanatory power by identifying trends over time and enabling extrapolation of system behavior into the future. Thus, dynamic MFA is used to understand the consequences of changes in a dynamic system over time and to analyze alternative scenarios with respect to future materials management. Overall, whether to use a dynamic or static model depends on the system under investigation, the data availability, and the goal and scope of the MFA. Both types of MFA provide quantitative decision support in the field of resource management and environmental pollution control with the question at hand and system of interest determining the complexity of the models.

3 Application of Material Flow Analysis

3.1 Resource Efficiency Evaluation

Material flow analysis can be used to evaluate the efficiency of resource recovery on the technology level (e.g., Laner and Rechberger 2007) as well as on the regional level (e.g., country level: Chen 2013; global level: Glöser et al. 2013). While various system definitions may be used to evaluate the efficiency of materials management (e.g., material required to provide a service/product or material intensity per monetary output (cf. Bringezu et al. 2003)), the focus of this section is on the efficiency of closing material cycles by recovering materials after use to reduce the consumption of virgin materials (e.g., Reck and Graedel 2012). MFA-based recycling efficiency evaluations often deal with metals, because of the high value of metals for society, the principally infinite recyclability, and the superiority of secondary to primary production from an environmental perspective. However, recycling is often inefficient due to social behavior, product design, recycling technologies, and the thermodynamics of separation. In this context, MFA provides a suitable means of investigating the potential for improvement in resource-efficient metal utilization and management.

Although recycling ratios are commonly used to evaluate resource efficiency, there is no generally accepted definition of recycling efficiency indicators. Recently, such an effort has been made by Chen (2013), who provided definitions for four groups of indicators related to the recycling of aluminum in the United States. The first group refers to indicators for measuring recycling efficiencies at the end-of-life (EOL) stage, including EOL collection rate, EOL processing rate, EOL recycling rate, and EOL domestic recycling rate. The second group consists of indicators for comparing generation or use of new with old scrap, including ratios of new, old, and new-to-old scrap. The third group is made up of indicators comparing the production or use of primary aluminum with secondary aluminum, and the fourth group comprises indicators used to identify the sinks (or losses) of the metal in the system (cf. Sect. 3.2). The first group and the second group of indicators, in particular, are often used to evaluate the recovery of secondary raw materials: the first group at the waste management stage and the second group at the production stage (cf. Fig. 7.4). Out of the first group, the EOL recycling rate is the most prominent indicator, because it depends on the collection rate and the subsequent processing efficiency. The formula for the domestic EOL recycling rate is given in Eq. 7.4, with EOL products (EOL-P) being the collected amount of metal in obsolete products and old scrap (OS) being the domestically recycled old scrap. It should be noted that the overall EOL recycling rate for a national budget can be calculated only by considering the import/export of EOL products and the import/export of scrap (included in flows I1, E1 and I3, E3, respectively, in Fig. 7.4). Because it is typically not possible to entirely distinguish new (processing or manufacturing scrap) and old (postconsumer) scrap from foreign trade statistics, the calculation of EOL RR is prone to assumptions about recycling efficiencies outside the region or country under investigation (Chen 2013). This has been extensively discussed by Buchner et al. (2014) for the old scrap ratio (OSR), a production-related recycling efficiency indicator, with respect to the national aluminum budget of Austria. The OSR designates the share of secondary aluminum produced from old scrap entering fabrication; see Eq. 7.5. The secondary aluminum produced from old scrap is the sum of the domestic old scrap (OS in Fig. 7.4) and the net old scrap import (old scrap balance for I1 and E1 in Fig. 7.4). The secondary aluminum to fabrication is the sum of the domestic old scrap (OS), the domestic new scrap (NS), the net scrap import (old and new scrap flows in I1 and E1), and internal scrap from fabrication to production (aggregated in the process "production and fabrication" in Fig. 7.4). For Austria, they find that the OSR is highly sensitive to assumptions about the share of old and new scrap in foreign scrap trade, resulting in a possible range for the OSR from 0 % to 66 %. The large range is the consequence of the high volume of foreign scrap trade and the limited data available about the composition (old vs. new) of imported and exported scraps:

$$\text{Domestic EOL RR} = \frac{\text{Old scrap}}{\text{EOL products}} \tag{7.4}$$

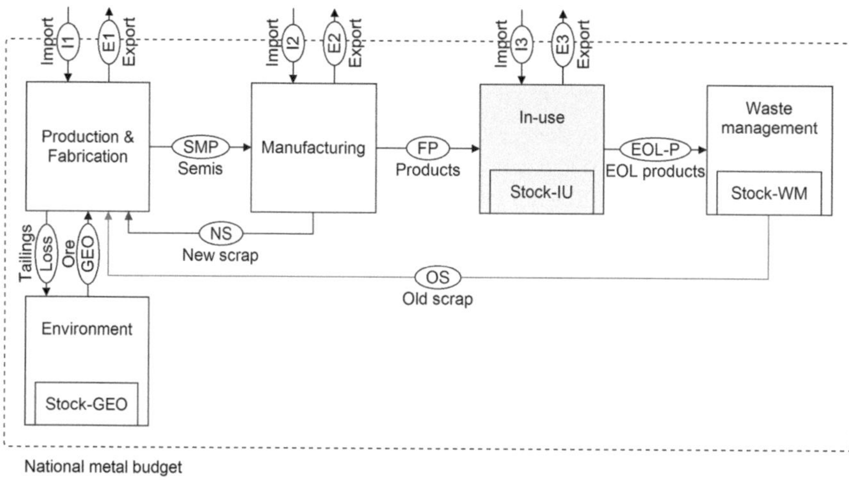

Fig. 7.4 Simplified, generic metal flow model to evaluate recycling efficiency on the national level (note: dissipative losses are not considered; scrap imports/exports are included in trade flows related to production)

$$\text{OSR} = \frac{\text{Secondary metal from old scrap}}{\text{Secondary metal entering fabrication}} \tag{7.5}$$

MFA is widely used to provide the quantitative basis for evaluating the recycling efficiencies of large-scale systems. However, comprehensive evaluation indicators are crucially dependent on rigorous definitions. In particular for national-level MFA, the resolution of foreign trade statistics in terms of pre- and postconsumer materials may be problematic for the calculation of robust recycling efficiency indicators. In such cases the inclusion of upstream and downstream processes may be necessary to generate a thorough understanding about the resource efficiency of the system under investigation, representing a potential link to supply chain modeling and network analyses. In the case of scarce metals, the poor data situation poses an additional challenge for the evaluation of recycling efficiency (cf. RPA 2012). For instance, the uncertainty about palladium flows in Austria has been shown to have a substantial impact on the evaluation of recycling efficiency (cf. Laner et al. 2015). System recycling efficiencies are also affected by the consideration (or non-consideration) of dissipative losses, as shown by Lifset et al. 2012, who conclude that dissipative releases of copper have a small but discernible effect on recycling rates and should be included in recycling efficiency evaluations. Furthermore, the distinction of in-use stock and hibernating stock may have a significant effect on the amount of recoverable material potentially available for recycling, as discussed by Daigo et al. 2007. These aspects of dissipative losses and the consideration of various types of stock are further elaborated on in the subsequent Sects. 3.2 and 4, respectively.

3.2 Evaluation of Resource Losses and Recovery: Statistical Entropy Analysis

A national economy can be conceived as a system that extracts highly concentrated resources from and returns low concentrated wastes and dissipative emissions to the environment (Georgescu-Roegen 1971; Stumm and Davis 1974). Take the example of copper: It is extracted from ore bodies, which have a typical concentration of 1 %. The ore is then concentrated by milling and flotation to some 30 %. The concentrate is smelted and blister copper of ca. 96 % copper content is produced, which is finally refined to over 99.99 % pure cathode copper, which is sold as semi-products (slabs, sheet, wire) to manufacturing and fabrication. These purification processes do not work 100 % efficiently, and wastes of minor copper content such as tailings and slag are shipped to the environment. The copper semis are then used to produce all kinds of products such as electrical and electronic equipment or alloys. This means that pure copper is combined with other materials, thereby mixed and consequently diluted. During use copper might be lost to the environment, mainly by wear and tear. Once the products become obsolete, they are either discarded as waste or become part of the hibernating stock. Waste usually means a mixture of a plentitude of products and materials. Therefore, the copper content in all types of wastes is usually below 1 %, with the highest fractions to be found in WEEE and EOL vehicles. If copper is to be recycled, this means that it has to be either separately collected as copper scrap; this happens, e.g., when buildings are deconstructed, or it is recovered from a waste stream, e.g., when bottom ash from municipal solid waste (MSW) incineration is processed to extract iron and nonferrous metals such as copper and aluminum. Both separate collection and scrap recovery represent a concentrating step.

The purpose of the above description of the life cycle of copper has been to demonstrate that such a life cycle can be conceived as a chain of concentration (flotation, smelting, refining, collecting, recovering) and dilution (producing all kinds of waste, mixing of materials, emissions) steps. Rechberger and Brunner (2002) used statistical entropy (SE) to describe such concentration and dilution phenomena. SE stems from Ludwig Boltzmann's formulation of entropy and was used by Claude E. Shannon to quantify information content (Shannon 1948). In statistics entropy is a means to describe distributions of any kind of characteristics. Often SE is calculated for probability (P_i) or frequency distributions (cf. Eqs. 7.6[1] and 7.7).

[1] Note: Statistical entropy (Shannon's entropy) is designated with the letter "*H*." In thermodynamics entropy is designated with "*S*" and "*H*" is used for enthalpy. "ld" denotes the logarithm to the base 2; the unit of *H* is then one bit. Usually *H* is normalized to H_{max} (not shown here), and the dimensionless relative statistical entropy (H_{rel}) is used.

$$H(P_i) = -\sum_{i=1}^{k} P_i \cdot ld(P_i) \tag{7.6}$$

with

$$\sum_{i=1}^{k} P_1 = 1 \tag{7.7}$$

Concentration can be conceived as the probability to find a substance in a good. Imagine a hypothetical process that turns MSW into only two outputs and we consider the substance cadmium. A reasonable concentration for cadmium in MSW is 10 mg/kg waste. One of the outputs contains no cadmium; the other contains only cadmium (i.e., Cd concentration is 100 %). The probability to find cadmium in the first flow is then zero and 100 % for the second flow. Such a distribution results in an SE value of zero. The other extreme case would be if all outputs have the same concentration. Then the entropy is maximal.

Inputs and outputs of a process for a substance are defined by concentrations and mass flows, i.e., the results of an MFA. Therefore, Eq. 7.6 has been transformed into a function of substance concentrations (c_i) and mass flows (m_i) (cf. Eq. 7.8):

$$H(c_i, m_i) = -\sum_{i=1}^{k} m_i \cdot c_i \cdot ld(c_i) \tag{7.8}$$

where

i: flow index
k: number of flows on the input or output side of a process/system

Equation 7.8 is then applied to both the input and the output of a process, and the difference of the SE values ($\Delta H = H_{Out} - H_{In}$) is considered. If ΔH negative, then the process is predominantly concentrating, otherwise predominantly diluting. Note that for gaseous and aqueous outputs (emissions), Eq. 7.8 is not applicable and more complex equations are adequate to consider the subsequent dilution in the environment (for details see Rechberger and Brunner 2002). Rechberger and Brunner (2002) applied statistical entropy analysis (SEA) to different historical states of MSW incinerators and showed that through technology development, incinerators became entropy-reducing facilities for substances such as Cd, Hg, Pb, and Zn. Rechberger and Graedel (2002) applied SEA to multiprocess systems and analyzed the European copper budget. They found that the entropy trend follows a "V shape" (cf. Fig. 7.5: 1→4) as had been qualitatively assumed by O'Rourke et al. (1996), Ayres and Nair (1984), and others. Throughout the life cycle of copper, the SE varies considerably and covers about 50 % of the possible range between total dissipation and maximal concentration of the total copper throughput. Nevertheless, the copper management did not show a clear entropy trend across its life cycle. The system as a whole neither dissipates nor concentrates copper significantly with regard to the original ore. The relatively limited impact of the contemporary waste management system on the entropy trend can be explained as follows: Even a more

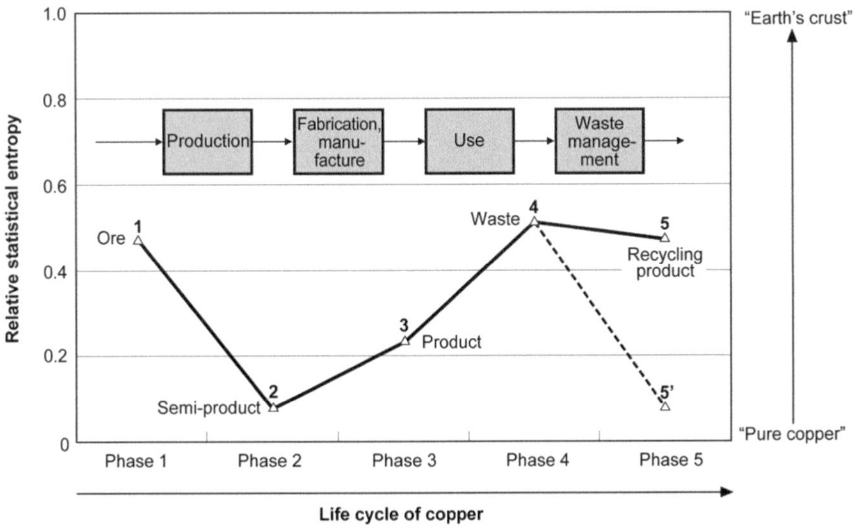

Fig. 7.5 Entropy trend of the European copper budget (Rechberger and Graedel 2002)

optimized waste management system with higher recycling efficiencies would not change the situation significantly because current copper flows into waste management are rather small compared to the consumption of copper. However, this limited impact may increase in the future when the infrastructure, which has been established over the last few decades, will be continuously renewed and replaced. As a result of these larger waste streams, a decreasing overall entropy trend might be realizable, provided efficient recycling technologies are applied. If one considers the variety and complexity of technologies that are required to produce semis out of ore (Fig. 7.5: 1→2), then it becomes clear that future waste management has to be based on similarly sophisticated technologies to fulfill its task, namely, to reduce the entropy at the end of the life cycle to resource level (Fig. 7.5: 4→5′).

Sobańtka and colleagues extended the SEA to chemical compounds (Sobańtka et al. 2012). So SEA is also applicable to systems in which the chemical speciation is of particular importance. This is often the case for carbon or nitrogen. In Sobańtka and Rechberger (2013), the extended SEA (eSEA) is applied to the nitrogen budgets of 13 Austrian wastewater treatment plants. There nitrogen appears in the following speciation: as NH_4^+ and N_{org} in the influent; NH_4^+, NO_3^-, N_{org}, and NO_2^- in the effluent; N_2 and N_2O in the gaseous emission; and N_{org} in the sludge. There are significant differences between the environmental impacts of these nitrogen compounds (e.g., the release of a certain amount of nitrogen either as N_2 or N_2O to the atmosphere), which could not be considered by the combination of classical MFA and SEA. It was shown that eSEA is a more sensitive and reliable metric than the usually applied nitrogen removal rate. As the plant-specific ΔH values express the nitrogen removal and transformation performance (benefit), they can be well combined with energy demand and costs and used for benchmarking of facilities and technologies.

SEA has so far proven to be a powerful tool for the evaluation of complex MFA results, because the final metric ΔH indicates if a process or the total system is a concentrating or diluting entity. It enhances the understanding of the metabolic characteristics of systems. On the other hand, SEA is rather complex and has therefore not experienced broad recognition and employment by the MFA community. Bai and colleagues (2015) employed SEA to assess the performance of a lead smelting process, Yue and colleagues (2009) applied SEA to the Chinese copper budget, and Kaufmann and colleagues (2008) used SEA in combination with life cycle impact assessment (LCIA) to compare landfilling and incineration with respect to carbon flows.

3.3 Sources, Sinks, and Final Sinks

Material flow analysis is used to analyze patterns of material use and to connect primary material sources (i.e., nature) to the sinks of materials in the environment (either intentional or unintentional releases to the natural environment). The need for appropriate sinks, where materials can be directed to without endangering humans and the environment, was recognized by Wolman (1965) and has received further attention with respect to the discussion about clean material cycles and the safe disposal of hazardous substances (cf. Brunner 2010; Kral et al. 2013). By definition, a final sink converts a problematic material into unproblematic materials (e.g., mineralization of organic pollutants in a waste incineration plant), or the material has a geologic residence time in the sink (i.e., thousands of years), for instance, as it may be the case for copper in soil. The relationship between sources of primary material, the (ideally circular) use of the material in society, and the intended and unintended material releases to sinks is illustrated in Fig. 7.6. With respect to hazardous substances introduced to anthropogenic material cycles, waste management plays a central role of directing them to appropriate sinks, such as incinerators for organic pollutants or properly engineered landfills for heavy metals (cf. Brunner and Tjell 2012). Hence, waste management aims at enhancing materials recycling without reintroducing problematic substances to anthropogenic material cycles via secondary raw materials. With respect to Fig. 7.6, it should be noted that a sink of anthropogenic materials may get a source again at a later time. For instance, old landfills, where hazardous but also valuable materials have been deposited in the past, are receiving increasing attention as potential sources of secondary raw materials nowadays within landfill mining initiatives (cf. Krook et al. 2012).

The evaluation of acceptable substance loads into regional sinks is based on an inventory of substance flows for the region under investigation and on the assessment of the (long-term) impact the loading of a sink may have on human health and the environment. The former is derived from material flow analysis investigating

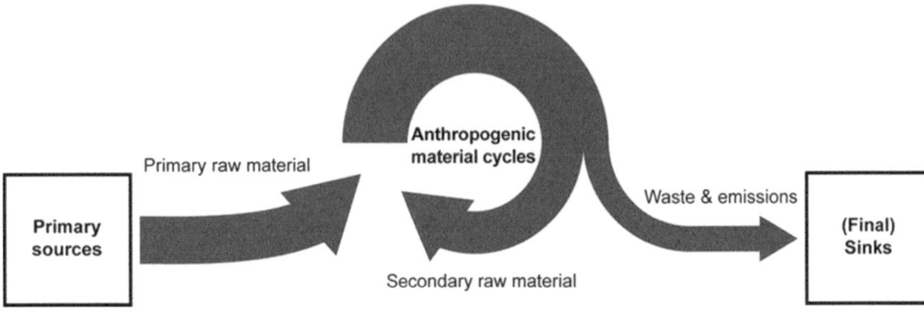

Fig. 7.6 Illustration of primary sources and sinks of materials in anthropogenic material cycles (Based on Kral et al. 2013)

the pathways of problematic materials through society to the receiving compartments in the environment. The latter corresponds to the substance flow's effect on the sink and may be expressed by accounting for the flow into the sink, in the simplest case, or by modeling the full cause-effect chain to derive the actual damage to human health and the environment, in the most complicated case (cf. Kral 2014). Based on an indicator to express the acceptable mass fraction of the total substance flows to regional sinks, Kral (2014) found that in Vienna around 99 % of the copper flows to sinks are acceptable. The remaining percent, however, may be problematic with respect to copper loadings of urban soil and of receiving waters. Using the same indicator, he identified 96 % of the perfluorooctane sulfonate (PFOS) flows in Switzerland to be acceptable, with the remaining 4 % not being evaluated due to data quality issues. In essence, the approach suggested by Kral (2014) resembles the evaluation underlying the ecological scarcity method, which is an impact assessment of life cycle inventories according to the distance-to-target principle (Frischknecht and Büsser Knöpfel 2013). The method of ecological scarcity is based on eco-factors which reflect the actual emission situation (in Switzerland) and the emission targets according to Swiss policy. The eco-factors are derived from the current flow expressing today's emission situation, the normalization flow as a reference quantity (typically equal to the current flow), and the critical flow representing the corresponding political target. Eco-factors have been widely applied in life cycle assessments (LCAs) to support decisions by companies as well as by authorities in Switzerland. The ecological scarcity method is applied also in LCAs related to other countries but ideally should be adapted to account for regional and societal differences between Switzerland and the country for which eco-factors are implemented (cf. Grinberg et al. 2012). Clearly, there is a direct link between material flow analysis to identify relevant pathways and sinks for the substances of interest and the regionally explicit impact assessment based on the ecological scarcity method in LCA. The fact that MFA and LCA are intertwined in several ways and that they offer complementary information for environmental decision support is further elaborated on in Sect. 4.

3.4 MFA for National Materials Accounting

If MFAs of a national system are routinely repeated, a national resource accounting scheme is obtained. Zoboli et al. (2015) established a retrospective accounting scheme for Austria and the resource phosphorus (P) by compiling yearly P budgets from 1990 to 2011 to demonstrate the feasibility of such a scheme. Their work delivered several important findings:

First, work load and number of budgets (years) are not linearly correlated. Most of the time had to be used to establish the system and identify the data sources. Once this is accomplished, the budgets for adjacent years were produced relatively fast.

Second, even in a relatively short and stable period of 22 years, the national P budget of Austria has undergone unexpected significant and partially abrupt changes, with important implications for environmental, resource, and waste management: For example, the agricultural use of phosphate mineral fertilizers has been significantly reduced. The use of meat and bone meal as animal feed was banned in 2001, giving place to a sudden and large loss of P in cement kilns. The bioenergy sector has gained increasing importance, affecting large flows of raw materials, wastes, and by-products. The percentage of available P in food that has been wasted has increased from 30 to 40 % in 20 years, but the ratio of P recovery in household wastes has considerably increased from 20 to 60 %. The ratio of recovery of the total waste P, however, declined from 55–60 % to approximately 40 %, because the generation of solid waste has grown more quickly than its recovery. The total loss of P in landfills has remained almost constant in time, although the P-carrying waste streams have largely varied, driven by new regulations regarding organic waste and sewage sludge, with important implications for future management possibilities. The analysis of the time series also revealed that more than 50 % of all flows changed more than 100 % compared to the initial year 1991. However, these changes did not necessarily happen gradually but sometimes occurred rather suddenly. Obviously, such developments could not be determined by a single-year analysis. The multiyear approach also improved the understanding of the system and helped making the model more comprehensive and more suitable to constitute the basis of material accounting and monitoring (cf. Sect. 2.1).

Third, from a methodological point of view, the multiple-year approach helped identifying systematic errors by analyzing the effect of data reconciliation on the original data set over time. After all some 10 % of the system's flows were systematically reconciled, that is to say, they were always altered in the same direction. This indicates the presence, even if small, of systematic errors and is an important feedback for data producers helping to continuously check and, if required, to improve data quality. The multiyear approach also allowed to identify inconsistencies in time series, which might occur from changes in the reporting procedure or changes in the accounting system and which might remain unrevealed by a classical single-year study.

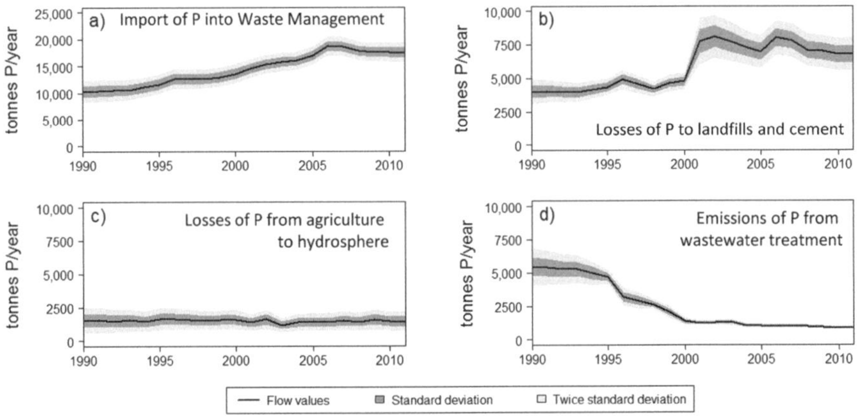

Fig. 7.7 (**a**) While the import of P into waste management increased constantly over the past years, the losses of P to landfills and cement increased even more, indicating a clear field for action. (**b**, **c**) The losses of P from agriculture to the hydrosphere remained rather constant indicating that efforts into optimized fertilizing practice have been rather inefficient. (**d**) Contrary, the emissions of P from wastewater treatment works could be reduced significantly, showing the effectiveness of technical solutions

Eventually, such resource accounting schemes are to be routinely maintained by national authorities. They could then provide the following services:

First, to identify negative trends of flows, stocks, or recovery rates. An example is given in Fig. 7.7 for the Austrian accounting scheme on P.

Second, to monitor the effectiveness of regulations and directives. For example, target values for certain flows can be expressed and the accounting scheme will give evidence if or when the target is or might be realized.

Third, the accounting scheme provides information on the required quantity and quality of data. This can lead into both directions: reduced data generation activities in areas that do not significantly contribute to the overall uncertainty of the system and increased activities where more or better data are required. So this could be used to optimize the total national spending on data generation.

3.5 MFA and Environmental Impact Assessment

MFA can be applied both as a tool for environmental impact assessment itself and as a basis for impact assessment methods such as life cycle impact assessment.

In Lederer and Rechberger (2010), five common options and one novel alternative to treat and dispose sewage sludge are compared with regard to environmental impact, resource recovery, and materials dissipation. The five common options are as follows: direct application of sewage sludge on agricultural soil (O1); mono-incineration of sewage sludge in a fluidized bed combustion and the ashes are spread on agricultural land (O2); like O2 but the ashes are landfilled (O3); sewage sludge is co-incinerated in a cement kiln (O4); and sewage sludge is co-incinerated in a coal-fired power plant (O5). With the exemption of O2, these options are frequently applied in Europe. Another option, direct landfilling of sewage sludge was not part of the study because this practice is going to be phased out as only nonreactive materials should be landfilled. Like options O2 and O3, the novel alternative technology (O6) includes a mono-incineration step, but unlike these options, the residual fly ash is further treated to recover the phosphorus, which concentrates in the ash and yields the main component for a fertilizer. For this purpose the ash is mixed with calcium or magnesium chloride and then thermally treated in a rotary furnace between 850 and 1000 °C. Under these conditions metal chlorides are formed which tend to evaporate. Thus, more than 90 % of cadmium, copper, lead, mercury, and zinc can be removed from the ash. In addition, the phosphorus availability for plants is increased through this treatment. The off-gas from the novel technology is cleaned by state of the art of the air pollution control (APC) equipment.

In order to describe the six options, MFAs were performed yielding the total mass balance and balances for As, Cd, Cr, Cu, Hg, Ni, Pb, Zn, and P for each option (a total of 60 budgets). These results were used to quantify the following environmental impacts: emissions to air, water, and soil applying LCA impact assessment; accumulation of heavy metals in the soil; extent of P recovery; and dissipation of heavy metals in the environment as a result of a SEA (cf. Sect. 3.2). These substance-related investigations were completed by a life cycle assessment using cumulative energy demand (CED) as a resource depletion indicator.

On the basis of the MFAs, the doubling times for the heavy metal contents for 1 ha (=10,000 m^2) of topsoil were determined under the assumption that the agricultural P demand of each option is covered by mineral fertilizer and the maximally recovered P amount from sewage sludge. This direct MFA-based impact assessment showed that option O1 is problematic with respect to Hg contamination of soil: The doubling time would be only 17 years indicating that the soil is considerably contaminated with Hg by such practice. Figure 7.8 shows the difference between options O1 and O6 with respect to P and metals' management. While both options achieve quite high recovery rates for P, the metal flows exemplified with Hg are very different. In option O1 the sink of the sludge-borne Hg concerns the soil and comparable high emissions to the hydrosphere. Option O6 transfers quantitatively all Hg into a safe underground disposal facility.

Another study, where MFA was applied in conjunction with LCA impact categories, deals with the recycling of cooling appliances such as refrigerators or freezers (Laner and Rechberger 2007). The motivation for this study was a then new ordinance on Waste Prevention, Collection and Treatment of Waste Electrical and Electronic Equipment, which raised the minimum recycling rate for cooling appliances to 75 %. The aim of the study was to find out whether the higher recycling rate would lead to better treatment practices for cooling appliances with respect to resource recovery and environmental protection. For this purpose two different treatment technologies, which achieved recycling rates between 80–90 % (T1) and

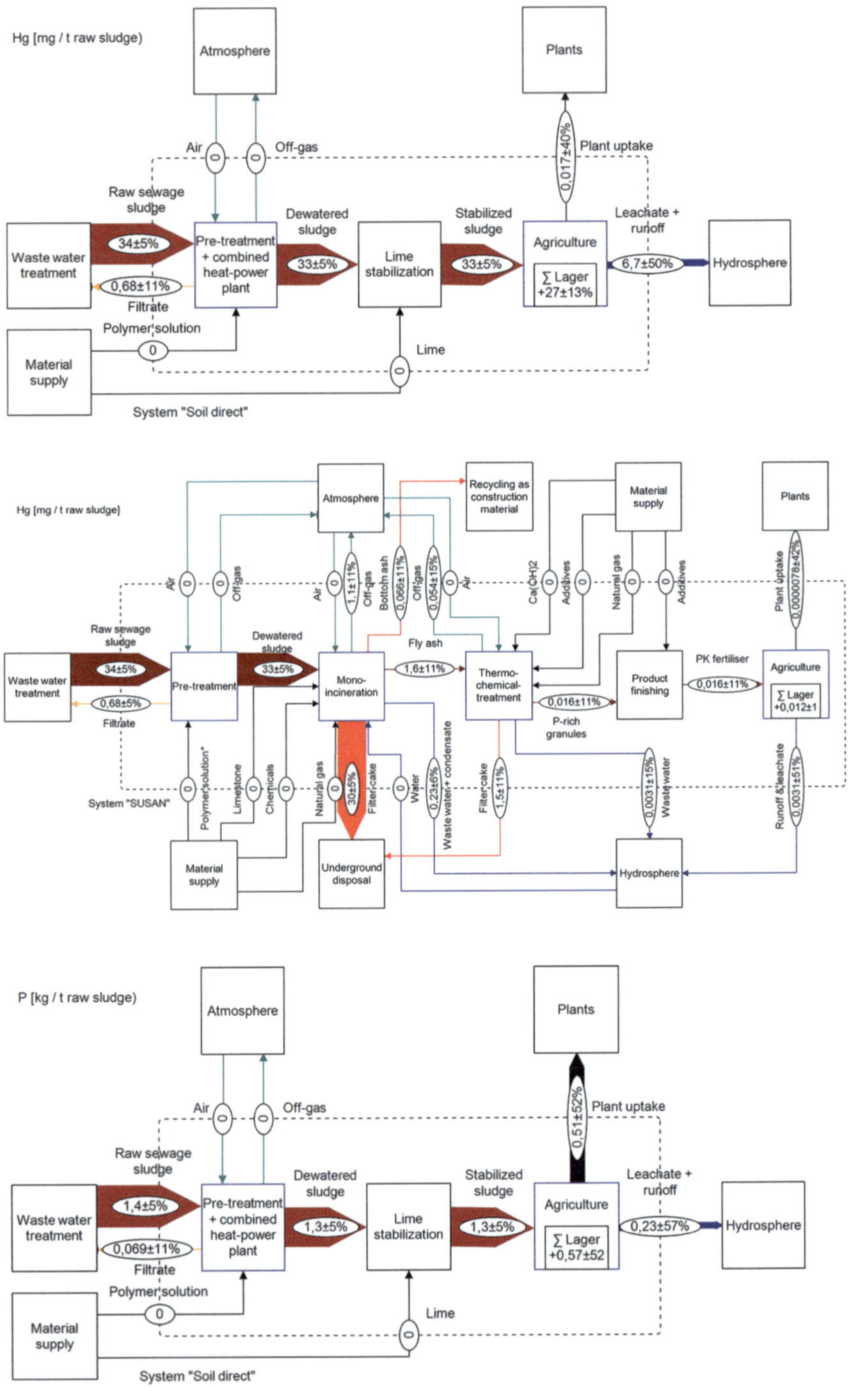

Fig. 7.8 Hg (*above*) and P budgets (*below*) of options O1 (*upper*: application to soil) and O6 (*lower*: fertilizer production)

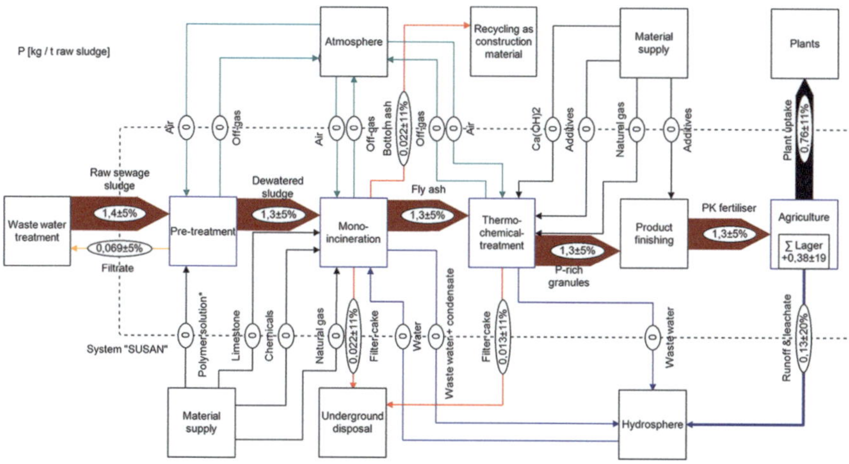

Fig. 7.8 (continued)

50–60 % (T2), respectively, were compared both for cooling appliances containing chlorofluorocarbons (CFCs) and volatile organic compounds (VOCs; not further addressed here). In order to analyze the environmental impact of the different technologies budgets for CFCs, carbon dioxide (CO_2), hydrogen fluoride (HF), and hydrogen chloride (HCl) were established and evaluated for their global warming potential (GWP), ozone depletion potential (ODP), and acidification potential (AP). However, the main purpose of the study was not the comparison of two technologies but to find out whether or not the predefinition of minimum recycling rates is suited to achieve goal-oriented treatment practices and, if not, which criteria could be formulated to provide for an optimal treatment of these appliances. This extended the relevance of this study beyond the technical question of how to recycle fridges optimally.

The first technology (T1) removes coolant and oil by a proprietary degassing unit. Refrigerant and oil are put into compressed gas cylinders, and the mixture is separated in a distillation facility. The distillation residual is incinerated, the oil recycled, and the CFCs are shipped to a reactor cracking facility, where they are cracked into HCl, HF, and residuals. In this first step, also the compressor, plate glass, cables, and possible mercury switches and capacitors are removed by manual dismantling. In the second step, the dismantled cooling appliance is shredded, and the liberated polyurethane (PUR) foam flakes are collected by air separation and are further ground. The CFCs are collected via activated charcoal filters and condensers. Then they are shipped to the cracking reactor, too. The PUR flour is bagged and recycled as an adhesive agent. Ferrous metals are separated magnetically. The rest is further treated in an advanced separation unit from which the following fractions are obtained: iron, aluminum, copper, polystyrene (for recycling), plastics of high calorific value for industrial combustion (plastics 2), plastics for advanced incineration (plastics 3), and residual material.

The first step of the other technology (T2) coincides widely with T1. The only difference is a more in-depth dismantling removing various plastic parts, cooling fins, and electr(on)ic components. In the second step, the disemboweled cooling appliances are shredded and combusted in a rotary kiln. The process gas is captured throughout the whole processing step and used as combustion air for the rotary kiln, where the CFCs are fully destroyed. The flue gas is treated in an advanced APC facility. Filter cake and fly ash are deposited in a special underground disposal facility. The bottom ash is collected and, after iron separation, landfilled. Figure 7.9 displays the total materials balances for both technologies.

The contributions of both technologies to resource conservation and protection of men and the environment were assessed applying the cumulative energy demand and the abovementioned impact categories. For CFC appliances, T1 achieved higher resource savings but resulted in a higher contribution to the ODP, because of substantial CFC losses. The impact of materials recycling to environmental impact and resource conservation was determined for both technologies and is shown as a function of the achieved recycling rate in Fig. 7.10 for technology T1. The maximum of resource conservation (CED) is achieved by the recycling of metals: aluminum, copper, and then iron. The recycling of the PUR as an adhesive agent is associated with higher environmental impacts (lower savings), because of dissipative emissions of CFCs from the PUR material. Therefore, the optimal recycling rate is a function of the composition of a product group (in this case cooling appliances) and the technology applied, both for recycling and for producing the primary raw materials replaced by the secondary ones. A high recycling rate per se does not automatically result in goal-oriented waste management if resource conservation is overcompensated by resulting emissions.

3.6 Scenario Analysis and Optimization

Material flow analysis generates a quantitative understanding of material flows in a system and thereby provides a basis for system optimization. Optimization of material flow systems is typically not done using formal optimization models, but based on the comparison of alternative management scenarios. Schaffner (2007) stated that scenarios should address sensitive parameters of the material flow model in order to investigate the most effective management measures. Following this principle, she analyzed various scenarios to reduce key nutrient flows in a river basin in Thailand and quantified the mitigation potentials associated with different measures. A similar procedure was used by Huang et al. (2007) to improve urban water management in Kunming City, where scenario analysis highlighted the effect of the implementation of best available technologies on the nutrient loads to receiving waters. Scenario analysis was also performed by Mastellone et al. (2009) to investigate alternative waste management strategies for the Campania region in Italy. They defined waste management scenarios based on the objectives and legislation for waste management and in consideration of the

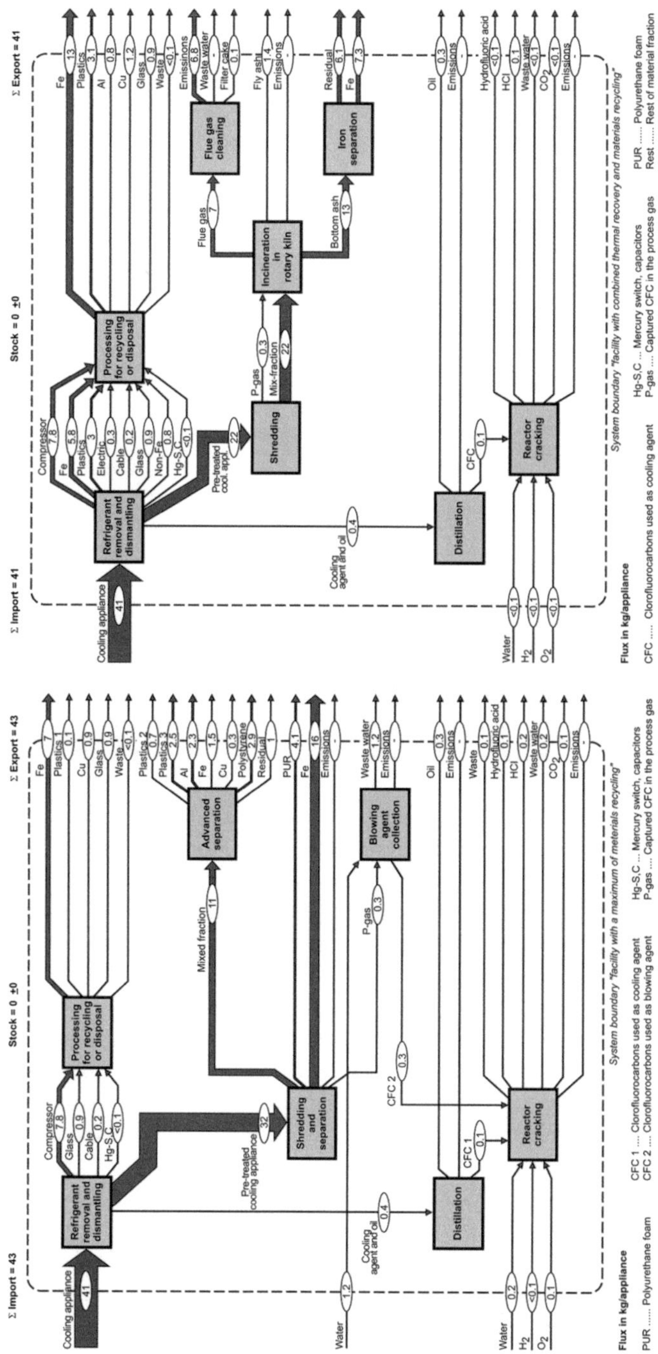

Fig. 7.9 Total materials budgets of technologies T1 (*left*) and T2 (*right*) (Laner and Rechberger 2007)

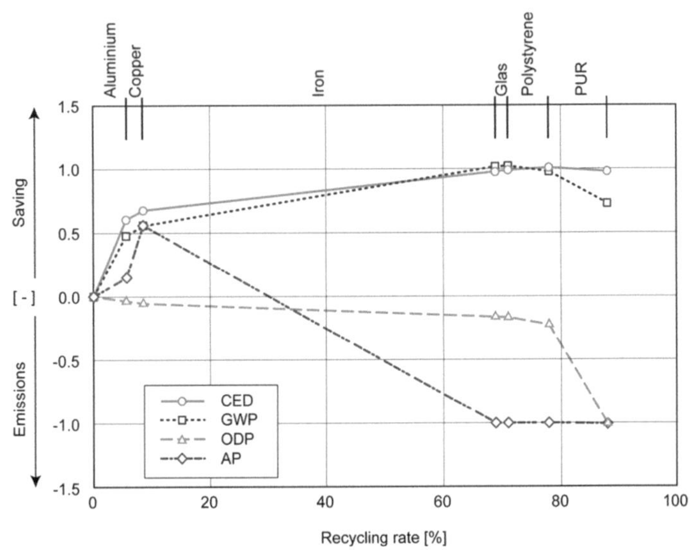

Fig. 7.10 Interrelation between materials recycling and investigated indicators. Each graph is normalized defining the maximum value as 1; therefore the figure does not provide information about the absolute relevance of the indicators (Laner and Rechberger 2007)

existing regional waste system. By comparing the contribution of the different scenarios to the goals of waste management, they derive recommendations for waste management policy in the Campania region. Another application of MFA-based scenario analysis to investigate alternative waste management solutions was presented by Laner et al. (2009), who compared different collection and treatment options for business waste management in Styria, a region in Austria. Based on the analysis of the status quo, they derive two sets of two different scenarios, each set comparing separate collection and treatment of business waste against mixed treatment together with residual household waste. The scenario sets differ with respect to the assumed composition of business waste from different sectors, because synthetic waste compositions had to be defined due to the lack of reliable waste sorting analyses (cf. Laner and Brunner 2008). The mass flows for one set of alternative business waste management scenarios are shown in Fig. 7.11, highlighting that the main difference between the alternatives is the installation of a splitting plant for separately collected business waste (Fig. 7.11a). In this case more of the business waste is directed to energy recovery processes than in the case of treating all business waste together with residual household in a mechanical biological treatment plant (Fig. 7.11b). The contribution of each scenario to the goals of waste management was evaluated based on the fraction of cadmium in the waste directed to appropriate sinks (cf. Sect. 3.2) and the environmental impact of the waste treatment scenario in terms of its contribution to climate change (global warming potential) as well as its cumulative primary energy demand (cf. Sect. 3.5). Using these criteria, the separate collection scenarios turned out to be preferable

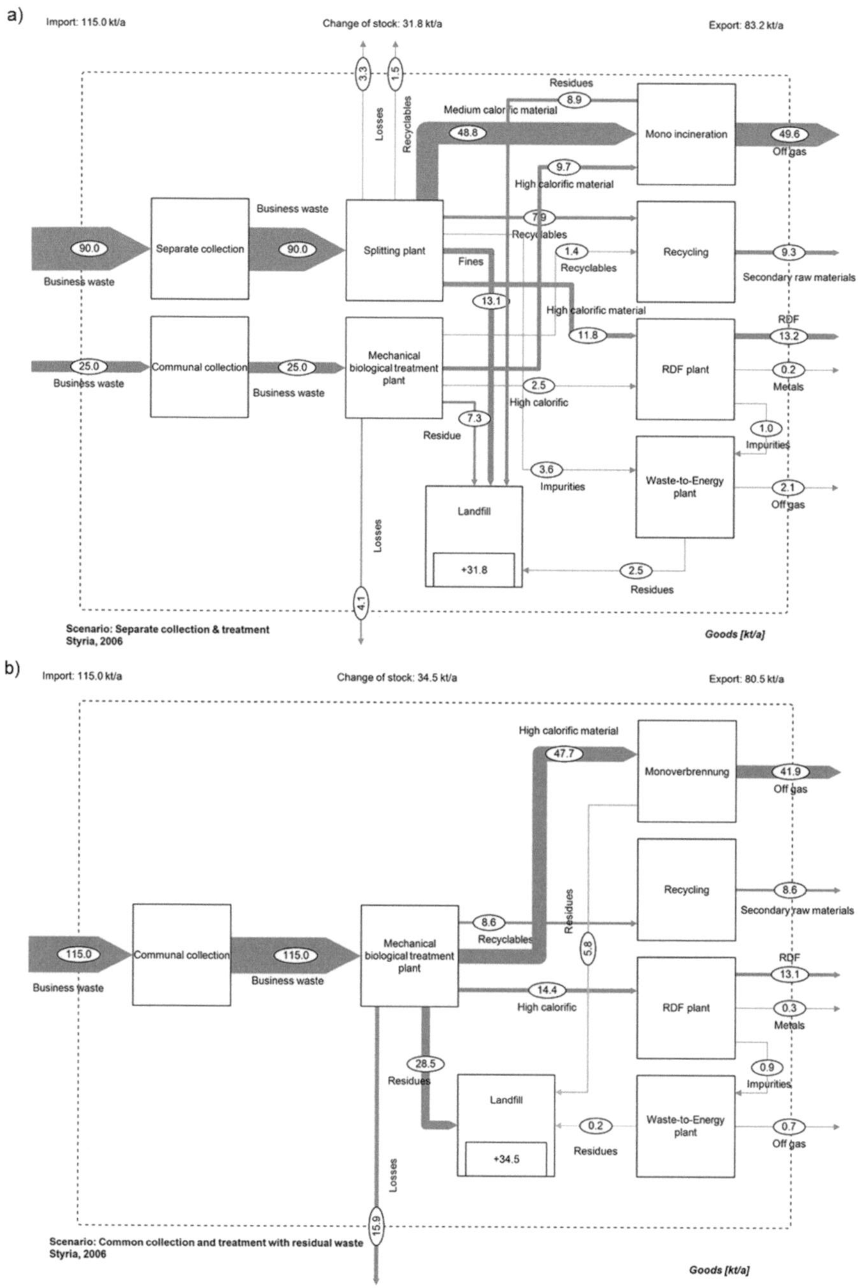

Fig. 7.11 Business waste treatment scenarios for Styria in the year 2006: (**a**) mass flows for the (largely) separate collection and treatment of Styrian business waste, (**b**) mass flows for the treatment of Styrian business waste together with residual waste from households (Based on Laner and Brunner 2008) Abbreviation: *RDF* residue-derived fuel

from an environmental point of view, mainly due to the more efficient utilization of the business waste's energy content in case of separate collection and treatment of most of this waste stream (i.e., approx. 80 % can be collected separately from residual household waste, while the remaining 20 % are collected together with household waste in any case; cf. Fig. 7.11).

The studies mentioned above have used scenario analysis to optimize material flow systems with respect to the magnitude of material flows into appropriate sinks as well as based on the assessment of environmental impacts associated with the scenarios. Indicators such as the global warming potential (GWP) and the cumulative energy demand (CED) are also widely used to assess the impact of life cycle inventories on the environment and the depletion of energetic resources (cf. Sect. 3.5). Thus, there is a natural link to LCA-based optimization, where systems are modified in a way to minimize environmental impact while providing a defined function – in the case of the mentioned MFA-based waste scenario assessments, this would be the treatment of a certain amount of waste generated in a specific region. However, instead of defining the scenarios up-front and then evaluating the environmental performance, one could aim at deriving optimal scenarios using a formal optimization model including various constraints such as waste characteristics, capacity limitations, or regulations. Such an approach has been put forward by Vadenbo et al. (2014b), who developed a multiobjective optimization model for waste and resource management in predefined networks. The model combines MFA, LCA, and formal optimization modeling in a framework for optimal resource management solutions. To illustrate the use of the model, they applied it to sewage sludge management in Switzerland (Vadenbo et al. 2014a). It turned out that the current situation of sewage sludge management could be improved in various ways with trade-offs between the different objectives for the investigated sewage sludge treatment options. This approach of combining MFA and LCA in a formal optimization framework provides a means of finding optimal solutions for environmental pollution reduction and resource management without predefining scenarios but based on formal objective functions to be minimized or maximized. Therefore, it represents an effective way to provide decision support in view of trade-offs between different environmental objectives and increasingly complex networks of resource utilization.

4 Combining Material Flow Analysis and Life Cycle Assessment for Environmental Decision Support

In this section, the use of MFA and LCA is discussed to highlight the synergies of combining MFA and LCA for decision support in environmental protection and resource management. Building on existing work, it is illustrated that MFA is essential to establish consistent inventory data sets in LCA, that MFA provides a basis for normalization in life cycle impact assessment (LCIA), and that the

MFA-based investigation of the anthropogenic stock offers the possibility of considering geogenic and anthropogenic stocks in resource depletion assessment in LCIA.

4.1 Consistent Material Flow Data and Waste LCA

LCA has once been criticized because of the careless use of inconsistent and sometimes impossible data in life cycle inventories (Ayres 1995). As erroneous data impairs objective environmental assessments of alternatives for providing a specific product or service, MFA, based on the fundamental principle of mass conservation, represents the basis for providing reliable and consistent inventory data for the systems under investigation. Therefore, LCAs should be founded on the analysis of material and energy flows of the system under investigation.

The overall goals of waste management are the protection of human health and the environment as well as the efficient use of resources. Therefore, MFA has been typically applied to investigate waste systems and their respective contribution to the safe disposal of hazardous substances and the recovery of valuable resources. The mapping of material flows and the consideration of mass balance constraints are used to generate inventory data for waste management processes and process chains, which often form the basis for consequent LCA to choose environmentally optimal solutions for waste management (e.g., Tonini et al. 2013; Vadenbo et al. 2014b). Detailed plant-level MFAs have been used in waste LCA studies to investigate substance and energy flows of new treatment technologies and to evaluate their environmental performance compared to other waste treatment alternatives (e.g., Andersen et al. 2010; Laner and Rechberger 2007; Lederer and Rechberger 2010). Similarly, MFA-based evaluations of waste systems use LCA databases and impact categories from LCIA to evaluate the environmental performance of waste treatment alternatives (e.g., Laner et al. 2009). In this context, MFA provides a means to generate transparent and consistent (with physical laws) inventory data, and LCA allows for considering environmental effects of waste management outside the main system of interest (upstream burdens of consumed materials, downstream processes, avoided production due to resource recovery), which are often decisive for the result of the environmental assessment (e.g., Vadenbo et al. 2013). Furthermore, impact categories from LCA are used in MFA studies of waste management to quantify environmental burdens or savings associated with the waste system.

4.2 MFA as a Basis for LCIA

In LCIA, classification, characterization, normalization, and valuation are distinguished as individual steps of the impact assessment, with the latter two being optional.[2] In the classification step, emissions are assigned to different impact categories based on their potential effects. In the characterization step, the contributions of the emissions to the individual impact categories are quantified. The normalization step serves to evaluate whether an environmental pressure is large compared to a reference situation. And the final valuation step is used to assign weights to the different impact potentials (cf. ISO 14044, 2006). The determination of material flows is of particular importance for the normalization of impacts, because the reference situation is typically defined as the emissions or resource extractions in a specific region or economy. MFA can serve as a basis to quantify the emissions or the material turnover in the reference system (e.g., a country) and is therefore a basic tool for normalizing impacts in LCA. This is also illustrated in the method of ecological scarcity, an impact assessment method based on a distance-to-target approach, where the current flows are determined for the annual pollutant emissions or resource extractions for Switzerland (cf. Frischknecht and Büsser Knöpfel 2013). The current flows are then compared to critical flows (based on policy goals or legal requirements) and used to calculate Swiss eco-factors. Due to the method's focus on Swiss conditions, the application of the ecological scarcity method in other countries should account for regional and societal differences between Switzerland and the country of interest via adapted eco-factors (cf. Grinberg et al. 2012). For this purpose, MFA represents the ideal tool.

4.3 Anthropogenic Material Stocks in LCA

In the past many MFA studies identified large material stocks as potential secondary raw materials (e.g., Graedel et al. 2004; Pauliuk et al. 2013). In the light of these large material stocks available in society, Schneider et al. 2011 introduced an extension for characterizing the depletion of abiotic resources in LCA. They acknowledge that anthropogenic material stocks have a significant effect on (potential) raw material availability and should therefore be considered as characterization factors for resource depletion. These extended characterization factors allow for a more realistic reflection of resource scarcity. Consequently, MFA to quantify anthropogenic material stocks allow for extending life cycle impact assessment (LCIA) for abiotic resource depletion by considering anthropogenic resources. However, so far only few efforts have been made to evaluate these anthropogenic stocks with respect to their value as a source of secondary raw materials and enable fair comparisons between primary deposits and anthropogenic stocks. Johansson et al. (2013) contributed to form the basis for such comparisons by classifying different anthropogenic stocks with respect to their availability and extraction type.

[2] See LCA Compendium, volume "Life Cycle Impact Assessment," editors Michael Hauschild and Mark Huijbregts, in particular chapter 14 and 15 (see Laurent and Hauschild 2015; Itsubo 2015).

They distinguish in-use mining, hibernation mining, dissipation mining, landfill mining, slag mining, and tailing mining. One step further, Lederer et al. (2014) proposed a method for the evaluation of anthropogenic resources, based on a stepwise procedure consisting of prospection, exploration, evaluation, and classification. The procedure was derived from existing classification approaches in economic geology. Thus, it provides a basis for comparing primary and secondary material stocks in a consistent way and ultimately for quantifying the actual amount of secondary raw materials which can be extracted from the anthropogenic stock. As the methodology has only been applied to phosphorus stocks in Austria so far, further case studies addressing various materials and types of stock are needed to demonstrate resource evaluations under different technological, environmental, and societal conditions (cf. Lederer et al. 2014). One concept of mining the anthropogenic stock is the so-called landfill mining, which is the recovery of wastes stored in landfills as secondary raw materials or fuels. This type of mining concepts has been evaluated with respect to various factors including environmental assessments using LCA (Frändegård et al. 2013). In summary, the connection between MFA to investigate anthropogenic stocks and LCA is twofold: On the one hand, the classification of anthropogenic resources based on MFA is important for the consistent evaluation of abiotic resource depletion in LCIA. On the other hand, LCA is an important tool to consider environmental effects of mining the anthropogenic stock, which need to be addressed within the evaluation of anthropogenic resources (Winterstetter et al. 2015).

5 Outlook

Due to the different perspectives of MFA and LCA on environmentally relevant systems, their application can be complementary to create a sound basis for decisions related to environmental protection and resource conservation (e.g., Lopes Silva et al. 2015). State-of-the-art environmental assessments in the field of waste management typically apply MFA and LCA in combination, to guarantee for consistent inventory data, on the one hand, and to environmental impacts of waste management outside the main system, on the other hand. In this context, frameworks for the optimization of waste and resource management systems have been put forward and are being further developed (e.g., Vadenbo et al. 2014a, b), which combine MFA, LCA, and formal optimization techniques. In general, due to increasingly complex systems to be considered in environmental and resource management, a sound data basis and the consideration of large process chains and networks are crucial for the objectiveness and reliability of decision support tools. The combination of MFA and LCA enables such analyses, and therefore, the application of these tools is expected to become more and more intertwined in integrated assessments of environmentally relevant systems. We therefore propose that both methodologies be jointly integrated in academic education, further reducing the number of explicit MFA and LCA experts.

References

Andersen JK, Boldrin A, Christensen TH, Scheutz C (2010) Mass balances and life-cycle inventory for a garden waste windrow composting plant (Aarhus, Denmark). Waste Manag Res 28:1010–1020. doi:10.1177/0734242x09360216

Andersson H et al (2012) WP4 background paper – identification of sources and estimation of inputs to the Baltic Sea. IVL Swedish Environmental Research Institute, Stockholm

Ayres RU (1995) Life cycle analysis: a critique. Resour Conserv Recycl 14:199–223. doi:http://dx.doi.org/10.1016/0921-3449(95)00017-D

Ayres RU, Nair I (1984) Thermodynamics and economics. Phys Today 37:62–71

Baccini P, Bader HP (1996) Regionaler Stoffhaushalt: Erfassung, Bewertung und Steuerung. Spektrum Akademischer Verlag, Heidelberg

Baccini P, Brunner PH (1991) Metabolism of the anthroposphere. Springer, Berlin

Baccini P, Brunner PH (2012) Metabolism of the anthroposphere: analysis, evaluation design, 2nd edn. MIT Press, Cambridge, MA

Bai L, Qiao Q, Li Y, Wan S, Xie M, Chai F (2015) Statistical entropy analysis of substance flows in a lead smelting process. Resour Conserv Recycl 94:118–128. doi:http://dx.doi.org/10.1016/j.resconrec.2014.11.011

Björklund A (2002) Survey of approaches to improve reliability in LCA. Int J Life Cycle Assess 7:64–72. doi:10.1007/BF02978849

Bonnin M, Azzaro-Pantel C, Pibouleau L, Domenech S, Villeneuve J (2013) Development and validation of a dynamic material flow analysis model for French copper cycle. Chem Eng Res Des 91:1390–1402. doi:http://dx.doi.org/10.1016/j.cherd.2013.03.016

Bringezu S, Moriguchi Y (2002) Material flow analysis. In: Ayres RU, Ayres LW (eds) Handbook of industrial ecology. Edward Elgar Publishing, Cheltenham

Bringezu S, Schütz H, Moll S (2003) Rationale for and Interpretation of economy-wide materials flow analysis and derived indicators. J Ind Ecol 7:43–64. doi:10.1162/108819803322564343

Brunner PH (2010) Clean cycles and safe final sinks. Waste Manag Res 28:575–576. doi:10.1177/0734242x10370987

Brunner PH, Rechberger H (2004) Practical handbook of material flow analysis. CRC Press LCC, Florida

Brunner PH, Tjell JC (2012) Do we need sinks? Waste Manag Res 30:1–2. doi:10.1177/0734242x11432367

Buchner H, Laner D, Rechberger H, Fellner J (2014) In-depth analysis of aluminum flows in Austria as a basis to increase resource efficiency resources. Resour Conserv Recycl 93:112–123

Cencic O, Frühwirth R (2015) A general framework for data reconciliation. Part I: linear constraints. J Comput Chem Eng 75:196–208

Cencic O, Rechberger H (2008) Material flow analysis with software STAN. J Environ Eng Manag 18(1):3–7

Chen W-Q (2013) Recycling rates of aluminum in the United States. J Ind Ecol 17:926–938. doi:10.1111/jiec.12070

Chen W-Q, Graedel TE (2012) Anthropogenic cycles of the elements: a critical review. Environ Sci Technol 46:8574–8586. doi:10.1021/es3010333

Daigo I, Igarashi Y, Matsuno Y, Adachi Y (2007) Accounting for steel stock in Japan. Tetsu-to-Hagane 93:66–70. doi:10.2355/tetsutohagane.93.66

Danius L, Burström F (2001) Regional material flow analysis and data uncertainties: can the results be trusted? In: Hilti LM, Giligen PW (eds) Sustainability in the information society. Metropolis Verlag, Marburg

Do NT, Trinh DA, Nishida K (2014) Modification of uncertainty analysis in adapted material flow analysis: case study of nitrogen flows in the Day-Nhue River Basin. Vietnam Resour Conserv Recycl 88:67–75. doi:http://dx.doi.org/10.1016/j.resconrec.2014.04.006

Dos Santos M, Spitzbart M, Weinlich M, Leitner T, Laner D, Cencic O, Rechberger H (2012) MoveRec: On-line tool for estimating the material composition of WEEE input streams. In: Electronics Goes Green 2012+ (EGG), 9–12 Sept., IEEE, Berlin, pp 1–5

Do-Thu N, Morel A, Nguyen-Viet H, Pham-Duc P, Nishida K, Kootattep T (2011) Assessing nutrient fluxes in a Vietnamese rural area despite limited and highly uncertain data. Resour Conserv Recycl 55:849–856. doi:10.1016/j.resconrec.2011.04.008

Dubois D, Fargier H, Ababou M, Guyonnet D (2014) A fuzzy constraint-based approach to data reconciliation in material flow analysis. Int J General Syst:1–23. doi:10.1080/03081079.2014.920840

Feketitsch J, Buchner H, Lederer J, Laner D, Fellner J (2013) Material flow analysis of plastic products in Austria: emphasis on data uncertainties in consumption sectors and solid waste management. Paper presented at the ISWA World Congress, 7–10 Oct 2013, Vienna, Austria, pp 1–11. ISBN: 978-3-200-03229-3

Fellner J, Aschenbrenner P, Cencic O, Rechberger H (2011) Determination of the biogenic and fossil organic matter content of refuse-derived fuels based on elementary analyses. Fuel 90:3164–3171. doi:http://dx.doi.org/10.1016/j.fuel.2011.06.043

Frändegård P, Krook J, Svensson N, Eklund M (2013) A novel approach for environmental evaluation of landfill mining. J Clean Prod 55:24–34. doi:http://dx.doi.org/10.1016/j.jclepro.2012.05.045

Frischknecht R, Büsser Knöpfel S (2013) Swiss eco-factors 2013 according to the ecological scarcity method. Methodological fundamentals and their application in Switzerland, vol 1330. Federal Office for the Environment, Bern

Georgescu-Roegen N (1971) The entropy law and the economic process. Harvard University Press, Cambridge, MA

Glöser S, Soulier M, Tercero Espinoza LA (2013) Dynamic analysis of global copper flows. Global stocks, postconsumer material flows, recycling indicators, and uncertainty evaluation. Environ Sci Technol 47:6564–6572. doi:10.1021/es400069b

Gottschalk F, Scholz RW, Nowack B (2010) Probabilistic material flow modeling for assessing the environmental exposure to compounds: methodology and an application to engineered nano-TiO2 particles. Environ Model Softw 25:320–332. doi:http://dx.doi.org/10.1016/j.envsoft.2009.08.011

Graedel TE et al (2004) Multilevel cycle of anthropogenic copper. Environ Sci Technol 38:1242–1252. doi:10.1021/es030433c

Grinberg M, Ackermann R, Finkbeiner M (2012) Ecological scarcity method: adaptation and implementation for different countries. Environ Climate Technol 10:9–15. doi:10.2478/v10145-012-0019-5

Udo de Haes HA, van der Voet E, Kleijn R (1997) Substance flow analysis (SFA), an analytical tool for integrated chain management. Paper presented at the ConAccount workshop, Leiden, The Netherlands, 21–23 January 1997

Hedbrant J, Sörme L (2001) Data vagueness and uncertainties in urban heavy-metal data collection water. Water Air Soil Pollut Focus 1:43–53. doi:10.1023/a:1017591718463

Huang D-B, Bader H-P, Scheidegger R, Schertenleib R, Gujer W (2007) Confronting limitations: new solutions required for urban water management in Kunming City. J Environ Manag 84:49–61. doi:http://dx.doi.org/10.1016/j.jenvman.2006.05.004

ISO 14044 (2006) Environmental management – life cycle assessment – requirements and guidelines. Geneva, Switzerland

Itsubo N (2015) Weighting. Chapter 15 "Life Cycle Impact Assessment" (Hauschild M, Huijbregts MAJ eds). In: LCA compendium – the complete world of life cycle assessment (Klöpffer W, Curran MA, series eds). Springer, Dordrecht, pp 301–330

Johansson N, Krook J, Eklund M, Berglund B (2013) An integrated review of concepts and initiatives for mining the technosphere: towards a new taxonomy. J Clean Prod 55:35–44. doi:10.1016/j.jclepro.2012.04.007

Kaufman S, Krishnan N, Known E, Castaldi M, Themelis N, Rechberger H (2008) Examination of the fate of carbon in waste management systems through statistical entropy and life cycle analysis. Environ Sci Technol 42(22):8558–8563

Kral U (2014) A new indicator for the assessment of anthropogenic substance flows to regional sinks. PhD thesis, Vienna University of Technology

Kral U, Kellner K, Brunner PH (2013) Sustainable resource use requires 'clean cycles' and safe 'final sinks'. Sci Total Environ 1:461–462. doi:http://dx.doi.org/10.1016/j.scitotenv.2012.08.094

Krook J, Svensson N, Eklund M (2012) Landfill mining: a critical review of two decades of res. Waste Manag 32:513–520. doi:http://dx.doi.org/10.1016/j.wasman.2011.10.015

Laner D, Brunner PH (2008) Kriterien zur Trennung von Siedlungsabfall aus Industrie und Gewerbe als Voraussetzung zur Zuordnung zu Behandlungsverfahren. Institut für Wassergüte, Ressourcenmanagement und Abfallwirtschaft, Wien

Laner D, Cencic O (2013) Comment on "Solid recovered fuel: materials flow analysis and fuel property development during the mechanical processing of biodried waste". Environ Sci Technol 47:14533–14534. doi:10.1021/es403403u

Laner D, Rechberger H (2007) Treatment of cooling appliances: interrelations between environmental protection, resource conservation, and recovery rates. Resour Conserv Recycl 52:136–155

Laner D, Pomberger R, Scherübl T, Brunner PH (2009) Voraussetzungen für eine zielorientierte Bewirtschaftung hausmüllähnlicher Gewerbeabfälle – eine Analyse am Beispiel der Steiermark. Müll Abfall 9:2–10

Laner D, Rechberger H, Astrup T (2014) Systematic evaluation of uncertainty in material flow analysis. J Ind Ecol 18(6): 859–870

Laner D, Rechberger H, Astrup T (2015) Applying fuzzy and probabilistic uncertainty concepts to the material flow analysis of palladium in Austria. J Ind Ecol 19(6):1055–1069

Lassen C, Hansen E (2000) Paradigm for substance flow analyses. Guide for SFAs carried out for the Danish EPA. Environmental Project No. 577, Danish Environmental Protection Agency, Copenhagen

Laurent A, Hauschild MZ (2015) Normalization. Chapter 14 "Life Cycle Impact Assessment" (Hauschild M, Huijbregts MAJ eds). In: LCA compendium – the complete world of life cycle assessment (Klöpffer W, Curran MA, series eds). Springer, Dordrecht, pp 271–300

Lederer J, Rechberger H (2010) Comparative goal-oriented assessment of conventional and alternative sewage sludge treatment options. Waste Manag 30:1043–1056. doi:10.1016/j.wasman.2010.02.025

Lederer J, Laner D, Fellner J (2014) A framework for the evaluation of anthropogenic resources: the case study of phosphorus stocks in Austria. J Clean Prod 84:1–848. doi:http://dx.doi.org/10.1016/j.jclepro.2014.05.078

Lifset RJ, Eckelman MJ, Harper EM, Hausfather Z, Urbina G (2012) Metal lost and found: dissipative uses and releases of copper in the United States 1975–2000. Sci Total Environ 417–418:138–147. doi:10.1016/j.scitotenv.2011.09.075

Lopes Silva DA, de Oliveira JA, Saavedra YMB, Ometto AR, Rieradevall i Pons J, Gabarrell Durany X (2015) Combined MFA and LCA approach to evaluate the metabolism of service polygons: A case study on a university campus. Resour Conserv Recycl 94:157–168

Mastellone ML, Brunner PH, Arena U (2009) Scenarios of waste management for a waste emergency area. J Ind Ecol 13:735–757. doi:10.1111/j.1530-9290.2009.00155.x

Montangero A, Belevi H (2007) Assessing nutrient flows in septic tanks by eliciting expert judgement: a promising method in the context of developing countries. Water Res 41:1052–1064. doi:http://dx.doi.org/10.1016/j.watres.2006.10.036

Müller D (2006) Stock dynamics for forecasting material flows—case study for housing in the Netherlands. Ecol Econ 59:142–156. doi:http://dx.doi.org/10.1016/j.ecolecon.2005.09.025

Müller E, Hilty LM, Widmer R, Schluep M, Faulstich M (2014) Modeling metal stocks and flows: a review of dynamic material flow analysis methods. Environ Sci Technol 48:2102–2113. doi:10.1021/es403506a

O'Rourke D, Conelly L, Koshland CP (1996) Industrial ecology: a critical review. Int J Environ Pollut 6(2/3):89–112

Ott C, Rechberger H (2012) The European phosphorus balance. Resour Conserv Recycl 60:159–172. doi:10.1016/j.resconrec.2011.12.007

Pauliuk S, Wang T, Müller DB (2013) Steel all over the world: estimating in-use stocks of iron for 200 countries. Resour Conserv Recycl 71:22–30. doi:http://dx.doi.org/10.1016/j.resconrec.2012.11.008

Rechberger H, Brunner PH (2002) A new, entropy based method to support waste and resource management decisions. Environ Sci Technol 34(4):809–816

Rechberger H, Graedel TE (2002) The European copper cycle: statistical entropy analysis. Ecol Econ 42(1–2):59–72

Rechberger H, Cencic O, Frühwirth R (2014) Uncertainty in material flow analysis. J Ind Ecol 18:159–160

Reck BK, Graedel TE (2012) Challenges in metal. Recycl Sci 337:690–695. doi:10.1126/science.1217501

RPA (2012) Study on data needs for a full raw materials flow analysis. Risk & Policy Analysts Limited, Brussels

Schaffner M (2007) Applying a material flow analysis model to assess river water pollution and mitigation potentials. PhD Thesis, University of Bern

Schaffner M, Bader H-P, Scheidegger R (2009) Modeling the contribution of point sources and non-point sources to Thachin River water pollution. Sci Total Environ 407:4902–4915. doi: http://dx.doi.org/10.1016/j.scitotenv.2009.05.007

Schneider L, Berger M, Finkbeiner M (2011) The anthropogenic stock extended abiotic depletion potential (AADP) as a new parameterisation to model the depletion of abiotic resources. Int J Life Cycle Assess 16:929–936. doi:10.1007/s11367-011-0313-7

Shannon CE (1948) A mathematical theory of communication. Bell Syst Tech J 27:379–423, 623–656, July, October

Sobańtka A, Rechberger H (2013) Extended statistical entropy analysis (eSEA) for improving the evaluation of Austrian wastewater treatment plants. Water Sci Technol 67(5):1051–1057

Sobańtka AP, Zessner M, Rechberger H (2012) The extension of statistical entropy analysis to chemical compounds. Entropy 14:2413–2426

Stumm W, Davis J (1974) Kann Recycling die Umweltbeeinträchtigung vermindern? In: Recycling: Lösung der Umweltkrise? Gottlieb Duttweiler-Institut für wirtschaftliche und soziale Studien (ed), Stuttgart, Zürich (in German)

Tonini D, Martinez-Sanchez V, Astrup TF (2013) Material resources, energy, and nutrient recovery from waste: are waste refineries the solution for the future? Environ Sci Technol 47:8962–8969. doi:10.1021/es400998y

Udo deHaes HA, van der Voet E, Kleijn R (1997) Substance flow analysis (SFA), an analytical tool for integrated chain management. Paper presented at the ConAccount workshop, Leiden, The Netherlands, 21–23 January 1997

Vadenbo CO, Boesch ME, Hellweg S (2013) Life cycle assessment model for the use of alternative resources in iron making. J Ind Ecol 17:363–374. doi:10.1111/j.1530-9290.2012.00543.x

Vadenbo C, Guillén-Gosálbez G, Saner D, Hellweg S (2014a) Multi-objective optimization of waste and resource management in industrial networks. Part II: model application to the treatment of sewage sludge. Resour Conserv Recycl 89:41–51. doi:http://dx.doi.org/10.1016/j.resconrec.2014.05.009

Vadenbo C, Hellweg S, Guillén-Gosálbez G (2014b) Multi-objective optimization of waste and resource management in industrial networks. Part I: model description. Resour Conserv Recycl 89:52–63. doi:http://dx.doi.org/10.1016/j.resconrec.2014.05.010

van der Voet E (2002) Substance flow analysis methodology. In: Ayres RU, Ayres LW (eds) A handbook of industrial ecology. Edward Elgar Publishing Ltd, Cheltenham

Winterstetter A, Laner D, Rechberger H, Fellner J (2015) Framework for the evaluation of anthropogenic resources: a landfill mining case study – Resource or reserve? Resour Conserv Recycl 96:19–30

Wolman A (1965) The metabolism of cities. Sci Am 213:179–190

Wu H, Yuan Z, Zhang Y, Gao L, Liu S, Geng Y (2014) Data uncertainties in anthropogenic phosphorus flow analysis of lake watershed. J Clean Prod 69:74–82. doi:http://dx.doi.org/10.1016/j.jclepro.2014.01.043

Yue Q, Lu ZW, Zhi SK (2009) Copper cycle in China and its entropy analysis. Resour Conserv Recycl 53(12):680–687

Zeltner C, Bader HP, Scheidegger R, Baccini P (1999) Sustainable metal management exemplified by copper in the USA. Reg Environ Chang 1:31–46. doi:10.1007/s101130050006

Zoboli O, Laner D, Zessner M, Rechberger H (2015) Added value of time series in MFA. The Austrian phosphorus budget from 1990 to 2011. J Ind Ecol. doi:10.1111/jiec.12381, open access, www.wileyonlinelibrary.com/journal/jie

Chapter 8
Life Cycle Assessment of Organizations

Julia Martínez-Blanco, Atsushi Inaba, and Matthias Finkbeiner

Abstract In order to protect the environment in a credible manner, organizations need to rely on stable schemes. The most applied and widespread approaches for environmental assessments at the organization level have only recently considered the full value chain and mostly concentrate on a single aspect. Carbon footprinting, for example, has shown that the environmental impacts beyond the walls of the organization can play an important role in the overall impact of an organization. While life cycle assessment (LCA) was originally conceived to be applicable only for products, its benefits and potential might also be extended for organizational assessment. The discussions on the carbon footprint of organizations and the development of the Scope 3-standard of the GHG (greenhouse gas) Protocol promoted the future use of a corporate approach. Several initiatives are on the way for the LCA of organizations: UNEP/SETAC Life Cycle Initiative proposes organizational LCA (O-LCA), using as a backbone ISO/TS 14072; moreover, the European Commission launched a guide for the Organisation Environmental Footprint (OEF). The main new elements of the methodology are at the scope and inventory phase, when the unit of analysis and the system boundary are defined, as well as the approach for data collection. LCA of organizations may represent a key element in the internal decision-making system of an organization, as it can provide insight on the organization and value chain and identify hotspots where action is more needed. It may also provide information and support the organization for voluntary or mandatory reporting to third parties and in its communication plan. This chapter aims to discuss the need and features of the LCA of organizations, present the several initiatives that currently exist, provide an overview of the technical framework and proposals to streamline the application of the methodology, and finally to illustrate each one with two case studies.

J. Martínez-Blanco (✉) • M. Finkbeiner
Chair of Sustainable Engineering – Institute of Environmental Technology, Technische Universität Berlin, Office Z1, Strasse des 17. Juni 135, 10623 Berlin, Germany
e-mail: julia.martinezblanco@campus.tu-berlin.de

A. Inaba
Department of Environmental and Energy Chemistry, Kogakuin University, 163-8677 Tokyo, Japan

M. Finkbeiner (ed.), *Special Types of Life Cycle Assessment*, LCA Compendium – The Complete World of Life Cycle Assessment,
DOI 10.1007/978-94-017-7610-3_8

Keywords Decision-making • Direct and indirect impacts • European commission • GHG protocol initiative • ISO 14040 • ISO 14044 • ISO/TS 14072 • LCA • Life cycle assessment • Multi-impact • OEF • OLCA • O-LCA • Organisation Environmental Footprint Sector Rules • Organisation Environmental Footprint • Organizational life cycle assessment • Performance tracking • Reporting flow • Reporting organization • Resource use and emissions • Suppliers • Sustainability strategy • UNEP/SETAC Life Cycle Initiative • Value chain

Acronyms

CDP	Carbon disclosure project
CSR	Corporate Social Responsibility
EMAS	Eco-Management and Audit Scheme
EMS	Environmental management system
GHG	Greenhouse gas
GRI	Global Reporting Initiative
IPCC	Intergovernmental panel on climate change
ISO	International Organization for Standardization
NACE codes	Nomenclature générale des activités economiques dans les communautés européennes
OBIA	Overall business impact assessment
OEF	Organisation Environmental Footprint
OEFSRs	Organisation Environmental Footprint Sector Rules
O-LCA	Organizational LCA (used by UNEP/SETAC Life Cycle Initiative)
OLCA	Organizational LCA (used by ISO)
OEF	Organisation Environmental Footprint
PEF	Product Environmental Footprint
SETAC	Society of environmental toxicology and chemistry
SMEs	Small and medium-sized enterprises (SMEs)
UNEP	United Nations Environment Program
USLP	Unilever sustainable living plan

1 Introduction

1.1 Outline of the Chapter

The chapter provides an overview of the state of the art of the life cycle assessment (LCA) of organizations. This is presented in four main sections. First, the need for the methodology and its connection with previous approaches is argued, along with the introduction of three existing initiatives addressing the LCA of organizations. In Sect. 2 the methodological framework is briefly presented. That is complemented, in Sect. 3, by proposals to streamline the implementation of the methodology and its outcomes. Finally, the application and potential of the LCA of organizations is illustrated with two case studies (Accor and Unilever) applying tailored approaches aligned with the one presented here.

Three different terms are used along the chapter to refer to the new methodology or approach presented. LCA of organizations is used as the generic term; however, when referring to the specific framework proposed by UNEP (2015) and ISO/TS 14072 (ISO 2014a), organizational LCA and O-LCA are used, while the term Organizational Environmental Footprint (OEF) is used within the framework of the European Commission (2013a). In most cases the terms are interchangeable – the main differences are explained in Sect. 1.4.4 – but we decided to stick to the original name in order to acknowledge the source and the different frameworks used.

1.2 Grounds for a New Approach

Organizations, companies, corporations, firms, public institutions, etc., have a key responsibility to reduce environmental impacts. The first step toward an improved environmental performance is the implementation of comprehensive schemes that frame the organization's strategy and decision-making, including environmental aspects – apart from technical and economic considerations. An approach that analyzes the whole organization (organizational approach), including not only the facilities of the organization but also its value chain (life cycle approach), and considers a set of relevant environmental aspects (multi-impact approach) can advance the integration of the environment in the organization's strategy and operation.

The environmental information required to support decision-making should provide guidance that is meaningful at the level the decisions are taken, i.e., the organizational level. Having sufficient understanding of a system is a prerequisite to design efficient strategies that can effectively improve its performance in the long term. The organizational approach reveals, among all the products and operations involved in the provision of the portfolio, the hotspots where the organization should focus energies and interventions. Understanding risks and impact reduction opportunities gives a solid ground to strategic decisions at different levels, for instance, decisions on technologies, investments, and new product lines.

Environmental burdens and risks are not restricted to the ones occurring within the organization's facilities; the decisions of the organization also affect the environmental impacts of the supplier network, as well as the use phase and end of life of the products in the portfolio. Indeed, life cycle's impacts could and usually do significantly contribute to the environmental performance of organizations (Downie and Stubbs 2011; WRI and WBCSD 2011; Makower et al. 2014). Focusing on internal operations is a good starting point, but it may not derive effective environmental improvements if most of an organization's impacts occur up and down the life cycle (UNEP 2015). Consequently, the organization may need to also intervene on the activities of the value chain. Understanding the risks and opportunities through the value chain drives more interest from companies on environmental issues. Additionally, the life cycle approach helps to ensure that organizations "do not benefit from 'outsourcing' steps/life cycle stages which are linked to high environmental burdens" (Pelletier et al. 2013).

Furthermore, while the assessment of specific, important environmental areas such as greenhouse gases or water has advanced the environmental awareness of organizations and society in general, a holistic approach is needed. This is essential to expose any potential trade-offs between different environmental issues and to help avoid unintended shifting of burdens (Pelletier et al. 2013). When multiple environmental aspects are considered, impacts are not only the result of emissions to air (as in GHG accounting); all the emissions should be included, to air as well as to soil and water. Furthermore, apart from emissions, the impacts are also produced due to the use/consumption of resources. By assessing multiple impacts, an organization has also more angles from which to assess how their operations, performance, and decisions affect different natural systems and how to improve them (Draucker 2013).

All organizations are key to reduce the pressure on the environment. Large corporations can play a promising role due to their relevant share on the global depletion of resources and emission of pollutants and toxic substances. For example, according to Unilever (2014a), the company purchases 12 % of the world's black tea supply, 1 % of cocoa, and has a relevant consumption of sugar, paper, oils, etc. Furthermore, large corporations have the right resources and influence to promote environmental tools and methodologies throughout their value chain. On the other side, small- to medium-sized enterprises (SMEs) have an important contribution to the world economy if addressed as a collective and thus to environmental impacts. SMEs represents more than 90 % of businesses and on average account for 50 and 60 % of the gross domestic product and employment, respectively, of all countries (UNIDO 2006). Very often SMEs produce components and services needed for producing the final products sold by larger organizations; in those cases it is also common that they should follow a list of specifications for production. Therefore, it may be more effective for them to focus efforts on the organizational, but not on the specific product level where they have less room for improvement of their environmental performance. Many of these SMEs are located in developing countries (where environmental tools are progressively more used) and partially prompted by bigger organizations that produce for developed countries with higher environmental standards. Not only private but also public organizations bear responsibility in the protection of the environment. Apart from reducing the impacts derived from their activities, public organizations have the mandate to act as an example and driving force of change.

1.3 The Way Toward Life Cycle Assessment of Organizations

The 2002 World Summit for Sustainable Development in Johannesburg called for a comprehensive set of programs focusing on sustainable consumption and production (UN 2002). Several methodologies, tools, and techniques are already available for organizations (Fig. 8.1). A referent approach for many organizations is the environmental management system (EMS); currently more than 300,000 organizations have a certified EMS according to ISO 14001 (ISO 2004) and a relevant number was certified with the European version, EMAS (Eco-Management and Audit Scheme) (European Commission 2009). They are mainly procedural tools, and when including a company ecobalance, they commonly analyze only gate-to-gate processes.

Carbon footprinting of corporations has been steered within the Greenhouse Gas Protocol initiative (WRI and WBCSD 2004) and incorporates the value chain since a few years ago through “scope 3” (WRI and WBCSD 2011). Using that initiative as a basis, the ISO/TR 14069 standardizes the quantification and reporting of GHG emissions of organizations (ISO 2013a). Other related schemes, not always including the value chain, are the Carbon Disclosure Project (CDP 2014), the Bilan Carbone (ADEME 2010), and the G4 Global Reporting Initiative (GRI 2013). According to their analysis, Pelletier et al. (2013) suggested that methodological guidance in current Organisation Environmental Footprint studies is still considerably less advanced than for product-level environmental footprint studies because of the complexity and variety of different organizations compared to a single product or even product category. In many of them, there is an apparent emphasis on high-level corporate sustainability reporting rather than on how to calculate organization environmental performance. Moreover, some approaches do not require considering the entire life cycle of the activities of the organization, and most focus on single indicators – both of which increase the likeliness of burden shifting. Existing methods vary in terms of their scope, requirements, and consistency; nevertheless their definition and the acquired experience promoted the future development of LCA of organizations (European Commission 2011).

In order to analyze the environmental performance of products, it has become standard to use a life cycle and multi-impact perspective to capture all impacts. The main standard documents for product LCA are ISO 14040 and ISO 14044 (ISO 2006a, b). While the LCA methodology was originally developed for products, its benefits and potential are not limited to that scope (ISO 2014a). However, life cycle thinking has not put much focus into the organizational perspective until recently. The first considerations concerning organizational footprinting were conducted in the 1990s. The Overall Business Impact Assessment (OBIA) was introduced by Clift and Wright (2000) and Taylor and Postlethwaite (1996), which applies the life cycle approach to explore the relationship between environmental impact and added economic value along the supply chain, and represented the first steps to the development of the environmental footprint of Unilever. Finkbeiner

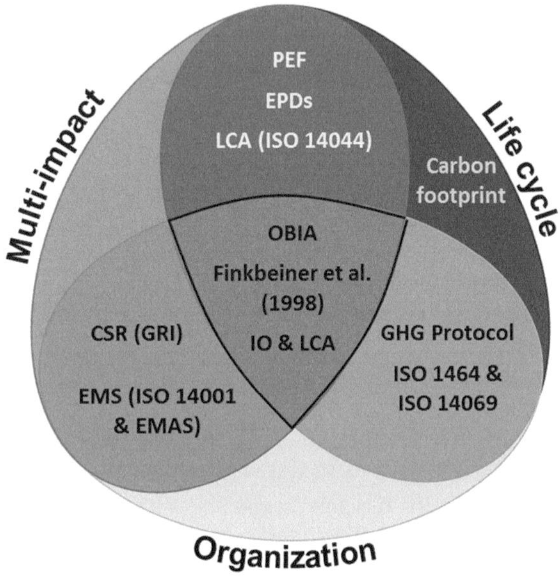

Fig. 8.1 Initiatives and reference works that laid the ground for the development of the LCA of organizations

et al. (1998) discuss the potential complementarity of the company, procedural approach of EMS, and the product oriented, analytical concept of LCA. The two approaches gained attention by science and industry, but it was not enough to be further developed into a broadly applied standard. Input-output analysis has been also proposed to be used along with LCA to evaluate environmental impacts of industry sectors and corporations (Lave et al. 1995; Huang et al. 2009a, b).

Now, the LCA community is in the process of adapting the life cycle concept at the organizational level, strengthened by the framework and acquired experience with product and spurred by the growing interest in environmental analysis tools. However, as stated in ISO (2014a), LCA of organizations may be more complex, as there is more than one product life cycle to follow and a large part of the environmental impacts can reside outside the organization's gate. It may be particularly challenging for large organizations operating in multiple sectors and/or countries (Pelletier et al. 2013).

1.4 Overview of Existing Initiatives for Life Cycle Assessment of Organizations

Several works exist or are underway on the development and agreement of approaches for the multi-impact assessment of organizations from a life cycle perspective (Fig. 8.2). The European Commission started a project in 2011 to develop a reference method for organization and product environmental footprinting. As a result in 2013, it launched the "Organization Environmental

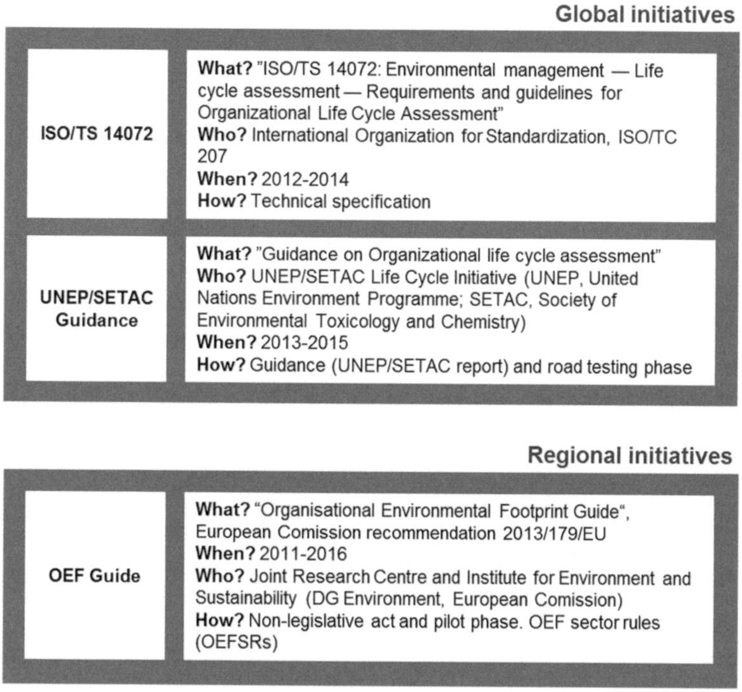

Fig. 8.2 Overview of the main existing initiatives for LCA of organizations

Footprint Guide" (European Commission 2013a), along with an equivalent guide for product footprinting (i.e., PEF Guide). In 2012, the International Organization for Standardization (ISO) started a project to harmonize the requirements and guidelines to apply life cycle thinking to organizations, ISO/TS 14072 (ISO 2014a). A year later, the United Nations Environment Programme (UNEP) and the Society of Environmental Toxicology and Chemistry (SETAC) partnership Life Cycle Initiative also launched the flagship project "LCA of Organizations," whose main outcome is the document "Guidance on Organizational Life Cycle Assessment" (UNEP 2015). These three initiatives that undertake LCA of organizations are further detailed in the following three sections and specific features and differences among them presented in Sect. 1.4.4.

1.4.1 "ISO/TS 14072: Environmental Management, Life Cycle Assessment, Requirements and Guidelines for Organizational Life Cycle Assessment" by the International Organization for Standardization (ISO/TS 14072)

The technical specification ISO/TS 14072 is dedicated to the application of LCA to organizations and proposes a framework for conducting organizational life cycle assessment (OLCA). It therefore extends the application of ISO 14040 and ISO

14044 to all the activities of an organization, which means that the reporting unit of the system allows coverage of different products and operations of the organization considered in the LCA study (ISO 2014a). ISO/TS 14072 provides additional requirements and guidelines for an easier and more effective application of the product LCA standards. It includes description of the advantages that LCA may bring to organizations, the system boundaries, and the limitations regarding reporting, environmental declarations, and comparative assertions. It is intended for any organization with interest in applying LCA. It is not intended for ISO 14001 interpretation and covers the goals of ISO 14040 and 14044.

It was prepared by the International Technical Committee ISO/TC207 Environmental management, Subcommittee SC5 "life cycle assessment," and Working Group WG10. The project was approved in February 2012, and the first meeting of the working group was in Bangkok in June 2012. As a technical specification, it was approved by two out of three of the members of the committee casting a vote. In total, ISO/TC207 includes 85 participating countries and 29 observing countries. One case study is included as an annex in the standard.

1.4.2 "Guidance on Organizational Life Cycle Assessment" by the UNEP/SETAC Life Cycle Initiative

The UNEP/SETAC Life Cycle Initiative (UNEP/SETAC 2014) launched in 2013 the flagship project FS1c "LCA of organizations" within its Phase III program. The primary goal of the project is to demonstrate that the benefits and the potential of the life cycle approach are not limited to the application to products and that the use in organizations is relevant, meaningful, and already possible. Two main outcomes are expected from the project: the UNEP (2015), hereafter UNEP Guidance, and its road testing that started by the end of 2015.

The document highlights the potential of an organizational perspective within life cycle thinking and especially strives to align with ISO/TS 14072 and thus with ISO 14040 and ISO 14044. At the same time, it complements the standard and aims to be a more detailed accompanying document. The document also builds on other existing works and initiatives on the assessment of the environmental performance of organizations, like the Greenhouse Gas Protocol initiative. The UNEP Guidance includes recommendations about the specific methodological issues to take into account when organizational life cycle assessment (O-LCA) is applied, but does not attempt to cover in detail those aspects of the methodology that are common with product LCA.

The UNEP Guidance is intended to be useful to organizations of all sizes, in all sectors, both public and private, and with diverse degree of experience on environmental management. Specific recommendations are provided for several pathways related to the organization's previous experience with environmental tools and for small and medium-sized organizations. Furthermore, the publication includes the experience of several case studies with the use of environmental multi-impact life cycle approaches, some of them making direct reference to the LCA framework.

A working group was established to support the lead authors with the document drafting. After several review rounds, face-to-face, and online meetings, an agreed draft was produced, which was consolidated by a broad group of feedback stakeholders. A total of 100 participants from all over the world and with different background cooperated in the UNEP Guidance preparation. In autumn 2015, the road testing of ten organizations started and first results are expected for summer 2016. Further detail about the flagship project and its activities can be found in the UNEP Guidance (UNEP 2015) and in Martínez-Blanco et al. (2015a).

1.4.3 "Organisation Environmental Footprint Guide" by the European Commission (OEF Guide)

The European Commission's Joint Research Centre (JRC IES) and other European Commission services have worked toward the development of a technical guide for the calculation of the environmental footprint of organizations, the OEF Guide (European Commission 2013a). They have worked in parallel on the methodological guide on Product Environmental Footprint (PEF). The two methodologies are tightly interlinked and have many elements in common.

The document provides guidance on how to calculate an OEF. It aims to increase reproducibility and comparability by emphasizing prescriptiveness over flexibility to ensure that the methodology is applied consistently (Pelletier et al. 2013). The methodology has been developed building on the *Reference Life Cycle Data System Handbook* (European Commission 2010a), as well as other existing methodological standards and guidance documents. The OEF Guide strives to align with existing or upcoming international methodological norms, including ISO 14069 (ISO 2013a) and GHG Protocol Scope 3 (WRI and WBCSD 2011), as well as the PEF Guide. Similarly, efforts have also been made to align insofar as possible with existing environmental management schemes (European Commission 2009; ISO 2014b).

At the European level, PEF and OEF are supposed to achieve an important goal, the implementation of LCA in European environmental policy. Other objectives claimed by PEF/OEF are the harmonization of methods to avoid proliferation, the cost reduction for business and increased applicability for SMEs, and credible communication to consumers avoiding confusion and mistrust (Pelletier et al. 2013; European Commission 2013b, c). This is necessary to produce methods that can be useful for application in the context of policy instruments at EU level.

The guide also explains how to create sector-specific methodological requirements via Organisation Environmental Footprint Sector Rules (OEFSRs), in order to further increase methodological harmonization, specificity, relevance, and reproducibility for a given sector. OEFSRs will furthermore facilitate focusing on the most important parameters, thereby also reducing efforts in completing an OEF study. The OEFSRs aim to support comparisons and comparative assertions between organizations (see Sect. 3.2).

The document was developed taking into account the results of a preliminary pilot phase, an invited expert consultation, and a consultation between European

Commission services. Currently, new pilots are being conducted on selected sectors in order to test further the methodology and prepare OEFSRs for the corresponding sectors.

1.4.4 Main Differences Between the Life Cycle Assessment of Organizations' Approaches

The three approaches differ from previous life cycle assessment techniques mainly because their object of study is the organization, rather than the product. The organization portfolio usually includes more than one product, thus the entire set of goods and services provided by the organization are assessed at the same time. They also differ from other organization-oriented methodologies by their approach, the life cycle, and from recent value chain tools because they are a multi-criteria environmental assessment. Additionally, some internal differences between the three approaches can be distinguished. These are summarized in Table 8.1.

The methodology depicted by ISO/TS 14072, organizational LCA (OLCA), is defined as a "compilation and evaluation of the inputs, outputs and potential environmental impacts of the activities associated with the organization as a whole or portion thereof adopting a life cycle perspective" (ISO 2014a). This definition was also adopted by the UNEP Guidance, but the acronym it uses includes a hyphen, i.e., O-LCA (see Martínez-Blanco et al. 2015a). The latter is the acronym used in this chapter. The European Commission proposed the term Organisation Environmental Footprint (OEF), defined as "a multi-criteria measure of the environmental performance of a product-providing organization from a life cycle perspective" (European Commission 2013a).

According to Finkbeiner (2013), OEF Guide has some requirements that do not align with life cycle standard principles (ISO 2006a, b, 2014a), which was confirmed for some of them and discussed for others by Galatola and Pant (2014). Some examples are the recycling formula for end of life and the default set of impact categories and methods. One of the key issues is whether comparative assertions intended to be disclosed to the public are supported. OEF Guide considers the option of comparative assertions intended to be disclosed to the public within the same sector and according to the OEFSRs (under development), while ISO/TS 14072 and UNEP Guidance clearly discourage this use. They argue that the comparability step is neither robust nor meaningful at this point in time, due to the lack of a consistent basis for comparison between organizations. Even within the same sector, the size, the location, the product segment, the vertical integration, the financial transactions, and overall business model can be significantly different (Finkbeiner and König 2013). Furthermore, the OEF Guide is communication driven while O-LCA is not.

Due to the several differences identified with the product LCA framework, Finkbeiner (2013) questioned whether the proposed OEF approach actually supports its own targets – like inclusion of LCA in environmental policy, harmonization of methods, comparability over flexibility, cost reduction for business, and

Table 8.1 Main features and differences of the existing initiatives for LCA of organizations. This does not aim to be a comprehensive comparison of the differences

	ISO/TS 14072	UNEP Guidance	OEF Guide
General	Organizational life cycle assessment, OLCA	Organizational life cycle assessment, O-LCA	Organizational Environmental Footprint, OEF
	Terminology is mainly based on ISO 14040 and ISO 14044. Few additional terms, like reporting unit, performance tracking, consolidation methodology, etc.	Terminology is mainly based on ISO 14040, ISO 14044, and ISO/TS 14072. Few additional terms, like reporting organization and flow, direct and indirect, etc.	Totally new terminology: resource use and emissions profile, environmental footprint, etc.[a]
Goal and scope	The unit of analysis is the reporting unit (defined as the quantified performance expression of the organization under study to be used as a reference)	The unit of analysis, the reporting unit, is broken down into reporting organization (includes the explanation of the subject of study, the consolidation approach, and the reference period) and reporting flow (measure of the outputs from the reporting organization)	The unit of analysis is defined by two elements: organization (unit of analysis) and product portfolio (type and amount of goods/services provided over the reporting interval)
	Consolidation method (operational control, financial control, or equity share) is selected during system boundary definition	As in ISO/TS 14072	Consolidation method is selected during system boundary definition. Only operational and financial controls are considered
	The idea of having two boundaries, for the organization and for the life cycle, is implied but not set as a requirement	Only one system boundary is defined, including both direct and indirect activities	The system boundaries shall include the organizational boundaries (in relation to the defined organization) and the OEF boundaries (which specify which aspects of the supply chain are included in the analysis)
	Cutoff allowed, based on mass, energy, environmental significance	As in ISO/TS 14072	Cutoff is not allowed, but any data gaps shall be filled using best available generic or extrapolated data. Such data shall not account for more than 10 % of the overall contribution to each impact category

(continued)

Table 8.1 (continued)

	ISO/TS 14072	UNEP Guidance	OEF Guide
Inventory	It shall be an iterative process	As in ISO/TS 14072	Screening step is recommended (based on readily available or generic data)
	Data sources and data quality assessment shall be carefully done. A list of criteria that should be specified for data quality assessment (based on ISO 14044) exists	As in ISO/TS 14072	List of criteria that shall be used for semiquantitative data quality assessment. Criteria shall be met by a study intended for external communication. Minimum data quality requirements are defined
	As for ISO 14044, data from specific sites or representative averages should be used for units that contribute the majority of the mass and energy flows, and are considered to have environmentally relevant inputs and outputs	The use of specific data is recommended, particularly for direct activities. Greater use of assumptions, extrapolations, and generic data is expected for indirect activities. It should collect higher quality data for priority activities (based on environmental significance, mass or energy, etc.)	Specific data shall be obtained for direct processes or activities and for indirect ones where appropriate. Generic data should be used only for indirect processes and activities
	Allocation procedures described in ISO 14044 apply. System expansion is not considered	As in ISO/TS 14072	Multifunctionality decision hierarchy proposed, similar to ISO 14044. System expansion is included
	Reuse and recycling are addressed separately, providing general principle of avoiding allocation (based on ISO 14044)	As in ISO/TS 14072	Recycling formula for end of life provided. Several ones have been proposed, still being discussed
Impact assessment	Classification and characterization are mandatory; normalization and weighting optional	As in ISO/TS 14072	Classification and characterization are mandatory, normalization is recommended, and weighting optional
	The selection of impact categories, category indicators, and characterization models shall be both justified and consistent with the goal and scope of the LCA. No default list	As in ISO/TS 14072	Default set of 14 midpoint impact categories and models

(continued)

Table 8.1 (continued)

	ISO/TS 14072	UNEP Guidance	OEF Guide
Interpretation	It shall not be used for comparative assertions intended to be disclosed to the public	As in ISO/TS 14072	It considers the option of comparative assertions intended to be disclosed to the public within the same sector and according to the OEFSRs. In the OEFSRs it is recommended to create benchmarks and classes of environmental performance for each sector (see Sect. 3.2)
Reporting and critical review	If OLCA results are communicated to a third party, a critical review should be performed according to ISO 14044 and ISO/TS 14071. Independent internal or external reviewer, with sufficient competencies (defined at ISO/TS 14071)	When O-LCA outcomes are communicated to a third party, a critical review shall be performed (mandatory as for external communication in ISO 14044) according to ISO 14044 and ISO/TS 14071. Independent internal or external reviewer, with sufficient competencies (defined at ISO/TS 14071)	Any OEF study intended for external communication shall be critically reviewed by at least one independent and qualified external reviewer. A scoring system exists: minimum necessary score to qualify as a reviewer

Source: European Commission (2013a), Pelletier et al. (2013), Finkbeiner (2013), ISO (2014a), Galatola and Pant (2014), UNEP (2015), Martínez-Blanco et al. (2015b)
[a]The European Commission is already reviewing this, as it received limited public support (Galatola and Pant 2014)

credible communication to consumers – in a substantial way or whether the proposal could have even an adverse effect on the policy targets it tries to achieve. According to Galatola and Pant (2014) during the 3-year pilot phase, there will be time to revisit the different elements of the method and the objectives of the pilot phase, and at the end necessary changes will be implemented in the OEF Guide.

Keeping in mind that both the ISO/TS 14072 and the UNEP Guidance build on product LCA standards and that a very similar methodological framework is provided, only few variations or adjustments on main requirements of ISO/TS 14072 were necessary on the text of the UNEP Guidance (see Table 8.1). In ISO/TS 14072, the unit of analysis used is the "reporting unit," while in the UNEP Guidance, that is broken down into two elements: the definition of the unit ("reporting organization") and the quantification of that unit ("reporting flow"). The OEF Guide draws the unit of analysis also according to two elements, which also respond to definition and quantification. Because consolidation methods affect both the definition of the reporting organization and the system boundary, the UNEP Guidance proposes to choose the consolidation method at an earlier stage rather than the other two methodologies. According to Martínez-Blanco et al. (2015b), "this is not in real conflict with ISO/TS 14072 but a matter of order on the introduction of the concepts."

1.5 Goals Served for Organizations

Some of the goals and opportunities that this approach may bring to organizations are listed in the following, based on the ones highlighted in ISO/TS 14072, OEF Guide and UNEP Guidance:

- Reduce pressure on the environment and avoid future negative effects on the organization.
- Gain insight about relationship, main actors, and impacts involved in internal operations and value chain.
- Identify environmental hotspots throughout the value chain for each of the environmental categories considered.
- Track environmental performance of the inventory and impacts of the organization over time.
- Get support to define which are the priority actions and targets at different levels.
- Make estimations of potential future scenarios due to different actions.
- Improve organizational procedures, for instance, in the gathering and managing of environmental data.
- Get the basis for voluntary or regulatory environmental communication with stakeholders and reporting.
- Show environmental awareness with marketing purposes.
- Foster suppliers in the value chain, consumers, and even competitors to adopt environmental friendly practices.

Motivations for LCA of organizations application may differ between large and small/medium organizations, as well as between organizations from developing and developed countries. In general, all of them would aim to get analytical results, though differences in the motivations behind could be identified. For instance, large organizations may have the objective to document their good practices, particularly when countries with poor environmental regulations are involved as suppliers, while one of those suppliers individually may decide to apply LCA of organizations to fulfill the requirements and standards of a large organization buying a big share of its products.

2 Main Methodological Issues of Organizational Life Cycle Assessment

In this section, the main features of the O-LCA are presented, organized as the four phases of an LCA: goal and scope, inventory, impact assessment, and interpretation (Sects. 2.2, 2.3, and 2.4). But first, in Sect. 2.1, main differences between organizational LCA and product LCA are identified and shortly explained. Additionally, Sect. 2.5 adds some recommendations for reporting and critical review. For the five sections, we focus on the framework defined by the UNEP Guidance, mostly aligned with ISO/TS 14072. The Guidance is not explicitly cited along the chapter in order to avoid repetition. Most of the recommendations and requirements stated here are similar for an OEF (Sect. 1.4.4).

This section does not aim to cover in detail those aspects of O-LCA that are common with product LCA, like the impact assessment, and much less to resolve gaps and unanswered questions for product assessment that are shared with the new organizational perspective; UNEP/SETAC (2012) and Finkbeiner et al. (2014) list some of these limitations. Thus, principles, requirements, or guidelines not specified either in the Guidance or in ISO/TS 14072 are, by default, equivalent to those for product LCA, and therefore ISO 14040 and ISO 14044 are the documents to check.

2.1 Main Differences with Product Life Cycle Assessment

As mentioned before, most of the principles and requirements of ISO 14040 and ISO 14044 for product LCA apply also for O-LCA with some minor terminology amendments, for instance, impact assessment, reporting, and review requirements. Others were partially adapted from ISO 14044, such as allocation procedures. See Finkbeiner and König (2013) for further discussion on the use of product LCA standards for the LCA of organizations. The major discrepancies between the two methodologies appear during scope definition and inventory. For further detail on the topic of this section, see Annex D in UNEP (2015) and Martínez-Blanco et al. (2015b). Section 2.1 mainly builds on these two publications.

The first difference between product and organizational LCA is the object under study. While product LCA is intended for the environmental evaluation of individual products, the latter aims to assess organizations. This influences the definition of the several elements of the goal and scope phase and of the life cycle inventory analysis. Figure 8.3 shows the three main elements of the scope definition of O-LCA, reporting organization, reporting flow and system boundaries, and the differences between the two approaches.

The need for a reference unit and consistent boundaries is common for both LCA of products and LCA of organizations, as they "firmly ground the analysis in concrete, physical relationships that connect upstream, organizational level, and

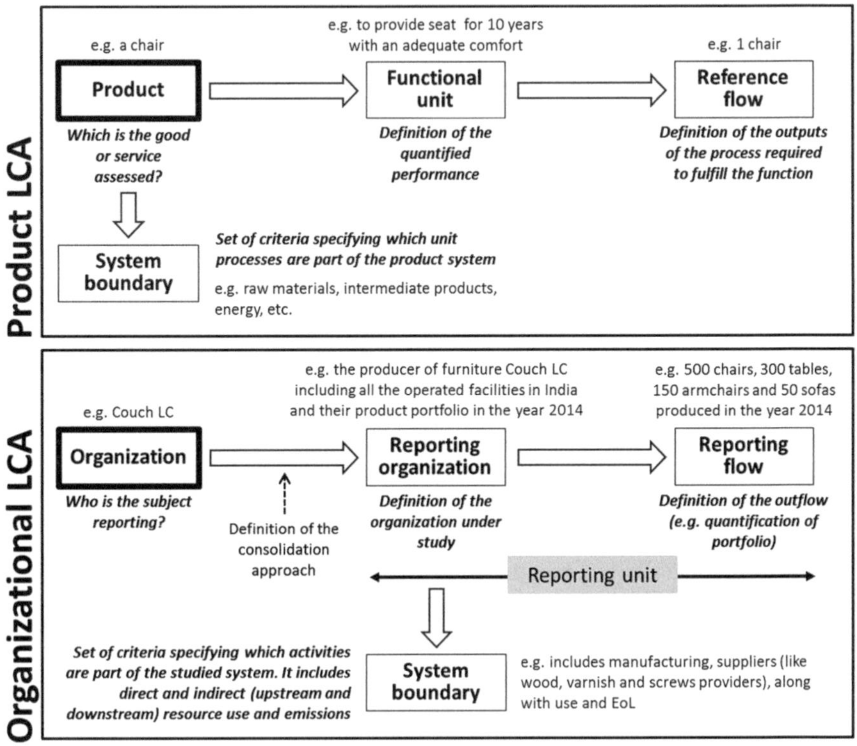

Fig. 8.3 Scope definition for product and organizational LCA

downstream processes in a coherent, internally consistent manner" (Pelletier et al. 2013). Scoping elements in product LCA aim to achieve comparability between different products. The definition of the scope in O-LCA is not intended to guarantee comparability (see Sect. 1.4.4); rather, it responds to reproducibility and aims to guarantee that environmental performance tracking over time is "based on the same time period, system boundaries and reporting unit" (ISO 2014a).

The functional unit and the reference flow (in product LCA) are defined in accordance to the main function/s of the product. In O-LCA, the reporting organization defines the organization per se (i.e., which parts of the organization are included), and the reporting flow ideally represents the quantification of its product portfolio (amounts, units, revenue, etc.). For product LCA the system boundary is derived from the type of product, whereas the definition of the reporting organization is the determining issue for stating system boundary in O-LCA. Furthermore, in O-LCA direct[1] and indirect[2] activities are differentiated. Main requirements and

[1] Activities from sites that are owned or controlled by the reporting organization (UNEP 2015).

[2] Activities that are a consequence of the operations of the reporting organization, but occur at sites owned or controlled by another organization (upstream or downstream) (UNEP 2015).

guidelines for system boundary in product LCA apply for O-LCA; however, the steps included are not only the raw materials, energy, intermediate products, etc., necessary for the production of one product (as in product LCA) but those organizations, processes, and activities involved in the production of the entire portfolio of the organization.

In any case, product LCA and O-LCA are complementary, due to the fact that they answer different questions. The two methodologies may either be implemented independently or be mutually complementary for the environmental management of the organization. One example of the former is when O-LCA is applied to identify environmental hotspots throughout the whole organization and value chain, and product LCA supplements the study with more detailed data for selected products or activities.

2.2 *Goal and Scope Definition*

The first step of an O-LCA is to describe the goal of the study. This should identify the reasons to conduct the study, the intended use, and the target audience. This strongly affects the following phases of O-LCA. A statement shall be also included that the results are not intended to be used in comparative assertions intended to be disclosed to the public (see Sect. 1.4.4). Examples of goals for an O-LCA are listed in Sect. 1.5.

The scope of O-LCA should be sufficiently well defined to ensure that the extent, granularity, type of data, and detail of the study can effectively fulfill the stated goals (ISO 2006b). Due to the iterative nature of LCA, the scope may have to be refined during the study. The elements to be described are organization to be studied; products, operations, facilities, and sites of the organization included in the reporting organization; the reference period considered; reporting flow; system boundary; allocation procedures; impact assessment methodology and types of impacts; interpretation to be used; data and data quality requirements; assumptions; value choices and optional elements; limitations; type of critical review, if any; and type and format of the report required for the study. Only reporting organization, reporting flow, and system boundary are defined here, as the recommendations and requirements for the other elements are equivalent to those of product LCA.

2.2.1 Reporting Organization

The reporting organization is "the organization under study to be used as a unit of analysis" (UNEP 2015). It should specify the organization under study, the consolidation method selected that defines the units included, and the reference period.

Subject of Study

The definition of organization by ISO/TS 14072 is a flexible term; it includes corporation but also institutions, firms, sole traders, etc., and it might be or not be a legal entity. In spite of full organizational assessment being recommended and encouraged, if properly justified, the application of the technical standard to segments or selected parts of an organization is foreseen (e.g., business divisions, brands, regions, or facilities). In any case, the subset selected should represent a clear unit of operation and shall be transparently justified and reported. For instance, several segments that Accor may have decided to assess are proposed in Fig. 8.4, apart from the real subject selected, the whole Accor group (see Sect. 4.1).

An example of an organization that may be interested on assessing only a segment is when this is understood as a pilot for a broad application in the future. Another alternative may be a subset of a bigger organization that has enough autonomy to pioneer the application of O-LCA. In some cases, specific divisions, brands, regions, or facilities may have a better knowledge and direct control over their operations and might obtain major cost savings and long-term sustainability of operations.

Consolidation Method

When a big and/or complex organization (i.e., including wholly owned operations, incorporated and non-incorporated joint ventures, subsidiaries, etc.) is assessed, it may not be straightforward to specify which facilities and activities are part of the subject of study. A systematic approach, referred to as consolidation methods, is proposed to specify which parts should be considered in the study (WRI and WBCSD 2004; ISO 2013a, 2014a), which include:

- Control approach: the organization includes units over which it has control. Control can be defined in either financial or operational terms. The organization accounts for 100 % of the units over which it has financial or operational control:
  - The organization has financial control over a unit if the former has the ability to direct the financial and operating policies of the latter with a view to gaining economic benefits from its activities.
  - The organization has operational control over a unit if the former or one of its subsidiaries has the full authority to introduce and implement its operating policies at the operation.
- Equity share approach: the organization includes units according to its share of equity interest, i.e., according to the organization's percentage ownership of each of the units.

Such consolidation methods reflect different perspectives. They should be chosen depending on the complexity of the organization and what is to be prioritized:

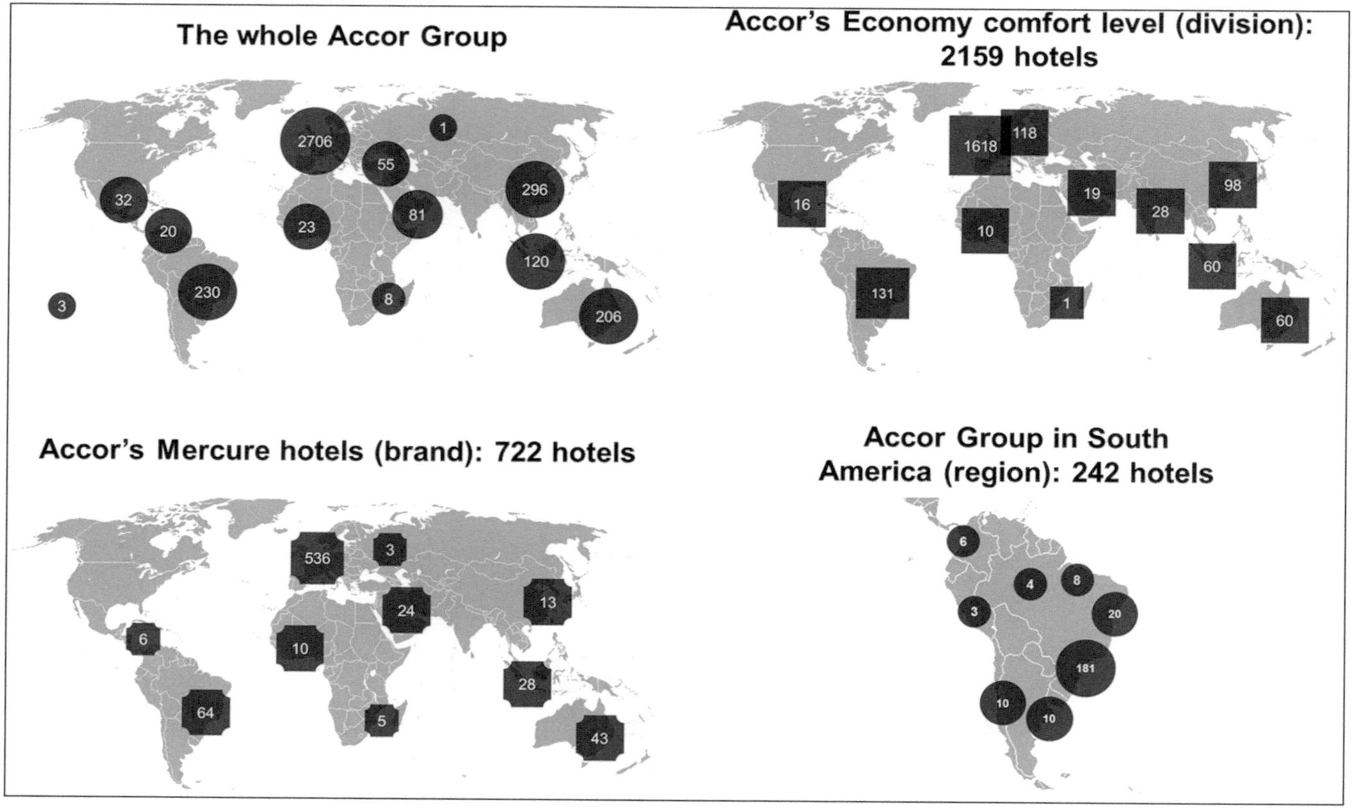

Fig. 8.4 Example of potential segments to be assessed in the case of Accor group (Source: own elaboration based on data from http://www.accorhotels.com/gb/booking/map-search.shtml). Note: In reality, the whole Accor group was assessed (see Sect. 4.1), which corresponds to the first option in the figure

risk or effective tracking and implementation of management policies (WRI and WBCSD 2004).

Reference Period

The reference period is the time period for which the organization is being studied, and it should be transparently stated. It is recommended to assess one operation cycle; 1 year is the preferred option, in accordance to financial and other reporting schemes.

2.2.2 Reporting Flow

The reporting flow is a measure of the outputs of the reporting organization during the reference period. Expressed in quantitative terms, it is aimed to provide a reference for the linkage between the different units in the value chain. It is recommended to match the reporting flow definition with existing records in the management control system of the reporting organization. The portfolio records of an organization are usually presented per unit of goods or services or in terms of weight or volume. In some cases, it may be useful to consider clusters or representative products, particularly when the organization provides a very wide portfolio. The organization may also choose to use nonphysical terms, such as economic revenue or number of employees, for example, when the complexity of the portfolio is dramatically high, or for providers of services or social functions that could find particularly challenging the identification and quantification of their portfolio.

Ideally, apart from the type of products and the amounts produced for each of them, the reporting flow should also include information about the quality and the durability of the products; in OEF Guide terms, it should answer the questions: What? How much? How well? How long? (European Commission 2013a). Incorporating quality and durability indicators within the definition of the portfolio can facilitate the interpretation of the results (for instance, more durable products may present higher energy use at the level of raw material acquisition). The level of detail on that will be subject to the resources available and the goal of the assessment.

2.2.3 System Boundary

The study shall define boundaries that properly define which social, financial, and physical relationships of the networks, where organizations are embedded, will be considered. The system boundary is defined as a "set of criteria specifying which activities are part of the studied system. It determines the direct and indirect resource use and emissions associated with the operations of the reporting organization" (UNEP 2015). Direct resource use and emissions are from sources owned or controlled by the reporting organization, while indirect take place throughout the value chain (upstream or downstream) linked to organization's activities. Requirements from ISO 14044 apply for O-LCA system boundary.

System boundary shall be documented and in accordance to the goal and scope of the study – reporting organization may particularly affect system boundary definition. It shall consider the complete life cycle to cover all inputs and outputs related to the reporting organization's activities and disclose and justify any exclusion. A complete cradle-to-grave assessment is recommended; it is mandatory to include upstream indirect activities and direct activities. In certain cases, downstream activities, like distribution, use phase, and end of life, may be excluded, particularly if the organization can argue that it has no direct influence on the use and end-of-life stage of its products (e.g., via product design or recycling campaigns). However, all the organizations are somehow able to influence upstream and downstream activities, so knowing whether the key drivers are downstream is always important. Downstream activities should be included if products use energy or generate emissions during use phase.

Environmental offsetting, i.e., discrete resource use or emission reductions used to compensate for resource use or emissions elsewhere, is not supported by the Guidance and shall not be aggregated with the organization's results (ISO 2013a). In any case, offsetting should be based on credible methods for each of the impact categories considered, which should be described in the study. Offsetting has been mainly used for climate change; within the context of a multi-impact approach, the organization should proof that offsetting makes sense for the considered impact categories. According to Curran et al. (2014), for instance, biodiversity offsetting policy leads to a net loss of biodiversity.

2.3 *Inventory*

The life cycle inventory is the phase of O-LCA that addresses data collection and modeling of the system. It is when inventory results are obtained based on the previous definition of the goal and scope and iteratively revised along with the other phases of O-LCA. The most laborious step in the inventory consists of effectively collecting data. The inventory should include the whole set of inputs and outputs from activities involved in the provision of the reporting flow and within the system boundary. All the inputs and outputs in the inventory should be expressed as elementary flows.

2.3.1 Direct and Indirect Activities

The activities considered within the system boundary can be divided in direct activities and in upstream and downstream indirect activities (see Sect. 2.2.3). A list of potential activities to consider is presented in Table 8.2, but additional ones

Table 8.2 List of most common direct and indirect activities

Indirect upstream activities	Direct activities	Indirect downstream activities
Business travel	In or from facilities, vehicles, and equipment owned or controlled by the reporting organization	Use of sold products
Capital equipment	Sourcing of energy	Franchises
Employee commuting	Physical or chemical processing	Leased assets
Leased assets	Waste disposal or processing	Processing and storage of sold products
Purchased electricity, fuel, and energy	Transportation and distribution	Transportation and distribution (of, e.g., sold product)
Purchased raw materials, goods, and services	Emissions and discharges from intentional or unintentional releases.	End-of-life treatment of sold products
Transportation and distribution (of, e.g., raw material, goods, fuel)	Consumption of natural resources	
Waste generated in operations		

Source: own elaboration based on WRI and WBCSD (2011), UNEP (2015)

may be defined. Also supporting activities – i.e., those activities of the organization that are not directly involved with the production of the products (like heating, cleaning, canteen services, commuting of employees, research and marketing activities, etc.) – should be taken into account.

2.3.2 Types of Data and Prioritization of Data Collection Efforts

Two general types of data can be used in the inventory quantification (European Commission 2013a):

- Specific data (also called site-specific or primary data) refer to directly measured or collected data representative of processes or activities at a specific facility or set of facilities.
- Generic data (also called secondary data) are not based on direct measurements or calculation for the respective specific process(es) or activity(ies), but rather sourced from a third-party life cycle inventory database or other sources.

Examples of specific data sources are bills and stock of consumables, emissions reported to authorities for legal purposes, and emission measurements. Examples of generic data sources include industry-average data from literature or scientific papers, life cycle inventory databases, or government statistics. Generic data can be either sector specific, i.e., particular of the sector being considered, or multi-sector.

It is desirable that all inputs and outputs of all the activities that are attributable to the product portfolio of the organization and included in the system boundary are addressed in the O-LCA study with specific and high-quality data. However, this may be neither feasible, due to time and resource constraint, nor necessary, as some inputs, outputs, or activities can be insignificant for the overall impacts. Organizations should focus on collecting data of sufficient quality to ensure that the inventory appropriately reflects the situation of the organization, supports its goals, and serves the decision-making needs (WRI and WBCSD 2013).

In general, specific data should be used for direct activities and for indirect activities identified as significant. Specific data are also recommended to model indirect activities, but higher use of assumptions, extrapolations, and generic data is expected for them.

ISO 14044 allows input and output prioritization by including a clause with the option to leave out of the system insignificant ones – defined by cutoff criteria. Several criteria can be used to apply cutoff: the first option is to prioritize according to environmental impacts, based on an initial estimation (screening) of the environmental impacts; when this is not possible, it is recommended to use a combination of other criteria, like mass or energy, spending or revenue, suppliers' closeness, risk, etc. (WRI and WBCSD 2011). Once the significant activities are identified, organizations should focus resources on the most significant ones, by using better data, or may exclude activities that are insignificant.

2.3.3 Data Collection Approaches

Two approaches can be used to collect the data of the inventory for O-LCA: bottom-up and top-down approaches. Additionally, a hybrid approach or intermediate approach may be pictured that is using both bottom-up and top-down data. Section 4.2 shows the tailored bottom-up approach used by Unilever; the top-down approach used by Accor is presented in Sect. 4.1. Although differences are expected between the outcomes of bottom-up and top-down approaches – particularly on the granularity of the results, the inclusion of supporting activities, and the comprehensiveness of inputs and outputs addressed – consistent outcomes should result from both approaches.

Bottom-up data collection approach entails adding the different LCAs of the products of the reporting organization, weighted by the amount of products that are produced during the reference period, together with the supporting activities (ISO 2014a). The organization may decide not to assess every individual product, but should guarantee representativeness and comprehensiveness. It can be done, for example, by defining clusters or families of products and identifying representative or proxy products within them. See a clustering example for Unilever in Fig. 8.12, based on Milà i Canals et al. (2010).

Top-down data collection approach considers the reporting organization as a whole and adds upstream (cradle-to-gate) models for all inputs of the organization and downstream (gate-to-grave) models for all outputs (ISO 2014a). Direct activities should be modeled with specific data, which could be done by direct measurement (using monitoring, mass balance, or stoichiometry) or calculation (using activity data and consumption/emission factors). For value chain activities, the organization may use generic data that model the entire value chain or seek specific data for the network of suppliers and other partners in the value chain.

2.3.4 Multifunctionality

There are situations when a process, activity, or unit delivers several outputs (i.e., products) and only one or some of them are included in the study. Two main situations in O-LCA may be identified as multifunctionality situations. First is when assessing impacts of suppliers with primary data and only some of the products in the supplier's portfolio are consumed by the reporting organization. See Finkbeiner and König (2013) for further definition of this situation. Second is when the reporting organization represents a part of a higher organization and shares facilities, activities, or processes with other organizations; see Sect. 2.2.1.

If resource use and emissions data are collected for the whole process, activity, or unit, the share of resource use and emissions attributable to the outputs considered in the study should be calculated. This should be done according to the hierarchy of solutions in Fig. 8.5. Organizations should avoid or minimize allocation and use it only when more accurate data are not available, as allocation adds uncertainty to the estimation of inputs and outputs. In general, system expansion should not be used at the organizational level, because of the concern of inconsistent or poorly representative substitution scenarios; hence it is not included in the hierarchy.

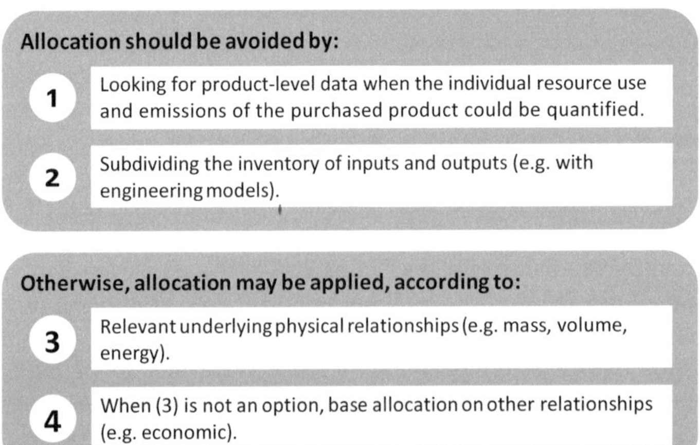

Fig. 8.5 Hierarchy to solve multifunctionality situations in organizational LCA (Source: based on ISO 2006a; WRI and WBCSD 2011)

2.4 Impact Assessment and Interpretation

The two last steps of O-LCA are presented here together, as the guidelines and requirements to be applied are basically the same as for the impact assessment and interpretation of product LCA, respectively. O-LCA should use one of the existing impact assessment methods (e.g., ReCiPe, CML 2002, EDIP, and LIME). Similarly, the challenges and limitations are also transferred to the new methodology – see, for example, Bare (2009). For further detail, please see, for instance, ISO (2006a, b) and European Commission (2010a, b).

Inventory-level indicators (e.g., waste produced, water consumed, or energy used) could be also considered in the results as these are important metrics for organizations, but it should be clearly acknowledged that they are not addressing environmental impacts. The organization may be also interested on single-score impact category indicators, as they make easier the interpretation of the results for non-LCA experts. However, they are based on value choices, add further uncertainty, and hide trade-offs between impact categories; hence both aggregated and disaggregated results should be provided, along with detail about the aggregation methodology behind.

The interpretation phase should indicate the consistency of the results according to all the aspects defined during the goal definition and scope phase. It is necessary to outline conclusions, explain limitations of the results, and provide recommendations. Furthermore, it should involve the iterative process of reviewing the scope of O-LCA, particularly the assumptions taken and the quality and sources of the data collected.

2.5 Reporting and Assurance

Environmental performance, performance tracking, and organization environmental strategy are elements largely reported by organizations. Organizational LCA may be a great source of information for supporting and communicating them to third parties (like policy makers, consumers, shareholders, etc.). O-LCA provides key environmental information on the performance of an organization that may be used for joining sustainability reporting and Corporate Social Responsibility (CSR) frameworks. If the O-LCA is communicated to a third party (i.e., interested party other than the commissioner of the study), a third-party report shall be prepared.

As required in ISO (2006a), the results and conclusions of the O-LCA study, along with the methods, assumptions, and limitations, shall be fully, accurately, and objectively reported and in accordance with the goals of the study. Particularly, it shall provide a clear definition of the reporting organization that is being assessed, the system boundary, and the data collection approach. Results can be presented for different levels of the organization depending on the granularity of the study results (for the whole organization or, for instance, business divisions, brands, regions, facilities, or activities).

The more accurate and coherent with the goal and scope of the study the data and assumptions are, the more valuable the outcomes of O-LCA. That is why it is useful to establish an assurance procedure. Accurate results make more likely that the organization can manage its environmental impacts effectively.

When O-LCA is communicated to a third party, a critical review shall be performed. It provides confidence to the stakeholders about the reported information and associated statements. Assessing the accuracy and completeness of the reported results, as well as the compliance with O-LCA principles, may be also voluntarily performed to improve the robustness and credibility of the results. The documents to follow to do so are ISO (2006a, 2014b), complemented by WRI and WBCSD (2004, 2011).

3 Practical Implementation of Life Cycle Assessment of Organizations

This section aims to provide further insight about the practical implementation of LCA of organizations and the use of its outcomes. The first two sections focus on strategies proposed to ease the implementation of the LCA of organizations: Sect. 3.1 explores how previous environmental assessments of an organization may streamline the LCA of organizations application, while Sect. 3.2 presents the EU proposal to define sector rules for LCA of organizations. Finally, how the organizations may use and integrate the outcomes from LCA of organizations in their own management control and decision systems is addressed in Sect. 3.3.

3.1 Experience with Environmental Tools: Pathways Proposal

Organizations that have applied other environmental tools and have available data may use that to facilitate the implementation of the new approach. Three pathways are described in UNEP (2015) that could steer an organization to conduct O-LCA; this section is a summary of the approaches proposed. The pathways correspond to the three main dimensions of O-LCA (Fig. 8.6). The organization may benefit from (1) existing on-site environmental data, (2) existing LCAs of products, and/or (3) existing assessments of the organization and its value chain for one sole aspect. Other pathways may be derived by the combination of these three or others.

Apart from the three pathways, organizations that have voluntarily embraced sustainability reporting schemes – like the Global Reporting Initiative (GRI), the Carbon Disclosure Project (CDP), and the United Nations Global Compact

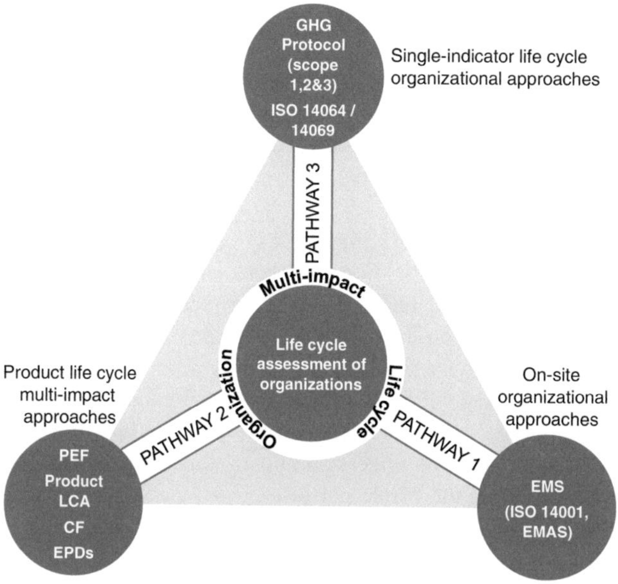

Fig. 8.6 Definition of the implementation pathways according to previous experience with environmental tools

principles – may also find a preliminary basis on the acquired knowledge. Information and experience obtained with sustainability reporting schemes may help for first description of organization's units, for identification of some important hotspots, for the collection of preliminary data of one or few indicators mainly at the inventory and gate-to-gate level, and for a first picture of the stakeholders inside and outside the organization's walls.

3.1.1 Pathway 1: On-Site Corporate Approaches

On-site organizational approaches are those that cover most of the processes that take place in the organization's sites. In LCA terms, it corresponds to a gate-to-gate scope. The most common and comprehensive framework for this pathway are corporate ecobalances and environmental audits according to EMS (ISO 2004) or EMAS (European Commission 2009) umbrella. These are typically used to determine the environmental aspects of an organization and as the baseline for decision-making.

An organization that has previously applied an on-site organizational approach may benefit by using the outcomes as ground for O-LCA. It is expected to mainly provide data for direct activities, but can also guide the identification of the targeted suppliers. O-LCA can complement and refresh the EMS of organizations mainly by broadening the horizon from on-site to value chain improvements.

3.1.2 Pathway 2: Multi-impact Life Cycle Product Approaches

The environmental schemes that may provide information about the environmental performance of products in the portfolio are product LCA and Environmental Product Declarations, EPDs (ISO 2006a, c). Also product single-indicator life cycle-based methodologies like carbon or water footprinting may contribute here (ISO 2013b).

When LCAs (or any of the other approaches) exist for most of the products in the portfolio of the reporting organization or at least enough representative individual LCAs are available, O-LCA may consist of the addition of the different LCAs weighted by the amount of products that are produced during the reference period, together with the supporting activities. Even if only a small share of the products in the portfolio were assessed with LCA, the outcomes obtained may roughly identify some important hotspots in the value chain that should be further assessed. O-LCA brings a more comprehensive understanding of the organization environmental performance by including the whole portfolio.

3.1.3 Pathway 3: Single-Indicator Life Cycle Organizational Approaches

This pathway refers to organizations that have assessed the environmental performance of the organization and its value chain for only one impact category indicator, like the well-known and largely used GHG Protocol Scope 3 Standard (WRI and WBCSD 2004, 2011) or the closely related ISO 14064 and ISO 14069 (ISO 2006d, 2013a).

The overall analytical framework and the data collection procedures and tools developed for the single-indicator assessment may be very useful to start the environmental multi-impact approach, particularly to define the scope of the study, for instance, a preliminary definition of the consolidation method and system boundary exists. Furthermore, previously created connections with different levels of management in the organization and with suppliers will ease the collection of the complete inventory. The environmental multi-impact approach of O-LCA will identify impacts beyond the specific single indicator, thus avoiding unintended trade-offs.

3.2 Specific Methodology by Sectors: OEFSRs Proposal

The Organisation Environmental Footprint Sector Rules (OEFSRs) are sector-specific guidelines derived from the OEF Guide. Currently, two OEFSRs are under development for the sectors retail and copper production within the context of the OEF testing phase. This section is based on European Commission (2013a, d). "The main aim of developing OEFSRs is to create consistent rules for the calculation of the environmental performance of organizations in a given sector. The information calculated on the basis of existing OEFSRs could then be used for communication purposes, enable year-on-year comparisons of the performance of a specific organization and, where appropriate, to enable comparisons and comparative assertions of similar organizations in the same sector (with a similar Product Portfolio)" (European Commission 2013d). When the OEF studies are not intended for public comparative assertions, they may be carried out without using OEFSRs.

OEFSRs are an extension of the OEF Guide (Sect. 1.4.3) and shall be developed according to the OEF Guide. An OEFSR shall further specify requirements from the OEF Guide and add new ones where necessary in order to focus the study on those aspects and parameters that are more relevant for the environmental performance of the sector. Their aim is to increase the reproducibility, consistency, and relevance of OEF studies and also possibly to achieve reduction on the time, efforts, and costs involved in completing an OEF study. Galatola and Pant (2014) stated that the use of OEF Guide along with OEFSRs may reduce between 30 and 50 % of the costs compared to the current situation.

The steps for the preparation of an OEFSR are described in the following, according to the European Commission (2013d). This is the process that is being followed to develop the first two OEFSRs, and it is foreseen as the main guidance for the development of future OEFSRs, after some adjustments derived from the pilot phase experience. The first step is to define the sector to which the rules refer, which shall be done based on NACE codes (Nomenclature générale des Activités Economiques dans les Communautés Européennes). The defined sector should be broad enough to include relevant organizations in the sector but guarantee that the sector is specific enough to allow the definition of meaningful screening and requirements. The second step is to choose a representative organization of the sector in the EU market, which may be real or be a proxy. Third, according to the model defined, a screening is applied to preliminarily identify most relevant steps of the life cycle, processes, impact categories, and definition of the benchmark for the sector. Based on this and stakeholder consultation, a draft of the OEFSR is to be prepared including key information like choice and description of system boundaries, relevant and irrelevant environmental aspects, how to model end of life, data quality requirements particular for activities or processes, rules for solving multifunctionality, etc. The draft of the OEFSR is then applied to supporting studies to test it. Finally, when comparability is considered meaningful within the sector, the OEFSR should include the particular benchmark, which corresponds to the environmental performance of the representative organization, along with five classes of environmental performance, being the benchmark the one in the middle. In case it is concluded that it is not meaningful, the OEFSR shall clearly state that the OEFSR cannot be used as a basis for comparative assertions. Major producers and stakeholders should be involved on the preparation of the OEFSRs, along with experts, and an independent third-party panel shall review each OEFSR.

3.3 Environmental Targets and Performance Tracking

LCA of organizations provides comprehensive information – along the value chain and for multiple impact categories – at the level at which decisions are taken and beyond the organization's walls. Through its results, the organization understands which are the risks and impact reduction opportunities and has strong arguments to elucidate which are the most effective actions to reduce the organization's environmental impacts. Once the LCA of organizations model is built, the organization may be interested in running different scenarios to assess the effect of proposed actions or measures. Furthermore, LCAs of organization results identify the hotspots for which further analysis may be necessary. An organization may decide to apply LCA of organizations in a regular basis and track environmental performance over time due to multiple reasons. For example, it may aim to set targets within organization environmental strategy. This section gives further guidance on the latter two applications, setting targets and tracking the environmental performance of the organization.

Common reasons for setting and tracking environmental targets include minimizing future risks and stimulating innovation, preparing for future regulations, and reporting, for instance (WRI and WBCSD 2004). A target should be defined as a quantified reduction to be achieved in a certain impact category in a target year on the basis of a reference year. It is measured either in absolute (e.g., kilograms of CO_2) or relative (e.g., percentage) terms and can be presented as an intensity or efficiency measure (e.g., per unit of revenue). Setting global targets for the whole organization and value chain is recommended. Moreover, setting specific targets for certain activities, products, business divisions, brands, regions, or facilities due to specific circumstances provides additional metrics. A combination of short- and long-term targets is recommended, to support the long-run strategy of the organization and at the same time measure the continuous progress (UNEP 2015).

Performance tracking of an organization is defined as the comparison of the performance of the same organization's products and operations over time, based on the same time period, system boundary, and reporting organization (ISO 2014a). A given tolerance is considered to state that two reporting organizations are the same, which should be according to the goal and scope of the study and should be quantified and transparently reported. When the organization undergoes structural changes (such as acquisitions, mergers, outsourcing, and divestments) or the methodological framework suffers relevant adjustments (like significant variations in system boundary, calculation methods, and improvements in data accuracy, or discovery of significant errors), the organization should recalculate historic impact performance or establish a new baseline period. Also changes in the portfolio of the organization, both in the amount and type of products, should be considered during the interpretation of the results and if very significant may trigger the recalculation of the baseline period. See WRI and WBCSD (2004, 2011) for more detailed guidance on target definition and performance tracking.

4 Case Studies

In order to support the explanations from the previous sections, two examples of companies that have applied organizational approaches for the environmental multi-impact assessment of organizations and their value chain are presented here. Both companies developed their own methodology to perform the assessment. The two examples were also included in UNEP (2015) as first mover stories for this kind of approaches. Here we have prepared, with the support of the companies, a complete and extended explanation of their experience.

The two companies are Accor and Unilever (Table 8.3). Accor is a French international hotel group, which operates in 92 countries with more than 3,500 hotels and a broad portfolio of hotel brands – Accor provides an extensive offer from luxury to budget. The company activities cover accommodation, restoration, and sale of food and beverages. It is a pure player in hotels and boasts a unique and universal business model as an owner, operator, and franchisor of hotels on all five continents. Unilever is an Anglo-Dutch multinational fast-moving consumer goods company with a wide-ranging portfolio in foods, household, and personal care products. Unilever owns more than 400 brands, including world-leading brands like Knorr, Ben and Jerry's, and Dove, alongside trusted local names such as Blue Band, Pureit, and Suave. Unilever's products are sold in over 190 countries and, on any given day, 2 billion consumers worldwide use them.

The examples are presented in Sects. 4.1 and 4.2. The text was drafted mainly using previous documents of the organizations; other complementary sources are cited on place. Accor section is based on an internal report (Accor 2011a), the public version of the report (Accor 2011b), and the company webpage (Accor 2014). Unilever section is based on several related publications (Unger et al. 2011; Unger and King 2013; Unilever 2013) and Unilever's webpage (Unilever 2014a).

Table 8.3 Characteristics of the two case studies "Accor" and "Unilever"

	Accor	Unilever
Sector	Services. Hotels and resorts	Consumer goods. Food, beverage, cleaning agents, and personal care
Headquarters	Europe (France)	Europe (UK)
International presence	Operates in 92 countries	Sells products in over 190 countries. 57 % of sales in emerging markets (like Brazil or India)
Employees	160,000 (Accor 2014)	174,000 (Unilever 2014a)
Brands	14 brands (like Mercure, Novotel, Ibis, and Sofitel)	400 brands (like Lipton, Knorr, Dove, Axe, Hellmann's, Omo, and Ben & Jerry's)

4.1 Accor

Accor's commitment to sustainable development dates back many years, with practical initiatives such as the creation of an environment department 20 years ago.[3] Promptly they kicked off a program called Earth Guest, working for people and for the environment, to create value for everyone – its customers, employees, and partners – in its hotels and in 90 different countries. Many other solutions have been adopted aimed at contributing to the development of local communities, optimizing water consumption and energy use, and reducing the hotels' environmental footprint.

The hotel group Accor performed its environmental footprint[4] in 2011 within the context of a CSR (Corporate Social Responsibility) strategic assessment, given Accor's desire to have a global view of its relevant environmental impacts. The study involved nearly a year of groundwork with the firm PwC. Accor's goal to quantify metrics on its global environmental impacts led to the creation of a specific methodology to provide accurate information about the real environmental issues of Accor's activity beyond CO_2 and for all Accor hotels. An update of Accor's footprint is currently under development.

Accor decided to base the study on the LCA method, which companies generally use to assess a product's full environmental impact. The environmental footprint is defined in this study as a technique to assess a company's environmental impacts, in terms of energy, water consumption and contamination, and waste. It is based on the consolidation of inputs and outputs at the level of the company, including upstream and downstream activities, for all the services that this company is providing.

The approach has been, therefore, largely inspired by the LCA principles. Accordingly, the three first subsections describe the four phases of LCA: goal and scope (Sect. 4.1.1), inventory (Sect. 4.1.2), impact assessment and interpretation (Sect. 4.1.3). Section 4.1.4 details how the company applied the outcomes of the study on its business and sustainable strategy.

4.1.1 Goal and Scope

The items required by product LCA to describe the goal and scope of the study were adapted and detailed to the corporate context as described in the following sections.

[3] The Sustainable Development Department of Accor supported the drafting of this section, particularly Arnaud Hermann and Pascal Fillon, the VP and Environment and ISO14001 manager, respectively. The text is based on (Accor 2011a, b, 2014).

[4] In this section we use Accor's terminology (see, for instance, Accor 2001b). It does not refer to the Organizational Environmental Footprint, OEF, presented in Sect. 1.4.3.

Goals

Accor had the following objectives:

- Obtain a comprehensive view of its environmental impacts.
- Identify the main environmental impacts over the entire life cycle and the key components of the supply chain impacting the group's environmental performance.
- Identify opportunities to improve the environmental performance and prioritize levers for action.
- Raise awareness among employees of the group's main environmental impacts.
- Communicate on Accor environmental impacts both internally and externally.

This represented the first ever environmental footprint study in the hospitality industry. Accor built a wealth of expertise and also aimed to share what it has learned with its peers, hoping that this information would spur new and more sustainable practices in the hospitality industry.

Unit of Analysis

The reporting organization can be defined as the worldwide international group, over 1 year, considering the operational control approach. Owned, operated, and franchised hotels are included in the study. In the case of Accor, operational control is considered to include franchised hotels because they are required to follow the brand's business model. The reporting flow is defined as the yearly number of overnight stays, breakfasts served, and meals served as representative of Accor's basic services offering. Accor has more than 450,000 rooms and estimates suggest it serves 56 million breakfasts a year.

System Boundary

Three main life cycle steps were considered on the life cycle of a hotel, namely, construction, use phase, and end of life. The use phase was the most significant one and included accommodation service, hotel restoration services, and hotel management. To ensure the comprehensiveness of the assessment, 100 % of Accor hotels were included in the scope of the study. The study covers 90 countries, split between three main areas: EMEA (Europe, Middle East, and Africa), Americas, and Asia-Pacific. The activities included in the system boundary were divided into 11 activities: water consumption and release, energy use on-site, hotel air conditioning, waste management, outside laundries, food services, construction and renovation, room furniture, housekeeping products, offices equipment and supplies, and employee travel. They are further described in Sect. 4.1.2.

For each activity, a life cycle perspective was considered. In general, every operation and by-product that contributes significantly to the group's impact was assessed. Some exclusion had to be made when there was not enough information or the contribution was negligible. Customer travel outside hotels was not considered because there were no reliable data and it was complex to derive a model from assumptions. Poorly documented operations such as organic and chemical compound processing in wastewater and meals for employees were also left out. Headquarters' energy use and water consumption could not be addressed. Accor did not include wastewater treatment in the scope, due to the lack of complete information about wastewater treatment in the 90 countries. Some environmental indicators were not assessed for some activities either because it was not feasible to assess the impacts of the activity to the indicator or because the activity did not represent a major contribution to the indicator.

Indicator Selection

Environmental indicators were selected according to their relevance to the accommodation services sector, to Accor's environmental program priorities, to their understandability to stakeholders, and to the availability of reliable assessment methods. Three inventory-level indicators and two impact categories were assessed:

- Total energy use (primary MWh). Quantity of energy resources extracted from the environment (petroleum oil, natural gas, uranium, wood, biomass, etc.), including both used energy for the process and feedstock energy.
- Water consumption (m^3).[5] All the resources taken from the environment apart from well water and rain water (not registered in hotel meters).
- Waste production (tonnes). Quantity of total waste generated along the total life cycle.
- Climate change (tonnes of CO_2 equivalent). Based on the method developed by the IPCC in 2008.
- Water eutrophication (tonnes of PO_4^{3-} equivalent). Based on the method developed by the CML (Guinée 2001).

Energy use, water consumption, and waste production are key indicators for Accor's environmental management and are already followed for on-site impacts through the environmental reporting. Climate change is a relevant impact for Accor as it is foreseen that action plans on energy use are likely to significantly reduce Accor carbon footprint. Water consumption monitoring and the quality of the water discharged into environment are also major issues. The group also has a major concern about the impact it could have on toxicity, eco-toxicity, and biodiversity; however these impacts could not be evaluated due to a lack of available indicators.

[5] Water consumption label was used by Accor and it is also used here, although it does not refer to the common definition of water consumption, including evaporative effect (see Berger and Finkbeiner 2010). The rationale behind this indicator refers to water use.

Reliability

Although rigorous, this study inevitably ran into methodological limitations. An overview of the overall reliability and accuracy of the "data sources used"; the "extrapolations, allocations, and main hypotheses"; and the "environmental factors" was provided. The level of reliability of each item was rated low, medium, or high, according to certain criteria (see some criteria in Table 8.4). Accordingly, an evaluation of the reliability of the calculated environmental impacts for each activity was provided using the same scale.

Critical Review

Accor asked a panel of experts to run this study and its findings through a critical review. Two French LCA specialists and an international hospitality industry expert spent 2 months analyzing the study's findings to ensure accuracy and transparency before result publication. Their input allowed Accor to fine-tune and expand a few issues and helped Accor to identify the environmental stakes in its operations even more accurately and reliably. A critical review report was established following the guidelines and contents of product LCA critical review reports and it is publicly available.

4.1.2 Inventory

The quantification of the inventory with a top-down approach (see Sect. 2.3.3) and the estimation of environmental impacts were performed for each of the 11 activity categories separately (Fig. 8.7). Within every category, Accor took into account key activity data referring to all the flows involved in Accor's operation. Depending on the available sources of information, activity data collection was handled with

Table 8.4 Examples of criteria proposed by Accor to be used to judge the reliability of the evaluations

Level of reliability	Data source	Extrapolations, allocations, and main hypotheses	Environmental factors
Good	Good data coverage, based on reporting system	Direct use of data and environmental factors	Accurate factors
Medium	Average data coverage	Estimations and extrapolations/ allocations	Estimated factors
Low	Low data coverage, low geographical coverage, data from one category of hotels	Large estimations and extrapolations/ allocations	Nonspecific factors

Source: Accor (2011a)

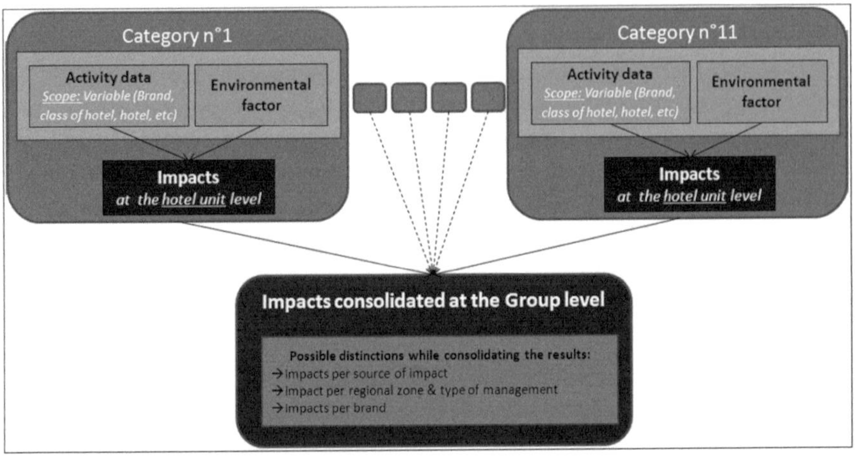

Fig. 8.7 Methodology used to estimate impacts at the Accor group level (Source: Accor 2011a)

different scopes: group, regional zone, class of hotel, brand, or hotel. Furthermore, Accor collected environmental factors (e.g., greenhouse gas emissions due to the consumption of fuel or m^3 of water per MJ of energy), which describe the intensity of environmental impacts due to activity data. The concept of hotel unit was used for the aggregation and assumptions.

Hotel Unit

Because the group includes around 3,500 hotels of different budget segments and situated in more than 90 countries, it was neither possible nor necessary for the goals of the study to provide specific data for each hotel. Thus the inventory and impact assessment were calculated at the hotel unit level. It is an artificial concept that represents the aggregation of all the hotels which have a common brand, management type, and country. The quantity of hotels considered in one hotel unit can vary from one to hundreds. A total of 359 hotel units were defined in the study; a small sample is presented in Fig. 8.8.

Activities Description

Description of the 11 activities included in the study:

- Water consumption and release. Public network water consumed by clients (baths, taps, and toilets) and for the hotel management (food preparation and laundry cleaning).
- Energy use on-site. It includes electricity and purchased steam, including use for heating purposes, and direct use of fuel from fossil origin (natural gas LNG, propane, butane, etc.).

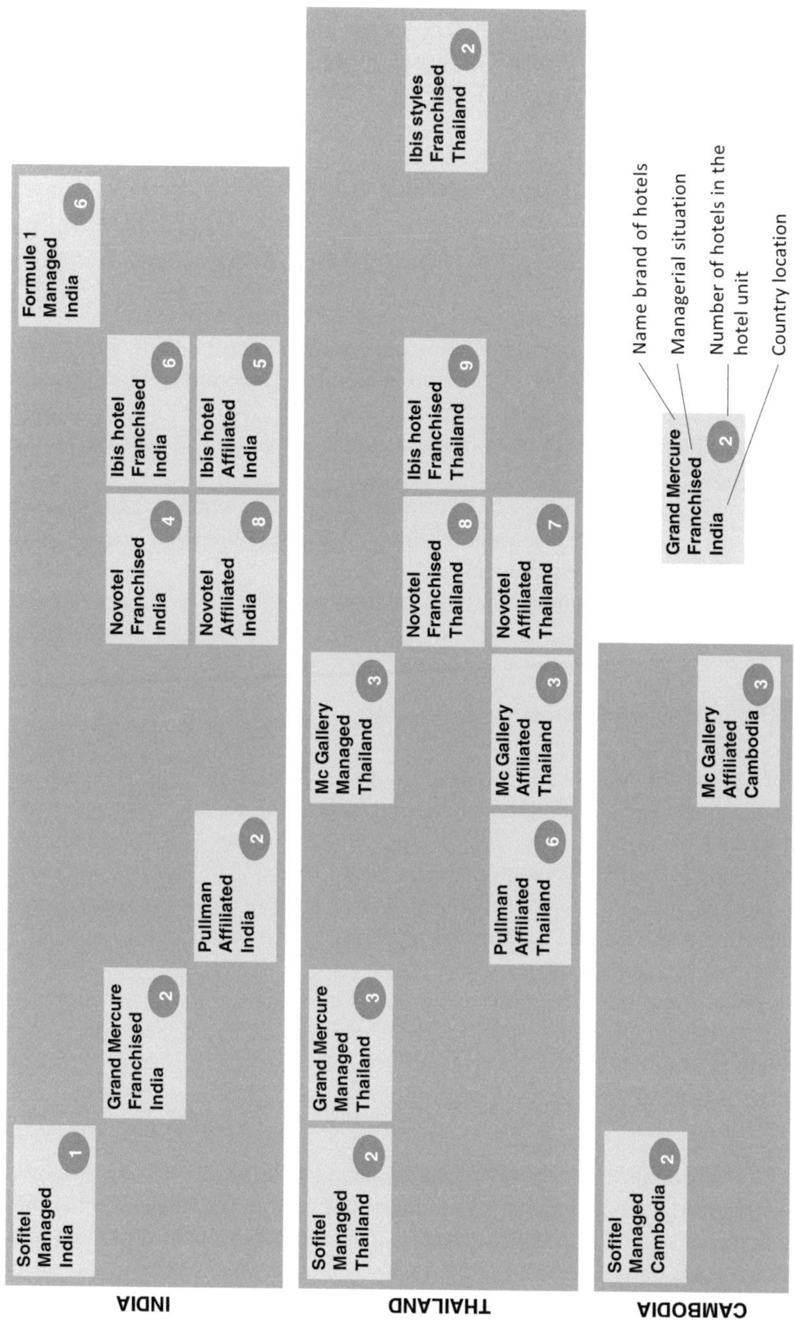

Fig. 8.8 Imaginary example of hotel unit clusters for Accor in India, Thailand, and Cambodia. Each box represents a hotel unit

- Hotel air conditioning. It includes leakages of refrigerants during the use of the air conditioning and the cooling systems installed in the rooms and general areas. Energy consumption is included at "energy use on-site."
- Waste management. Production of waste in the hotels (kitchens, bedrooms, offices, reception, corridors, and parking lots) and the treatment of waste produced.
- Outside laundries. It includes laundry detergent production (internal and external laundry) and external laundry cleaning activity (other impacts of internal laundry are included with water consumption and release and energy use on-site).
- Food services. Catering activity (for both breakfasts and meals) for clients. Upstream impacts for most food and beverage have been included.
- Construction and renovation. Hotel building structures and building materials.
- Room furniture. It includes TV equipment, guest amenities, linen, plastic glasses, and toilet paper.
- Housekeeping products. Production of cleaning chemicals consumed in the hotels.
- Office equipment and supplies. It includes paper for external and internal printing (for corporate communication), telecommunication equipments (phones), and computer equipments.
- Employee travel. Commute and professional travel.

Data Collection

Two types of activity data were collected: global and environmental. The former includes, for instance, the number of hotels, number of rooms, total area, and number of meals and breakfasts. They were useful for extrapolations and allocations and mostly obtained through corporate departments and Accor's reporting system. The specific data sources for environmental activity data were the procurement department, Accor's environmental reporting system, hotel census, and specific suppliers' data collection.

The activity was either from primary or secondary sources. Primary data refer to direct measurements about the specific organization's life cycle. Secondary data refer to external sources that are not specific to the organization, but rather represent an average or general measurement of similar activities. The main sources of data used were:

- Hotel Census. Source: Accor Corporate Finance Department.
- OPEN reporting. Annual global environmental reporting for managed and affiliated hotels of several activity data. It was an invaluable source of information in the completion of the study.
- Inventory of Accor's external purchases based on Country Mappings 2009 includes purchases of energy, water, food, equipment, waste, manpower, etc. Source: Accor Procurement department.

- Carbon footprint studies.
- Interviews with Accor staff and suppliers.

Apart from the activity data, environmental factors (describing the intensity of environmental impacts due to activity data) were collected from several databases, studies, and literature not specific to Accor. They provide the link that converted the quantities (activity data) into the resulting environmental impacts (for instance, the kg of CO_2 eq. emitted per kWh of energy used), depending on the indicators selected.

Assumptions

As a conclusion we can highlight that one of the main difficulties during the study was to obtain activity data at the hotel unit level. Indeed, the level of data availability was not always the same within Accor's network. Especially, there was less accurate data available for franchised hotels. As a consequence extrapolations and allocations had to be made on the basis of certain level of reference (on the basis of the number of rooms, the number of hotels, the area covered, the hotel brand, or the regional zones). Some of the assumptions were:

- The OPEN reporting only covers managed and affiliated hotels. Data from the franchised hotel units were extrapolated to cover the entire Accor scope. The extrapolation used data from OPEN reporting for hotels in the same country and/or brand.
- There were no consolidated hotel data available (for all the hotel units) concerning the occupancy rates and the attendance indexes. When data were not available, the average values for the company were used.
- It was considered that the Inventory of Accor's external purchases accounted for 90 % of the real purchases. The non-referenced 10 % was extrapolated.

4.1.3 Results and Interpretation

The impacts were consolidated at the group level in order to assess Accor's impacts (Figs. 8.9 and 8.10). Additionally, obtaining impacts per hotel unit offered Accor the possibility of different levels of assessment, for instance, the impacts of a brand, of Accor's activities in a specific country, or to compare the impacts of the hotels according to their management type. According to the number of hotels represented in each hotel unit, the impacts at the whole company level were consolidated. The results are discussed in the following sections broken down by indicators.

The environmental results were complemented with an indicator of reliability (low, medium, or high), in order to reflect the robustness of the data used, methods, hypothesis, etc. (see Figs. 8.1 and 8.9). For instance, data sources for the calculation of energy use on-site were considered highly reliable, while waste management was noted to the least reliable.

	ENERGY CONSUMPTION (primary MWh)	CLIMATE CHANGE (tonnes of CO_2 equivalent)	WATER CONSUMPTION (m^3)	WATER EUTROPHICATION (tonnes of PO_4^{3-} equivalent)	WASTE PRODUCTION (tonnes)
Total	18,200,000	3,660,000	544,000,000	3,180	1,250,000
Water consumption and release			62,000,000	120	
Energy consumption on-site	13,800,000	2,420,000	6,960,000		332,000
Hotel air-conditioning		73,900			
Waste management	16,200	75,800	38,900	4	65,600
Outside laundries	1,200,000	48,000	6,750,000	68	6,340
Food services	1,120,000	495,000	467,000,000	2,990	
Construction and renovation	810,000	165,000	806,000		848,000
Room furniture	353,000	75,100	371,000	5	6,020
Housekeeping products	5,660	678	7,510		118
Office equipment and supplies	48,400	11,900	35,500	6	849
Employee travel	892,000	303,000			

VERY RELIABLE: the data gathered, conversion factors, assumptions and extrapolation/allocation rules are considered reliable.
FAIRLY RELIABLE: several aspects of the data gathering process, conversion factors, assumptions and extrapolation/allocation rules are considered reliable.
UNRELIABLE: the data gathered, conversion factors, assumptions and extrapolation/allocation rules are by and large considered unreliable.

Fig. 8.9 Accor's footprinting results including reliability (Source: Accor 2011b)

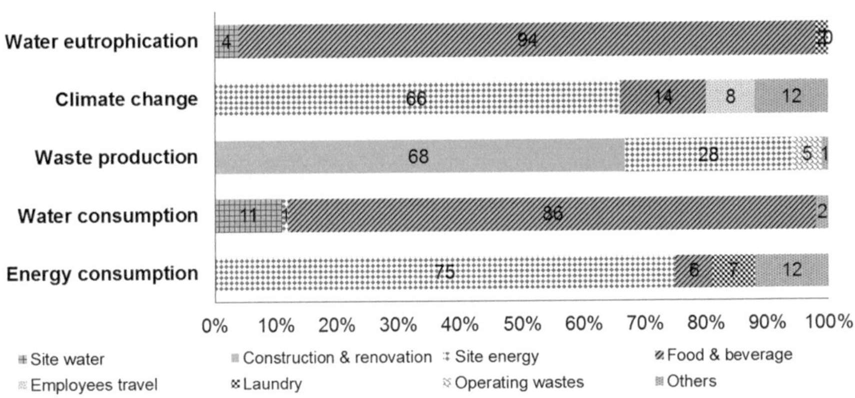

Fig. 8.10 Contribution of the main activities of the hospitality business (Source: data from Accor 2011b)

Water Consumption

Accor consumed 544 million cubic meters of water in 2011 (Fig. 8.9). A full 86 % of that water came from irrigation systems feeding crops and livestock feeding products, especially beef (Fig. 8.10). A closer look at those figures reveals that farm-produce "water equivalents" vary a lot according to how long they take to grow and how much water they need to do so. Fruit and vegetables by and large need less water than livestock. Direct water consumption in hotels (showers, kitchens, laundries, swimming pools) added up to some 60 million cubic meters of water a year, an 11 % of the total consumption.

Eutrophication

Based on the available data, estimates suggest that the eutrophication generated by Accor around the world was approximately 3,180 tonnes of PO_4^{3-} (see Fig. 8.9). As shown in Fig. 8.10, 94 % of that came from fertilizer used to meet Accor's food-service requirements. Wastewater that hotels release into deficient sewerage systems may add to this impact but this impact has not yet been measured.

Energy Use

Most of the roughly 18 billion kWh of primary energy that Accor used in 2011 (75 % of the total) went to direct use in hotels (Figs. 8.9 and 8.10). Laundries working for the group accounted for 7 % of that energy use. The other sources inter alia included electricity consumption in the laundries that work for group hotels, necessary consumption in farming operations to produce food, and group employee fuel consumption to travel. As shown in Fig. 8.11, luxury and upscale hotels were the ones with higher use of energy per available room and per day.

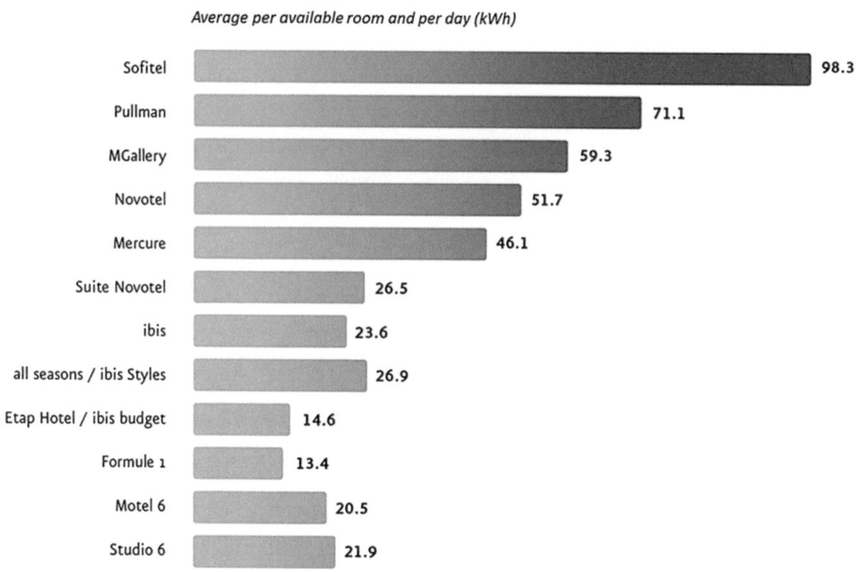

Fig. 8.11 Hotel on-site energy use in 2010 (Source: Accor 2011b)

Climate Change

In 2011, CO_2 emissions added up to nearly 3.7 million tonnes of CO_2 equivalents. About two-thirds of that came from energy use (electricity mainly). Food services ranks second with 14 % of total emissions. These emissions are mainly due to the transport and distribution circuits and from meat and dairy cow rumination. Lastly, employee travel (to work, for work, etc.) accounted for 8 % of the group's emissions. Estimates suggest that employees fly about 120 million kilometers in total every year. Measurements show that liquid-refrigerant leaks are rare, and, at the end of the day, they only contributed a minor roughly 2 % of the group's carbon footprint (see Figs. 8.9 and 8.10).

Waste

Accor produced 1.25 million tonnes of waste. About 70 % of that waste came from hotel building and refurbishing work. Most of it was "inert" waste (concrete, rubble, and the like), which appears at the end of a hotel's life cycles (100 years on average) or during refurbishing work. Some of it can be sold and reused (e.g., concrete can be recast and used to level land and build roads). And about one-quarter of that impact was due to energy-related waste (extracting and preparing fuel). Waste volumes bulged fast in countries that use a lot of coal to generate energy, like China, the USA, and Australia, where Accor runs large hotel networks (about 1,300 hotels in total). Hotel operations generated comparatively little waste in relation to other aspects of Accor operations (a 5 %).

Sensitivity Analysis

Sensitivity analyses were conducted in order to assess the influence of certain hypotheses on the results. When possible, better hypotheses were defined and results recalculated. The sensitivity analysis examined four issues:

- The exclusion of on-site wastewater discharge contribution to the water eutrophication impact, on the wastewater treatment, which is a very local issue. The calculated contribution of wastewater discharges was relatively high and changed significantly the results. It underlines the importance for Accor hotels to be connected to a wastewater treatment plant.
- Building material composition regarding the hotel categories. The consequences of allocating more assets to the hotel luxury and upscale and midscale were tested. The increase of the impacts was quite negligible, above all if they were considered at the macro level, never higher than 0.5 %.
- Life expectancy value of hotel buildings shorter than 100 years. Depreciation times were tested for 80 and 50 years. There was a real sensitivity of the results to building depreciation times, but the influence on total results was low.

- Occupancy rates and attendance indexes values were calculated under several assumptions. Several options were tested. It may have an impact in terms of repartition of the impact between all brands, but did not have a strong influence on the consolidated results.

4.1.4 Operationalizing the Results

Accor underlined that the main results and lessons of the study were in line with the set objectives. They have put Accor in a position to map out new down-to-earth action plans and to enhance the group's sustainable development strategy for years to come. The results of the environmental footprint were a valuable input for Accor to define the main Corporate Social Responsibility (CSR) issues and the action plan for the following years. Accor's CSR evolved with the ambition to be a value differentiator for the whole hotel group. Results are also being used by Accor in order to raise awareness among its guests and employees on the relevant environmental impacts.

The complete findings of this environmental footprint shaped Accor's new sustainable development strategy, "PLANET 21," and its related action plan. The strategy defines 21 commitments and ambitious goals for achievement in 2015 and includes a program to inform guests and employees and encourage them to contribute to reinventing hotel sustainability. The impacts gauged also have economic and financial consequences, and managing those consequences is also pivotal to the group's sustainable development. From these three key angles, Accor aims to build up environment policy on a solid, factual, documented foundation in line with its business performance objectives.

The study itself was part of the awareness-raising drive. Presenting findings already provided the opportunity to train more than 300 employees. The technical departments, purchasing, brand teams, and operations factored the findings into their 2012 action plans. This initiative also blends beautifully into the Accor group's aim, of innovating, inventing new approaches to break down barriers, and imagining hospitality tomorrow.

In accordance to the main outcomes of the study, Accor defined specific measures to improve its environmental performance from a life cycle perspective, for instance:

- Promote the reporting to OPEN beyond the 2,000 hotels that were doing so in 2011, and accommodate the specific variables in Accor's business on a regular basis.
- Running a continuous-improvement drive to use more energy-efficient systems and bear that in mind when refurbishing projects, equipment purchases, maintenance investments, and installation operation.
- Curbing water consumption, especially in areas under water stress (for instance, with wastewater recycling systems and rainwater recovery systems).

- Find pointers for progress in food-processing companies that have opportunities to break new ground with less polluting options.
- Gravitate toward promoting more balanced and smarter menus in its restaurants, for its guests and for the planet.
- Cut waste at source, including using less packaging for transport and using more economical packaging for toiletries, cleaning products, and food.
- Stepping up its recycling channels and selecting efficient service providers.

4.2 Unilever

Unilever has a long-standing reputation on sustainability, which goes back to the two founding parties of the business, Margarine Unie and Lever brothers.[6] First activities were mainly socially aimed, but were followed by many more. Product LCAs have been performed in Unilever for over 20 years. They are generally performed on a case by case basis and often on a specific product in a market. Although this contributed to the improvement of the particular products, it provided an incomplete picture of the global business and was of limited value at the company or category levels. As was mentioned in Sect. 1.3, in the late 1990s, the OBIA (Overall Business Impact Assessment) approach was introduced and represented the first steps to the development of the environmental footprint of Unilever (Taylor and Postlethwaite 1996; Clift and Wright 2000). In parallel to "footprint" activities, Unilever was deeply involved in SETAC's LCA activities.

In 2008, Unilever started an ambitious initiative to assess its global footprint including GHG (greenhouse gas) emissions, water use, consumer waste, and sustainable sourcing. This was the first time the full life cycle GHG impact of the company's portfolio was assessed consistently across all product categories. The aim was to obtain a picture of its global business and to support business strategy and decision-making at the various organizational levels. The assessment informs the Unilever Sustainable Living Plan, which was launched in 2010, and is now updated and reported annually. In 2010, Unilever set the target to double the size of the business by 2020 while reducing its environmental footprint and increasing its positive social impact.

The Unilever footprinting methodology[7] comprises three main phases: a business data extraction phase covering sales, product specifications, and consumer habit information, a footprint measurement phase that combines the business data with environmental information, and an interpretation/reporting phase. The goal

---

[6] The Safety & Environmental Assurance Centre of Unilever supported the drafting of this section, particularly Dr. Henry King, the team leader of Science & Technology – Sustainability. The text is based on Unger et al. (2011), Unger and King (2013), and Unilever (2013, 2014a).

[7] In this section we use Unilever's terminology (see, for instance, Unger and King 2013). It does not refer to the Organizational Environmental Footprint, OEF, presented in Sect. 1.4.3.

and scope of Unilever's footprint methodology and general framework for the inventory are described in Sect. 4.2.1, while Sect. 4.2.2 provides specific detail for each of the indicators assessed. Section 4.2.3 presents the main results of the footprinting and Sect. 4.2.4, the value and application of the results by Unilever.

4.2.1 Methodological General Description

Goal and Scope

The global environmental footprinting methodology was designed specifically for the external reporting of Unilever's footprint and for internal management needs. It aims to allow the establishment of a baseline situation for each indicator and the tracking of the progress, by a measurement process that was repeatable. Although it was not specifically designed for the measurement and external communication of product-level data, with the appropriate caveats, the insights can be used to guide product development and assess future category innovation plans.

The unit of analysis of the footprinting analysis can be defined as the sales of products in 14 countries in 1 year (which represents over 70 % of global annual sales) characterized by over 2,000 representative products. The boundary includes the steps of the life cycle from cradle to grave, though this varies by environmental indicator depending upon the availability of data and the relevance of the management plan.

Four environmental indicators were considered; the scope specificities for each of them are pointed out in Sect. 4.2.2. The motivation for each indicator, based on Unilever (2013), was:

- Greenhouse gases. Climate change also has a significant impact on Unilever's business and consumers. Changes in weather patterns will affect the sourcing of agricultural raw materials due to increases in energy and food prices and extreme weather events that displace communities.
- Water. Water shortages are already affecting many parts of the world and the consumer's access to water is key to the use of many Unilever products. Around 70 % of available freshwater is used for agriculture, which provides the large part of Unilever's manufacturing inputs.
- Waste. Packaging plays a key role in protecting Unilever's products. But it increases resource scarcity and can also end up as waste in landfill, dumping grounds, or litter, particularly in developing markets with less developed infrastructure to manage packaging waste.
- Sustainable sourcing. Half of Unilever's raw materials come from farms and forests, and the decisions made on the sources and the collaboration with producers can have profound implications on global resources, climate change, and farmer livelihoods.

Inventory: Business Data Extraction Phase

A key challenge for a business such as Unilever is its diversity and size. Unilever sells a wide portfolio of products in over 190 countries; half of those sales are in developing and emerging countries. The wide diversity of products and their use, as well as the size of the company, made a bottom-up conventional product-based footprint approach, including every single product LCA globally, impractical. Therefore, the footprinting process was streamlined by defining a representative set of countries and products.

A total of 14 countries were deemed representative: Brazil, China, France, Germany, India, Indonesia, Italy, Mexico, the Netherlands, Russia, South Africa, Turkey, the UK, and the USA. The selection factors were both related to business (e.g., annual sales, coverage of all product categories, and consumer habits) and environment (e.g., country infrastructure and environmental profile, like carbon intensity of the electricity grid, degree of water scarcity, and waste management infrastructure). Unilever's product portfolio was grouped into clusters of similar products in each country (see Fig. 8.12). From each cluster a representative product was selected for subsequent measurement. A key challenge in this clustering exercise was to strike a balance between the level of detail necessary to guaranteeing representativeness of the results versus the input demand (e.g. data, time, and resources). Previous product portfolio assessments of Unilever brands, e.g., on Ben and Jerry's Europe or the global Knorr brand (Garcia-Suarez et al. 2008; Milà i Canals et al. 2010) provided important insights and understanding into this process. Currently over 2,000 representative products are footprinted in the 14 countries and this represents about 70 % of Unilever's global sales.

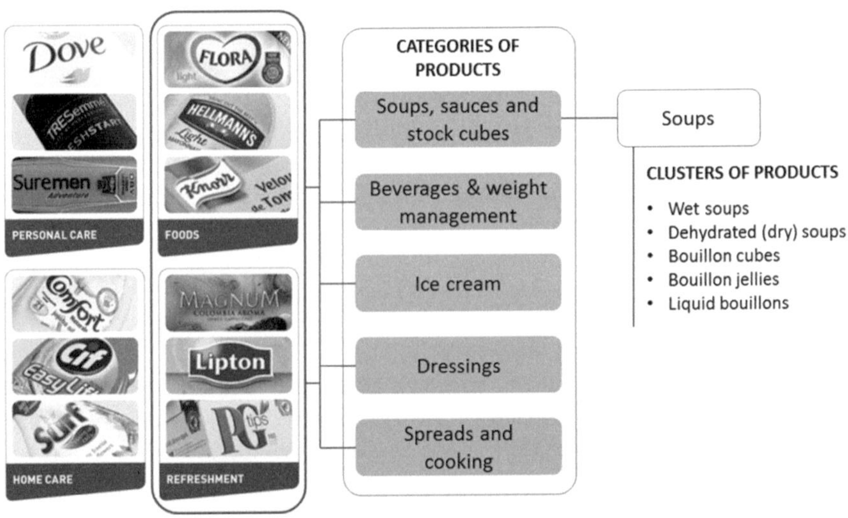

Fig. 8.12 Example of clustering for soups and bouillons (Source: pictures from Unilever 2012)

For each representative product, Unilever analyzed sourcing and ingredient information, packaging, manufacturing impacts, and data on consumer habits (which often vary by country). Apart from business data, secondary data were used due to the wide variety of ingredients and processes involved. Internal data were allocated as a rule by mass. However, life cycle-based processes from various sources are used and therefore they are used according to the allocation method applied in the corresponding source.

There are ongoing efforts in Unilever to systematize and improve the quality and efficiency of the annual measurement. The Unilever's footprint has now been performed three times and it is planned to repeat it annually. In 2012, Unilever invested in an automated process to improve the speed and accuracy of the footprint calculations, which is measured on a rolling basis from 1 July to 30 June. This made possible the development of bespoke data validation and reporting tools that hold and manage data from the different business IT systems. The first footprint took approximately 18 months to complete, while time was reduced to 12 and 8 months in the subsequent yearly repetitions. Not only time was saved, there have been significant improvements in data quality and increased granularity, and the number of representative clusters was expanded, enabling greater specificity and brand-level assessments and reporting (Unger and King 2013).

Critical Review

Over 2010–2011 Unilever invited an external panel of environmental life cycle assessment experts to review the footprinting approach. This aimed to provide assurance of the robustness of the approach, including the way in which Unilever's data were collected and compiled, a scientific review of the individual metrics, and assurance that the results and conclusions were fit for purpose, including the scope of the data and how the results were communicated. In 2013, PwC, the assurers for the Unilever Sustainable Living Plan, undertook the assurance of Unilever's GHG footprint measurement process and results for the first time, and this will be extended to all metrics in subsequent years.

4.2.2 Specificities for Each Metric

As it was aforementioned, Unilever assesses four environmental indicators or metrics. Due to data availability and the importance given to each of the issues, a differentiated scope, inventory, and impact assessment (if any) were developed for each of them. Specific detail is presented in the following.

GHG Emissions

The GHG emissions was the more comprehensively assessed indicator; it considers a 100-year time horizon for the CO_2 equivalent impacts. The footprint includes the assessment of GHG emissions for all life cycle stages from cradle to grave; it means from raw materials, to manufacturing to consumer use and disposal (Fig. 8.13). GHG emissions from land use associated with agricultural ingredients were partially included where data were available. GHG emissions from the biodegradation of petrochemical ingredients following discharge to the sewerage system or environment were excluded as it was very difficult to differentiate the feedstock of various ingredients from the existing product specification data and retrieval process. The aim is to address this area in future assessments. Finally, fixed assumptions were used for phases of the life cycle that contribute a small proportion to the product life cycle (e.g., a single distance/mode of transport was used to describe product distribution to the retailer).

After product clustering and selection of representative products, the necessary specification data and other relevant life cycle data (e.g., consumer habits data) were extracted from the various business IT systems. Published GHG data were used where possible, but if there were no specific data (e.g., for an ingredient or process), expert choice was applied. Two generic life cycle models were then built with the commercial life cycle assessment software GaBi: one for food products (see example in Fig. 8.13) and one for household and personal care products. Each

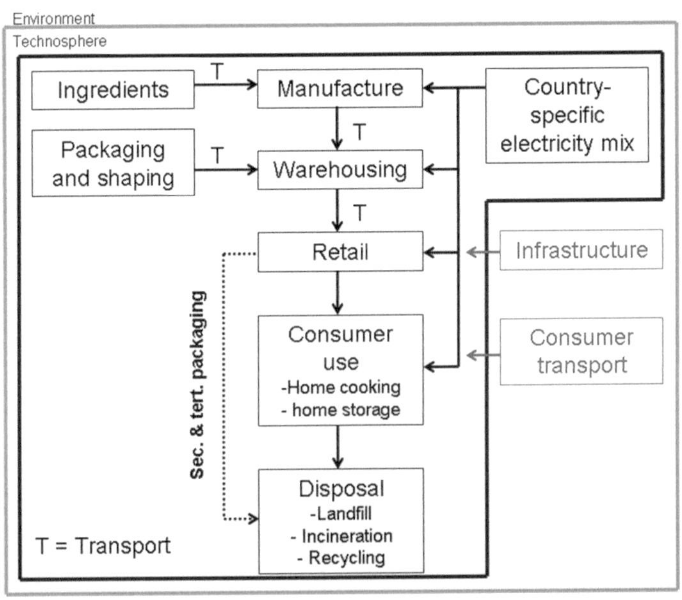

Fig. 8.13 Example of schematic outline of system boundaries for the food model (Source: Unger et al. 2011)

of them contains all possible ingredient inventories for the two types of products. This helped to ensure consistency in the measurement process. Later, simpler tools were created to automate representative product data entry. GHG calculations were performed on a per consumer use basis, and the results scaled up with the underlying sales figures of the cluster to give the total impact of all clusters in each country. The average Unilever GHG impact (per consumer use) was then calculated.

Water Use

Unilever uses water resources both directly (in the factories both as an ingredient in the products and during the manufacturing process) and indirectly (suppliers of agricultural raw materials for the growing of crops (Hillier et al. 2011) and consumers when using products to do their laundry, showering, cleaning, and cooking). The current water metric of Unilever considers the water added to the product and the water used by consumers. In the definition of domestic water scarcity, Unilever evaluated how many people in each country experience physical water scarcity as well as the number of people who have access to an improved water source. Of the 14 USLP (Unilever Sustainable Living Plan) countries, Unilever chose to focus on those seven countries that were defined as water scarce. These are China, India, Indonesia, Mexico, South Africa, Turkey, and the USA, representing around half of the world's population. The steps taken for water assessment were similar as for GHG emissions in terms of business process (i.e., clustering, habits, specific data, etc.). The only difference was that the calculation of the product category units for water is conducted in the central tool rather than in GaBi.

Although the metric excludes some steps of the life cycle, Unilever aims to consider water across the full value chain and has conducted a number of specific studies to understand its full water footprint. Water used in manufacturing operations is captured separately, as part of Unilever's eco-efficiency program[8] and has been reported regularly since the 1990s. In 2012, using data from the Water Footprint Network, Unilever completed a groundbreaking assessment of the amount of irrigation water used to produce their key agricultural raw materials in all the water-scarce countries their source is from. This included a detailed assessment of the key agricultural materials (around two-thirds of Unilever's volumes) and consideration of a further 30 materials. Calculations only included physical water scarcity, as access to an improved water source is not relevant for growing crops. The study findings indicate that approximately 85 % of Unilever's water footprint in water-scarce countries is associated with the consumer use phase.

[8] http://www.unilever.com/sustainable-living-2014/reducing-environmental-impact/eco-efficiency-in-manufacturing/

Waste and Packaging

Waste is defined as the amount of product left in the pack and the amount of packaging (i.e., paper, board, metals such as aluminum and steel, glass, and mixed material laminates) that ends up either in landfill or as litter; it means the amount of packaging that is not recycled, reused, or recovered for energy recapture. The waste footprint is measured in the grams of packaging and product leftovers that have not been reused, recycled, or recovered, on a per consumer basis. Local recycling context is considered, published national indices for recycling and recovery are used, or at least own estimates where these are not available. As for water, waste generated by Unilever's manufacturing operations is taken into account in the Unilever's eco-efficiency program and therefore it is not included in the footprint exercise. Solid waste from across the wider life cycle and activities such as electricity generation is also excluded.

Sustainable Sourcing

Finally, Unilever developed a metric for assessing the sustainable sourcing of raw materials, by quantifying the verifiable sustainable renewable sources (% by weight). The criteria for sustainable sourcing cover the three pillars of sustainability and focus on the agricultural and packaging production. The sustainable sourcing program either uses external certification where possible or self-assessment and verification against the Unilever Sustainable Agriculture Code (Unilever 2014a). The Code is applicable to all raw materials; it details the standards to be adopted and indicates the need for improvement over time. The assessment first concentrates on the top ten agricultural raw materials (like palm oil, soy bean, tea, cocoa, sugar, eggs, and tomatoes), which account for around two-thirds of Unilever's volumes. This indicator is not following a life cycle approach; however, it deals with the supply chain assessment.

4.2.3 Results and Interpretation

The Unilever's footprint is measured at an individual representative product level across the life cycle (as defined for each impact). The results are aggregated at several levels, depending on the intended use, product cluster (e.g., ready-to-use liquid bouillons), category (soup, sauces, and stock cubes), country (India), and company level, and are expressed per consumer use (e.g., the GHG impact of preparing a serving of soup or the water needed for one hair wash with shampoo) and as absolute totals (e.g., the waste associated with the product of the brand Ben and Jerry's in 1 year).

Product, category, and company footprint details provide valuable insights. Each category can be analyzed in detail to understand how much each product format contributes to the category footprint and understand the countries with the most consumer uses and the drivers for the impact. In addition, looking across the results of all categories and contributing life cycle aspects help identifying the key contributors. The footprinting activity initiated a number of research projects to address science gaps identified.

In the following, results at different levels are presented for GHG emissions, waste, and water. Sustainable sourcing outcomes are presented in Sect. 4.2.4, as it followed a different approach than the other indicators.

GHG Emissions

Figure 8.14 shows that less than 2 % of the product life cycle GHG impacts occur in Unilever's own operations – the main contributions occur either with the suppliers of raw materials or in the consumer use phase. Thus, the largest reduction opportunities exist across the value chain, in particular in the consumer use stage (68 % of the impact).

When reviewing the results by business category, additional insights can be obtained. The product categories which make the largest contribution to the greenhouse gas footprint are soaps, shower gels, and skin care products. The top contributor of the total footprint is the category "skin cleansing and care," just below 50 %. The second highest contributor is "hair care" with 14 % (Table 8.5). This reflects the contribution from sales in countries such as the USA where the consumer habit is to use hot water showers. In contrast in many developing countries, washing is often performed with water at ambient temperature. The contribution from laundry is 11 %, although the majority of sales exist in countries with handwash or ambient washing conditions, which do not contribute much to GHG emissions. In the case of ice cream, while the impact per consumer use is relatively high, the total number of servings (individual uses) is much smaller than the individual uses of say shower or shampoo products.

Water Use

The water used to produce Unilever's agricultural ingredients is about 15 % of the total footprint, and about 85 % relates to water used by consumers (see Fig. 8.2).

Fig. 8.14 Greenhouse gas footprint breakdown according to life cycle stages for the Unilever portfolio for 2013 (Source: Unilever 2014b)

Table 8.5 Indicators breakdown according to product category for the Unilever portfolio for 2013 (% contribution to Unilever's footprint by metric and by category)

	Metrics (%)		
Category of product	GHG emissions	Water (consumer use)	Waste (packaging)
Skin cleansing and care	46	29	14
Laundry	11	41	12
Hair care	14	8	9
Household care	5	13	5
Savory (soups, sauces, and stock cubes)	6		16
Beverages and weight management	5		11
Ice cream	6		9
Oral care	1	9	3
Dressings	2		8
Deodorants and fragrances	2		8
Spreads and cooking	3		4
Total	≈100	≈100	≈100

Source: Unilever (2014a)

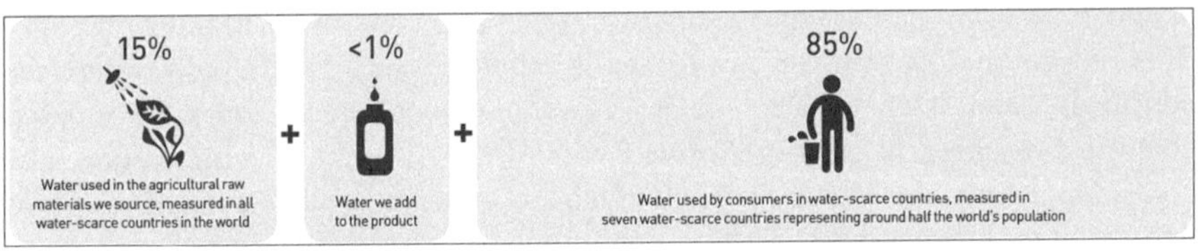

Fig. 8.15 Water footprint breakdown according to life cycle stages for the Unilever portfolio for 2012 (Source: Unilever 2013)

The water used by the consumers in washing and cleaning is more than seven times greater than the water embedded in the agricultural raw materials Unilever buys. The priority water-intensive crops for Unilever are tomatoes and sugar cane.

Table 8.5 shows the water associated with the consumer use of Unilever's products. Around 41 % of the water footprint comes from the laundry process. A significant proportion of this is washing laundry by hand in the developing world. In water-scarce countries, around 38 % of domestic water is used to clean clothes, often by hand. A further 37 % of the footprint occurs when people use Unilever's soaps, shower gels, and shampoos during showering, bathing, and washing hair and 13 % from household care, largely washing dishes by hand (Fig. 8.15).

Waste and Packaging

Based on Fig. 8.16, primary packaging represents the larger share, while transport packaging contributes 19 % and product leftovers are 23 % of the total weight of

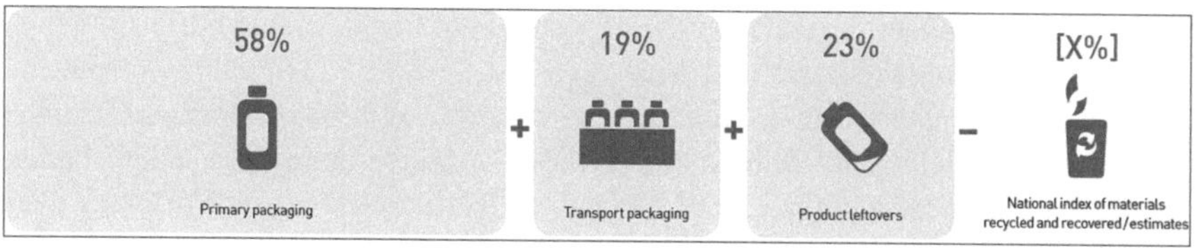

Fig. 8.16 Waste footprint breakdown according to life cycle stages for the Unilever portfolio for 2012. Potential amount of waste as defined before national adjustment (Source: Unilever 2013)

material that could potentially end up as waste. The total amount of waste is calculated by adjusting the potential waste footprint by the waste management infrastructures of each country for the various packaging materials considered.

The analysis has helped Unilever to see which categories of the product portfolio generate more waste than others and which could therefore yield the biggest opportunities for reductions. Food packaging is one of the biggest contributors to the waste footprint, with 48 % of the waste footprint (see Table 8.5). Tea bags make a significant contribution to the overall product leftovers because a small amount of the material used to enclose the tea itself is not compostable. Measured by material type, paper and board, flexible laminates, and glass make up the majority (two-thirds) of the waste footprint.

4.2.4 Operationalizing the Results

The results of the global environmental footprint of Unilever were communicated for the first time at the launch of the Unilever Sustainable Living Plan (USLP) in 2010. While understanding the impact of the company's portfolio is important, it is only when there is a management reduction plan that there will be benefits to the environment. Only by considering a wider range of aspects, potential trade-offs and synergies can be recognized and addressed. The footprint baseline has enabled Unilever to understand its key impacts by category, life cycle phase, and business.

The baseline was invaluable in getting buy-in from senior business leaders and guided the development and enhancement of reduction programs and targets and tracking of its achievement, which have been communicated internally and externally in the USLP. The project team included members from marketing, R&D, supply chain, packaging, IT, finance, as well as environmental experts. Key to completing the baseline was training and awareness raising, and this was often made more challenging due to the different levels of expertise and career backgrounds in each category and also the global spread of the project team.

Unilever Sustainable Living Plan

Launched in November 2010, the Unilever Sustainable Living Plan sets out to decouple the growth of Unilever from the environmental impact while increasing positive social impact. It spans the entire portfolio of brands, all countries in which Unilever sells products, and it applies across the whole value chain – from the sourcing of raw materials to the company's factories and the way consumers use the products. Three pillars are defined in the USLP – improving health and well-being, reducing environment impact, and enhancing livelihoods – which contain over 50 public, time-bound goals specified across nine themes. The pillar related to environment includes four themes: greenhouse gases, water use, waste and packaging, and sustainable sourcing.

The themes were chosen because of their scientific relevance and scale of impact for Unilever's portfolio, in accordance to assessments previously conducted by the company. The relevance of themes to external stakeholder expectations and the company's ability to quantify the metrics were also taken into account. Objective measurement techniques were established for each of the targets, including appropriate estimates and assumption, and for the environmental-related goals, Unilever environmental footprint (presented in Sects. 4.2.1, 4.2.2, and 4.2.3) represents the backbone.

Unilever's Targets and Measures Taken

In 2010, Unilever established a big goal to achieve by 2020 in the environmental pillar: "By 2020 Unilever's goal is to halve the environmental footprint of the making and use of its products as it grows Unilever's business." The environmental targets are expressed on a "per consumer use" basis. Each environmental, and other, theme was broken down into roadmaps with actionable steps using nonexpert friendly terms. Unilever uses a code to define whether each of the targets has been "achieved by target date," "on-plan for target date," "off-plan for target date," and "% achieved by target date." A range of approaches have been explored by Unilever to reduce the environmental impact and achieve the targets of 2020.

The contribution analysis highlights that it is not (only) the inherent design of the product that drives the impact but also the number of sold consumer uses, local infrastructure (e.g., grid electricity), and the typical consumer behavior. Reducing the contribution from consumer phase is a significant challenge since this life cycle phase is not under the direct control of the organization but has a high contribution in GHG emissions, water use, and waste and packaging. Some opportunities are innovation-led reductions, consumer habit change, and advocacy on relevant public policy areas such as low carbon energy, machine efficiency, and building standards, for example, the rollout of dry shampoo under many brands such as Dove, Suave, and TRESemmé that prevent the use of warm water. Unilever published the behavior change model "Five levers for change" (Unilever 2014a), comprising of a set of key principles, which, if applied consistently to behavior change interventions, increases the likelihood of having an effective and lasting impact reduction.

In addition, Unilever will continue to take actions to reduce impacts upstream of its manufacture, like promoting drip irrigation with their tomato suppliers. Furthermore, the Cool Farm Tool (Hillier et al. 2011, 2012) was commissioned by Unilever from the University of Aberdeen. It is a greenhouse gas calculator for farming. It is user friendly and gives instant results that invite users to try out alternatives and run scenarios for low carbon farming.

Additionally, the footprint assessment has been used as the basis of a new approach to assessing the environmental impact of internal product renovations and innovations within the innovation process management systems. It aides thinking about new business models and enables ways of delivering a service which might be outside the current business practice. Data and the underpinning life cycle models enable future scenarios. Unilever plans to use this, combined with the project management process, to systematically challenge and reduce the environmental impacts of Unilever's future product launches.

Target Achievement

To ensure that Unilever is progressing toward its target of halving the average per consumer use impact, regular updating of the footprint is essential and planned for the following years. Between 2010 and 2012, the GHG footprint per consumer use could be reduced by about 6 %; however it was increased again by 5 % on 2013 (Table 8.6). The increase was due to the evolution of the portfolio, in particular the Personal Care business expanded in hair and shower products via the Alberto Culver acquisition (which accounts for three percentage points of the GHG increase). However, Unilever has made good progress in those areas in their control, for instance, associated with own manufacturing operations (e.g., CO_2 from energy has been reduced by 32 % since 2008 compared to 1995 by 62 %).

In 2013, Unilever's water impact per consumer use increased by around 15 % since 2010 (see Table 8.6). This is again because the biggest impact comes from the water used by consumers, where the company has less control. Laundry business experienced high growth from bars in India which, while very affordable for people on low incomes, are also associated with a more water-intensive washing habit than other laundry handwash formats. However, Unilever is making progress in some parts of the organization through product innovation.

Unilever is on track to meet waste and packaging 2020 commitment. The total footprint from packaging waste to landfill has reduced by 11 % (see Table 8.6). Efficient pack designs and innovative use of materials, as well as the disposal of sauce brands with large waste footprints, have been the main drivers. Unilever is also working with others to stimulate recycling infrastructures. This can take time but is crucial for the long term. Likewise, the manufacturing teams exceeded in 2013 their 2020 target, reducing waste by 66 % per tonne of production since 2008.

Regarding sustainable sourcing sustainably, the progress has been strong. The commitment of Unilever was to source 10 % of Unilever's agricultural raw materials sustainably by 2010, 30 % by 2012, 50 % by 2015, and 100 % by 2020. In 2013, Unilever was close to achieve the target for 2015. Specific targets are set for each of the Unilever's top ten agricultural raw materials. The company is on track against the 2020 goal for eight of the raw materials.

Table 8.6 Environmental indicators target achievement for the Unilever portfolio 2011–2013

Year	GHG emissions	Water use	Waste and packaging	Sustainable sourcing
Unit	Impact per consumer use increased (+) or decreased (−) since 2010			% of our agricultural raw materials sustainably sourced
2011	Unchanged	Unchanged	Unchanged	14 %
2012	−6 %	Unchanged	−7 %	36 %
2013	+5 %	+15 %	−11 %	48 %

Sources: Unilever (2013, 2014a)

Reporting

Unilever has a long-standing commitment to reporting progress on sustainability performance each year, and it has been doing so since 1996. There are two main elements for reporting on the commitments set out in the Unilever Sustainable Living Plan as well as wider topics of interest to the stakeholders: Annual Report and Accounts and Unilever Sustainable Living Report. Additionally, Unilever annually assesses its progress against three indices: UN Global Compact, GRI (Global Reporting Initiative), and UN Millennium Development Goals (more information in Unilever (2014a)). A device called "Product Analyzer," publicly available in its webpage, shows the environmental impacts of a selection of Unilever's products across their life cycle, in order to exemplify the results of the footprint. This provides the GHG, water, or waste impacts of a representative food, home, or personal care product on a "per consumer use" basis.

5 Conclusion and Outlook

LCA of organizations has significant potential to be used by organizations, corporations, authorities, institutions, and other organizations in the society in their efforts to improve their environmental performance. Organizations need credible information at the level the decisions are taken and beyond its walls. Apart from moving to more resource-efficient and less polluting practices in their sites, a life cycle perspective supports identifying how to implement improvement opportunities in a more efficient way by different actors in the value chain. Many different types of inputs and outputs are involved in the provision of goods and services, thus many different types of environmental impacts occur.

Therefore, LCA of organizations offers a complete picture and supports decisions that find the right balance between environmental impacts. It offers insight into the organization and value chain and identifies hotspots where action is more needed. LCA of organizations also provides a structure for tracking the performance of the organization and the achievement of the targets defined within the organization environmental strategy. Furthermore, the strengths provided by the results are very adequate to support reporting and communication to third parties.

Currently, two main initiatives work on the development and testing of the LCA of organizations: O-LCA proposed by UNEP Guidance (UNEP 2015), which lines up with and complements the ISO/TS 14072, and the OEF by the European Commission (European Commission 2013a). They have a different scope and goals. While the former aims to adapt LCA framework to the organizational level and promotes the complementarity of the two methodologies, the main principles of the latter are increased reproducibility and comparability by producing a new methodology that harmonizes preceding approaches and that is policy oriented. OEF Guide does not align with some LCA principles. One of the main differences is that OEF Guide foresees comparability at the sector level, while UNEP Guidance discourages the comparability statements intended to be disclosed to the public.

An overview of the methodological framework proposed by the UNEP Guidance was included in the chapter. Most of the principles and requirements for product LCA apply also for organizational LCA with some minor terminology amendments. Major discrepancies between the product and organizational LCA are during the definition of the unit of analysis and the associated system boundaries and for the completion of the inventory. During the implementation of the methodology, an organization may benefit from existing environmental assessments, like EMS, product LCA, corporate carbon footprints, etc. The data and experience generated can streamline the application of LCA of organizations by providing inventory data for some activities or indicators, as well as guide the definition of the scope, identify hotspots and targeted suppliers, and facilitate the communication between different departments of the organization or with suppliers.

Some organizations are already working in obtaining the full picture of the environmental performance of the organization. Two of these stories were included in the chapter, Accor and Unilever. A complete overview of the case studies was presented, including the motivations for the application of the organizational approach, the methodological and implementation framework behind, the results obtained, and the implementation of the outcomes of the study within the companies' sustainability strategy and reporting scheme. They illustrate both the potential and the challenges of the LCA of organizations.

There is an identified need for checking and promoting the application of LCA of organizations in SMEs, as collectively they have an important role on global environmental impacts and particularly in developing countries. LCA of organizations may overcome some of the barriers for the implementation of LCA in developing countries. First are the high costs of LCA application since LCA of organizations provides an overall idea of the environmental performance without having to perform independent LCAs for many products. Another barrier is the threat of selecting against non-best available technologies (that could be overcome when discouraging comparison for communicating purposes). A simplified version of the Guidance could be considered in the future for those target organizations.

At this point of development, existing initiatives for the LCA of organizations focus on environmental impacts; however, the organizational level seems also a promising approach for the assessment of social aspects. Social performance is determined by how the organization conducts toward its stakeholders, rather than by the processes involved to provide the product. Therefore, an organization-related assessment may overcome some of the methodological and practical challenges of the social life cycle assessment (Martínez-Blanco et al. 2015c).

The technical framework of both methodologies (OEF and O-LCA) is already available to be applied by the international community. Both initiatives for the LCA of organizations are on the process of road testing the methodological approaches and the application of the reference documents. The UNEP Guidance is being road tested by at least ten organizations that volunteered to take the lead. The pilot phase of the OEF initiative currently focuses on the development of two sectorial guides, the OEFSRs, and aims to develop further ones in the future for other sectors. Finally, the authors want to encourage the readers to contribute on the road toward the worldwide use of the organizational approach within the context of life cycle sustainability assessment.

References

Accor (2011a) Accor group's environmental footprint (confidential). Pricewaterhouse Coopers Advisory, Paris

Accor (2011b) The Accor group's Environmental Footprint. Accor Sustainable Development Department, Paris, http://www.accor.com/fileadmin/user_upload/Contenus_Accor/Developpement_Durable/pdf/earth_guest_research/2011_12_08_accor_empreinte_environnementale_dp_bd_en.pdf

Accor (2014) Accor Group website. http://www.accor.com

ADEME (2010) Bilan carbone companies and local authorities version. Methodological guide version 6.1: objectives and principles for the counting of greenhouse gas emissions. French Agency for the Environment and Energy Management. http://www.associationbilancarbone.fr/sites/default/files/guide_methodologique_v6_euk-v.pdf

Bare JC (2009) Life cycle impact assessment research developments and needs. Clean Techn Environ Policy 12:341–351

Berger M, Finkbeiner M (2010) Water footprinting: how to address water use in life cycle assessment? Sustainability 2:919–944

CDP (2014) Carbon disclosure project – climate change program guidance website. https://www.cdp.net/en-US/Pages/guidance-climate-change.aspx

Clift R, Wright L (2000) Relationships between environmental impacts and added value along the supply chain. Technol Forecast Soc Chang 65:281–295

Curran M, Hellweg S, Beck J (2014) Is there any empirical support for biodiversity offset policy? Ecol Appl 24:632. doi:10.1890/13-0243.1

Downie J, Stubbs W (2011) Evaluation of Australian companies' scope 3 greenhouse gas emissions assessments. J Clean Prod 56:156–163

Draucker L (2013) GHG protocol: moving Corporate Accounting Beyond GHGs. Abstr. B. SETAC North Am. 34th annual meeting, Nashville

European Commission (2009) Regulation (EC) No 1221/2009 of the European Parliament and of the Council of 25 November 2009 on the voluntary participation by organisations in a Community eco-management and audit scheme (EMAS), repealing Regulation (EC) No 761/2001 and Commission Deci. Off J Eur Union

European Commission (2010a) International reference life cycle data system (ILCD) handbook – general guide for life cycle assessment – detailed guidance. European Commission – Joint Research Centre – Institute for Environment and Sustainability. http://bookshop.europa.eu/en/international-reference-life cycle-data-system-ilcd-handbook-general-guide-for-life cycle-assessment-detailed-guidance-pbLBNA24708/

European Commission (2010b) International reference life cycle data system (ILCD) handbook – analysing of existing environmental impact assessment methodologies for use in life cycle assessment. European Commission – Joint Research Centre – Institute for Environment and Sustainability, http://eplca.jrc.ec.europa.eu/uploads/ILCD-Handbook-LCIA-Background-analysis-online12March2010.pdf

European Commission (2011) Analysis of existing environmental footprint methodologies for products and organisations: recommendations, rationale, and alignment. European Commission – Joint Research Centre – Institute for Environment and Sustainability. http://ec.europa.eu/environment/eussd/pdf/Deliverable.pdf

European Commission (2013a) Organisation environmental footprint guide. European Commission – Joint Research Centre – Institute for Environment and Sustainability. http://eur-lex.europa.eu/JOHtml.do?uri=OJ:L:2013:124:SOM:EN:html
European Commission (2013b) Commission recommendation of 9 April 2013 on the use of common methods to measure and communicate the life cycle environmental performance of products and organisations (2013/179/ EU). Off J Eur Union
European Commission (2013c) Building the single market for green products facilitating better information on the environmental performance of products and organisations. Communication COM/2013/0196 final
European Commission (2013d) Guidance for the implementation of the EU OEF during the EF Pilot Phase – Ver 3.1. European Commission – Joint Research Centre – Institute for Environment and Sustainability. http://ec.europa.eu/environment/eussd/smgp/pdf/Guidance_organisations.pdf. 20 May 2001
Finkbeiner M (2013) Product environmental footprint – breakthrough or breakdown for policy implementation of life cycle assessment? Int J Life Cycle Assess 19:266–271
Finkbeiner M, König P (2013) Carbon footprint and life cycle assessment of organizations. J Environ Account Manag 1:55–63
Finkbeiner M, Wiedemann M, Saur K (1998) A comprehensive approach towards product and organisation related environmental management tools. Int J Life Cycle Assess 3:169–178
Finkbeiner M, Ackermann R, Bach V, Berger M, Brankatschk G, Chang Y-J, Grinberg M, Lehmann A, Martínez-Blanco J, Minkov N, Neugebauer S, Scheumann R, Schneider L, Wolf K (2014) Challenges in life cycle assessment: an overview of current gaps and research needs. Chapter 7 "Background and Future Prospects in Life Cycle Assessment" (Klöpffer W ed). In: Klöpffer W, Curran MA, (eds) LCA compendium – the complete world of life cycle assessment. Springer, Dordrecht, pp 207–258
Galatola M, Pant R (2014) Reply to the editorial "Product environmental footprint—breakthrough or breakdown for policy implementation of life cycle assessment?" written by Finkbeiner M (Int J Life Cycle Assess 19(2): 266–271). Int J Life Cycle Assess 19:1356–1360. doi: 10.1007/s11367-014-0740-3
Garcia-Suarez T, Sim S, Mauser A, Marshall P (2008) Greenhouse gas assessment of Ben & Jerry's ice-cream: communicating their "Climate Hoofprint." In: Proceedings of the 6th international conference on LCA agri-food sect, Nov 12–14, Zurich, pp 341–351
GRI (2013) Sustainability reporting guidelines, G4. Global reporting initiative. https://www.globalreporting.org/reporting/g4/Pages/default.aspx
Guinée JB (2001) Life cycle assessment: an operational guide to the ISO standards. Part 1 and 2. Ministry of Housing. Spatial Planning and Environment (VROM) and Centre of Environmental Science (CML). The Netherlands
Hillier J, Walter C, Malin D, Garcia-Suarez T, Mila-i-Canals L, Smith P (2011) A farm-focused calculator for emissions from crop and livestock production. Environ Model Software 26:1070–1078. doi:10.1016/j.envsoft.2011.03.014
Hillier J, Brentrup F, Wattenbach M, Walter C, Garcia-Suarez T, Mila-i-Canals L, Smith P (2012) Which cropland greenhouse gas mitigation options give the greatest benefits in different world regions? Climate and soil-specific predictions from integrated empirical models. Glob Chang Biol 18:1880–1894. doi:10.1111/j.1365-2486.2012.02671.x
Huang YA, Lenzen M, Weber CL, Murray J, Matthews HS (2009a) The role of input-output analysis for the screening of corporate carbon footprints. Econ Syst Res 21:217–242
Huang YA, Weber CL, Matthews HS (2009b) Categorization of scope 3 emissions for streamlined enterprise carbon footprinting. Environ Sci Technol 43:8509–8515
ISO (2004) ISO 14001: environmental management systems – requirements with guidance for use. International Organization for Standardization, Geneva
ISO (2006a) ISO 14044: environmental management – life cycle assessment – requirements and guidelines. International Organization for Standardization, Geneva

ISO (2006b) ISO 14040: environmental management – life cycle assessment – principles and framework. International Organization for Standardization, Geneva
ISO (2006c) ISO 14025: environmental labels and declarations – type III environmental declarations – principles and procedures. International Organization for Standardization, Geneva
ISO (2006d) ISO 14064-1: greenhouse gases – part 1: specification with guidance at the organization level for quantification and reporting of greenhouse gas emissions and removals. International Organization for Standardization, Geneva
ISO (2013a) ISO/TS 14067: greenhouse gases – carbon footprint of products – requirements and guidelines for quantification and communication. International Organization for Standardization, Geneva
ISO (2013a) ISO/TR 14069: greenhouse gases – quantification and reporting of greenhouse gas emissions for organizations – guidance for the application of ISO 14064-1. Geneva
ISO (2014a) ISO/TS 14072: environmental management – life cycle assessment – requirements and guidelines for organizational life cycle assessment. International Organization for Standardization, Geneva
ISO (2014b) ISO/TS 14071: environmental management – life cycle assessment – critical review processes and reviewer competencies: additional requirements and guidelines to ISO 14044:2006. International Organization for Standardization, Geneva
Lave LB, Cobas-Flores E, Hendrickson C, McMichael FC (1995) Life cycle assessment: using input-output analysis to estimate economy-wide discharge. Environ Sci Technol 29:420–426
Makower J, Mattison R, Salo J, Kelley D (2014) State of green business 2014. GreenBiz Group and Trucost. http://www.trucost.com/_uploads/publishedResearch/SOGB2014.pdf
Martínez-Blanco J, Inaba A, Finkbeiner M (2015a) Activities of guidance of organizational LCA by UNEP/SETAC WG. Jpn J Life Cycle Assess 11(2):97–103
Martínez-Blanco J, Inaba A, Finkbeiner M (2015b) Scoping organizational LCA – challenges and solutions. Int J Life Cycle Assess 20(6):829–841
Martínez-Blanco J, Lehmann A, Chang Y-J, Finkbeiner M (2015c) Social organizational LCA (SOLCA) – a new approach for implementing social LCA. Int J Life Cycle Assess 20 (11):1586–1599
Milà i Canals L, Sim S, García-Suárez T, Neuer G, Herstein K, Kerr C, Rigarlsford G, King H (2010) Estimating the greenhouse gas footprint of Knorr. Int J Life Cycle Assess 16:50–58
Pelletier N, Allacker K, Pant R, Manfredi S (2013) The European Commission Organisation Environmental Footprint method: comparison with other methods, and rationales for key requirements. Int J Life Cycle Assess 19:387–404
Taylor AP, Postlethwaite D (1996) Overall business impact assessment (OBIA). In: Proceedings of the 4th SETAC case study symposium, Brussels
UN (2002) Report of the world summit on sustainable development. Johannesburg, South Africa, 26 Aug–4 Sept 2002. United Nations, New York. http://www.un.org/jsummit/html/documents/summit_docs/131302_wssd_report_reissued.pdf
UNEP (2015) Guidance on organizational life cycle assessment. Life-Cycle Initiative, United Nations Environment Programme and Society for Environmental Toxicology and Chemistry, Paris, http://www.lifecycleinitiative.org/resources/reports/
UNEP/SETAC (2012) Greening the economy through life cycle thinking: ten years of the UNEP/SETAC life cycle initiative. Life Cycle Initiative, United Nations Environment Programme and Society for Environmental Toxicology and Chemistry, Paris, http://www.lifecycleinitiative.org/wp-content/uploads/2013/03/2012_LCI_10_years_28.3.13.pdf
UNEP/SETAC (2014) Life cycle initiative website. http://www.lifecycleinitiative.org/
Unger N, King H (2013) Use of the life cycle approach on company, category and product level to support the greenhouse gas strategy in a multinational company. 6th int. conf. life cycle manag – LCM 2013. Gothenburg, Sweden. ISBN NO: 978-91-980973-5-1, Gothenburg, pp 63–66
Unger N, King H, Calvert S (2011) How to measure and manage the life cycle greenhouse gas impact of a global multinational company. In: Finkbeiner M (ed) Towards life cycle sustainability management. Springer, Netherlands, pp 207–215

UNIDO (2006) Responsible trade and market access: opportunities or obstacles for SMEs in developing countries? United Nations Industrial Development Organization, Private Sector Development Branch. http://www.unido.org/fileadmin/user_media/Publications/Pub_free/Responsible_trade_and_market_access.pdf

Unilever (2012) Unilever sustainable living plan. Progress report 2011. Unilever. http://www.unilever.com/images/uslp-Unilever_Sustainable_Living_Plan_Progress_Report_2011_tcm13-284779.pdf

Unilever (2013) Unilever sustainable living plan. Progress report 2012. Unilever. http://www.unilever.co.uk/Images/USLP-Progress-Report-2012-FI_tcm28-352007.pdf

Unilever (2014a) Unilever website. http://www.unilever.com

Unilever (2014b) Unilever sustainable living plan 2013: Making progress, driving change. Unilever. http://www.unilever.com/images/slp_Unilever-Sustainable-Living-Plan-2013_tcm13-388693.pdf

WRI, WBCSD (2004) GHG protocol corporate accounting and reporting standard. World Resources Institute and World Business Council for Sustainable Development. http://www.ghgprotocol.org/files/ghgp/public/ghg-protocol-revised.pdf

WRI, WBCSD (2011) Corporate value chain (Scope 3) accounting and reporting standard – supplement to the GHG protocol corporate accounting and reporting standard. World Resources Institute and World Business Council for Sustainable Development. http://www.ghgprotocol.org/files/ghgp/public/Corporate-Value-Chain-Accounting-Reporing-Standard_041613.pdf

WRI, WBCSD (2013) Technical guidance for calculating scope 3 emissions – supplement to the corporate value chain (Scope 3) accounting & reporting standard. World Resources Institute and World Business Council for Sustainable Development. http://www.ghgprotocol.org/feature/scope-3-calculation-guidance

Index

M. Finkbeiner (ed.), *Special Types of Life Cycle Assessment*, LCA Compendium – The Complete World of Life Cycle Assessment,
DOI 10.1007/978-94-017-7610-3

MIX
Papier aus verantwortungsvollen Quellen
Paper from responsible sources
FSC® C105338
FSC
www.fsc.org

If you have any concerns about our products,
you can contact us on
ProductSafety@springernature.com

In case Publisher is established outside the EU,
the EU authorized representative is:
Springer Nature Customer Service Center GmbH
Europaplatz 3, 69115 Heidelberg, Germany

Printed by Libri Plureos GmbH
in Hamburg, Germany